MODULATION OF HOST GENE EXPRESSION AND INNATE IMMUNITY BY VIRUSES

Modulation of Host Gene Expression and Innate Immunity by Viruses

Edited by

PETER PALESE
Mount Sinai School of Medicine, New York, NY, USA

Springer

A C.I.P. Catalogue record for this book is available from the Library of Congress.

ISBN 978-90-481-6823-1 (PB)
ISBN 978-1-4020-3242-4 (e-book)

Published by Springer,
P.O. Box 17, 3300 AA Dordrecht, The Netherlands.

Sold and distributed in North, Central and South America
by Springer,
101 Philip Drive, Norwell, MA 02061, U.S.A.

In all other countries, sold and distributed
by Springer,
P.O. Box 322, 3300 AH Dordrecht, The Netherlands.

Printed on acid-free paper

Contents

Chapter 1
VIRUSES AND THE INNATE IMMUNE SYSTEM
Megan L. Shaw
Peter Palese

1. INTRODUCTION 1

2. HISTORICAL PERSPECTIVE 1

3. WHAT WE HAVE LEARNED SO FAR 2
   3.1 Interferon signaling pathways 2
   3.2 Viral anti-interferon activities 4

4. EXPLORATION OF NEW FRONTIERS 7
   4.1 The influence of host genetics on viral disease outcome 7
   4.2 The role of Toll-like receptors in the antiviral response 8
   4.3 Function of ISG products and the specificity of interferons 9
   4.4 Applications – antivirals and vaccines 9

Chapter 2
HOW VIRUSES ELICIT INTERFERON PRODUCTION
David E. Levy
Isabelle J. Marié

1. INTRODUCTION 19

2. TRANSCRIPTIONAL CONTROL OF IFN GENE EXPRESSION 20
   2.1 Regulation of the IFNβ gene 20
   2.2 Complex regulation of IFNα genes 22

3. THE VIRUS-ACTIVATED SWITCH 23
   3.1 Virus-activated kinases 24
   3.2 Virus-induced signals 25
   3.3 IFN induction through an RNA helicase switch 27

3.4 Additional mechanisms of IFN induction 28

4. CONCLUSIONS 29

Chapter 3
GENES MODULATED BY INTERFERONS AND DOUBLE-STRANDED RNA

Saumendra N. Sarkar
Gregory A. Peters
Ganes C. Sen

1. INTRODUCTION 35

2. GENE INDUCTION BY VIRUSES, INTERFERONS AND DOUBLE-STRANDED RNA 36
2.1 Characteristics of gene induction 36
2.2 Signaling pathways 38

3. FUNCTIONS OF IFN-INDUCED PROTEINS 40
3.1 Proteins induced by IFN and dsRNA 40
3.2 2-5 (A) Synthetases 41
3.3 RNase L 45
3.4 PKR 46
3.5 P56 family of proteins 52

4. CONCLUSIONS 56

Chapter 4
IMMUNOEVASIVE STRATEGIES: HOST AND VIRUS

Markus Wagner
Shahram Misaghi
Hidde L. Ploegh

1. INTRODUCTION 65

2. AVOIDANCE OF IMMUNE RECOGNITION BY PASSIVE STRATEGIES 67

3. CHANGE OF THE ANTIGENIC REPERTOIRE (ANTIGENIC VARIATION) 68

4. ACTIVE MANIPULATION OF HOST CELLS AND IMMUNE FUNCTIONS BY VIRAL PROTEINS ... 69
4.1 Interference with cytokine and chemokine functions and related intracellular signaling ... 69
4.2 Impairment of extracellular recognition of viruses by the humoral immune response ... 70
4.3 Impairment of host mediated killing of infected cells ... 70
4.3.1 Inhibition of apoptosis in infected cells ... 70
4.3.2 Impairment of class I and II MHC mediated antigen presentation in infected cells and protection from CTL killing ... 71
4.3.3 Evasion from natural killer (NK) cells ... 77

5. IMPAIRMENT OF THE FUNCTION OF IMMUNE CELLS BY DIRECT INFECTION ... 83
5.1 Functional paralysis of DCs after CMV infection ... 83
5.2 Impairment of macrophage function after infection with MCMV ... 84
5.3 Infection and inhibition of NK cells ... 84
5.4 Down-regulation of CD4 and the TCR/CD3 by infection of T-cells ... 84

6. CONCLUSIONS ... 85

Chapter 5
INTERFERON ANTAGONISTS OF INFLUENZA VIRUSES
Adolfo García-Sastre

1. INTRODUCTION ... 95

2. INFLUENZA VIRUSES ... 95

3. TRIGGERING OF THE TYPE I IFN RESPONSE BY INFLUENZA VIRUSES ... 97

4. THE NS1 PROTEIN MEDIATES EVASION OF THE TYPE I IFN RESPONSE BY INFLUENZA A VIRUSES ... 99

5. INTERFERON ANTAGONISTS OF INFLUENZA B VIRUS AND OF THOGOTO VIRUS ... 104

6. MULTIPLE FUNCTIONS ASSOCIATED WITH THE NS1 PROTEIN OF INFLUENZA A VIRUS 106

7. ROLE OF THE INFLUENZA A VIRUS NS1 PROTEIN IN VIRAL PATHOGENICITY 106

8. CONCLUSIONS 107

Chapter 6
THE ANTI-INTERFERON MECHANISMS OF PARAMYXOVIRUSES
Nicola Stock
Stephen Goodbourn
Richard E. Randall

1. INTRODUCTION 115

2. THE PARAMYXOVIRUSES 115

3. GENERAL CONCEPTS ARISING FROM THE INTERACTION OF PARAMYXOVIRUSES WITH THE INTERFERON SYSTEM 117
   3.1 IFN evasion strategies of paramyxoviruses 117
   3.2 Consequences of IFN antagonism for viral pathogenesis 119
   3.3 IFN antagonism and the development of novel vaccines and antiviral drugs 119

4. MECHANISMS OF IFN EVASION BY PARAMYXOVIRUSES 120
   4.1 Rubulavirus: Targeted degradation of STATs via the V protein 121
      4.1.1 Requirement for STAT proteins in the degradation process 121
      4.1.2 DDB1 and the "V degradation complex" 123
      4.1.3 Mumps V and RACK1 124
      4.1.4 Suppression of IFN production by Rubulaviruses 125
   4.2 Respirovirus: STAT sequestration, hypo/hyper-phosphorylation and degradation 126
   4.3 Morbillivirus: IFN antagonism via the V and C proteins 128
   4.4 Henipavirus: Binding and sequestration of STAT proteins 129
   4.5 Pneumovirus: IFN resistance via the NS1 and NS2 proteins 131

4.5.1 Role of NS1 and NS2 in IFN resistance 131

5. CONCLUSIONS 133

Chapter 7
THE STRATEGY OF CONQUEST
Sunil J. Advani
Bernard Roizman

1. INTRODUCTION 141

2. THE DEGRADATION OF PML BY THE ICP0 – AN E3 UBIQUITIN LIGASE 143

3. MODIFICATION OF CELL CYCLE PROTEINS BY HSV-1 144

4. PREVENTION OF SHUTOFF OF PROTEIN SYNTHESIS BY ACTIVATED PROTEIN KINASE R (PKR) 147

5. HSV BLOCKS APOPTOSIS INDUCED BY EXOGENOUS AGENTS OR VIRAL GENE PRODUCTS 150

6. IMMUNE EVASION: HSV BLOCKS MHC CLASS I AND II PATHWAYS 153

7. CONCLUSIONS 155

Chapter 8
IMMUNOMODULATION BY POXVIRUSES
James B. Johnston
Grant McFadden

1. INTRODUCTION 163

2. SELECTIVELY DELETING POXVIRUS GENES 165
2.1 Manipulating the poxvirus genome 165
2.2 Constructing knock-out poxviruses: Current and future strategies 166
2.3 Animal models of poxvirus pathogenesis 168
2.4 A caveat 169

3. OVERVIEW OF POXVIRUS IMMUNOMODULATORY STRATEGIES 169

4. VIROSTEALTH - PLAYING HIDE AND SEEK 170

5. VIROTRANSDUCTION - A FAILURE TO COMMUNICATE. 171
5.1 Inhibition of apoptosis 171
5.1.1 Modulation of caspase activity 172
5.1.2 Other anti-apoptotic strategies 174
5.2 Modulation of the interferon response 175
5.3 Uncoupling cell signaling events 175

6. VIROMIMICRY - SEEING A FAMILIAR FACE 176
6.1 Viroceptors 176
6.1.1 TNF viroceptors 176
6.1.2 IFN viroceptors 177
6.1.3 IL-1β viroceptors 178
6.2 Virokines 179
6.2.1 Viral IL-18 binding proteins 179
6.2.2 Viral IL-10 179
6.2.3 Viral growth factors 180
6.2.4 Viral semaphorins 181
6.2.5 Viral anti-inflammatory serpins 181
6.2.6 Modulation of chemokine function 182
6.2.7 Control of the complement system 183

7. CONCLUSIONS 184

8. APPLICATIONS OF KO VIRUS TECHNOLOGY 186

Chapter 9
INTERFERON ANTAGONISTS ENCODED BY EMERGING RNA VIRUSES
Christopher F. Basler

1. INTRODUCTION 197

2. THE IFNα/β RESPONSE 197
2.1 Viral interferon antagonists as virulence factors 200

3. FILOVIRUSES 201
3.1 Immune and cytokine responses during the course of Ebola virus infection 201
3.2 Ebola virus and IFN 202
3.2.1 IFNα/β treatment does not prevent Ebola virus disease in non-human primates 202
3.2.2 The role of the IFN response in restricting Ebola virus infection in mice 202
3.2.3 Evidence for Ebola virus evasion of the IFNα/β system . 204
3.2.4 Production of IFN during the course of Ebola virus infection 205
3.2.5 Identification of VP35 as an Ebola virus-encoded IFNα/β-antagonist 205
3.2.6 VP35 inhibition of IRF-3 206
3.2.7 Use of "reverse genetics" to further clarify roles in viral replication 206

4. HENIPAVIRUSES 207
4.1 Henipavirus IFN antagonists 207
4.2 Impact of Nipah virus V and W proteins on induction of IFNα/β production 209

5. BUNYAVIRUSES 209

6. FLAVIVIRUSES 212
6.1 Dengue virus 212
6.2 Other flaviviruses 213

7. ALPHAVIRUSES 213

8. CONCLUSIONS 214

Chapter 10
VIRAL PATHOGENESIS AND TOLL-LIKE RECEPTORS
Susan R. Ross

1. INTRODUCTION 221

2. TOLL-LIKE RECEPTORS AND SIGNALING 222

3. INTERACTION OF VIRUSES WITH DIFFERENT TLRS 225

3.1 Viral proteins as TLR ligands 225
3.1.1 RSV, TLR4 and induction of antiviral immunity 225
3.1.2 MMTV, TLRs and subversion of the host adaptive immune response 227
3.1.3 Measles virus, TLR2 and up-regulation of CD150 229
3.1.4 Human cytomegalovirus and TLR2-mediated cytokine production 230
3.2 Viral nucleic acids and TLR-mediated cell activation 231
3.2.1 Viral dsDNA and TLR-mediated antiviral responses 231
3.2.2 Viral RNA and TLR-mediated responses 234

4. CONCLUSIONS 236

Chapter 11

DIGESTING ONESELF AND DIGESTING MICROBES

Montrell Seay
Savithramma Dinesh-Kumar
Beth Levine

1. INTRODUCTION 245

2. INTRODUCTION TO THE MOLECULAR AND CELL BIOLOGY OF AUTOPHAGY 247
2.1 Formation and structure of autophagosomes 249
2.2 Regulation of autophagy 250
2.3 Autophagy genes 251
2.3.1 Protein kinase signaling 251
2.3.2 Lipid kinase signaling 252
2.3.3 Ubiquitin-like conjugation reactions 253
2.3.4 Atg protein recycling 254

3. INTRODUCTION TO THE BIOLOGICAL FUNCTIONS OF AUTOPHAGY 254
3.1 Role of autophagy in protection against nutrient starvation 256
3.2 Role of autophagy in differentiation and development 257
3.3 Role of autophagy in cell growth control 257
3.4 Role of autophagy in lifespan extension 259
3.5 Role of autophagy in preventing diseases associated with protein aggregates 260

4. AUTOPHAGY AND INNATE IMMUNITY TO VIRUSES 261
4.1 PKR-dependent autophagy degrades cellular and viral components 262
4.2 Mammalian autophagy genes play a role in host antiviral defense 265
4.3 Plant autophagy genes play a role in the plant innate immune response 267

5. EVASION OF AUTOPHAGY BY VIRUSES 268

6. SUBVERSION OF AUTOPHAGY BY VIRUSES 270

7. CONCLUSIONS 272

Chapter 12
GENETIC VARIATION IN HOST DEFENSES AND VIRAL INFECTIONS
Eunhwa Choi
Stephen J. Chanock

1. INTRODUCTION 281

2. HOST GENETIC VARIATION AND HIV-1 INFECTION 283

3. ASSOCIATION OF GENETIC POLYMORPHISMS AND HEPATITIS C INFECTION 287

4. ASSOCIATION BETWEEN *HLA* ALLELES AND HBV INFECTION 287
4.1 *HLA* class I and II alleles 288
4.2 Vaccine response 289

5. HOST GENETICS IN SEVERE RSV BRONCHIOLITIS 289

6. CONCLUSIONS 291

Index 295

Chapter 1

VIRUSES AND THE INNATE IMMUNE SYSTEM

Answers and yet more questions

MEGAN L. SHAW and PETER PALESE
Department of Microbiology, Mount Sinai School of Medicine, New York, NY, USA

1. INTRODUCTION

Infection of a naïve (non-immune) host with a virus elicits an immediate response which results in a cascade of changes in the host, including an interferon response (innate immunity). The outcome of this interaction is influenced by the genes of the virus as well as the genes of the host. Interestingly, different viruses affect this response in different ways. Not only is there a plethora of mechanisms used by the invading organisms, but the host has also evolved a great variety of redundant and robust countermeasures. This interplay of host and virus represents one of the most significant frontiers in biology today. A clearer understanding of the mechanisms involved will arm us with better strategies to deal with viruses, including emerging pathogens and potential bioterrorism agents.

2. HISTORICAL PERSPECTIVE

Almost fifty years ago, Isaacs and Lindenmann[1] found that addition of heat-treated influenza virus to pieces of chicken chorioallantoic membrane induced an antiviral factor, which they called interferon (IFN). Interestingly, the use of live, untreated influenza virus followed by heat inactivated virus inhibited the induction of IFN[2]. Most likely, expression of the NS1 protein (an IFN antagonist) by the live, untreated virus induced a sufficient amount of anti-IFN activity to counteract the synthesis of IFN in

P. Palese (ed.), Modulation of Host Gene Expression and Innate Immunity by Viruses, 1-18.

these cells. Viral transcription in cells infected by heat-inactivated influenza virus was sufficient to induce a vigorous IFN response but not enough functional NS1 protein was made to neutralize the antiviral response of the infected cell. Thus, early on, the principles (but not the precise mechanisms) of the antiviral response of an infected cell and the antagonist function of live virus had been recognized.

Also, many decades ago, the potential of IFN as a therapeutic agent was clearly foreseen by Jan Vilcek[3] and Kari Cantell[4] and the subsequent cloning and expression of human leukocyte IFN by Charles Weissmann's laboratory revolutionized the field[5]. The last 10-15 years have brought another renaissance to the IFN field, which was driven by the growing interest of molecular biologists in unraveling the signal transduction pathways of type I and type II IFNs[6]. This exploration of cellular pathways was complemented by the realization that viruses can counteract the antiviral strategies of the infected cell. First, the large DNA viruses were found to express proteins or small RNAs that have antagonist activity directed against the antiviral program of the cell[7-9]. Subsequently, it was demonstrated that RNA viruses also possess IFN antagonist activity[10], suggesting that most if not all viruses have ways of overcoming, or at least limiting the antiviral response of the host cell.

3. WHAT WE HAVE LEARNED SO FAR

3.1 Interferon signaling pathways

Interferons are generally classified into two families. In humans, the type I IFNs (IFN-α/β) include 13 IFN-α species and a single IFN-β species. These IFNs all bind to the same receptor and are secreted by almost all cell types. The type II IFN family consists of one member, IFN-γ, which is synthesized primarily by immune cells in response to IL-12 production. Production of IFN-α/β is triggered in direct response to virus infection and thus the IFN system constitutes one of the earliest (innate) phases of the host antiviral immune response. The importance of the IFN response to host defense against viral infection has been demonstrated by the fact that mice lacking specific components of the IFN signaling pathway, including the IFN-β gene itself, are acutely sensitive to virus infection even though their adaptive immune system remains intact[11-13].

The establishment of the IFN-mediated antiviral response within an infected cell can be broadly broken down into three signaling pathways (Figure 1). The first pathway (IFN production pathway) involves the transcriptional upregulation of the IFN-β gene and the secretion of IFN-β from the infected cell. The precise signaling events that are initiated upon virus infection and result in the activation of the IFN-β promoter are an active line of enquiry and are discussed in detail in Chapter 2. Suffice it to say that the transcription factors, nuclear factor κB (NF- κB) and interferon regulatory factor 3 (IRF3) are both activated in response to virus infection, and this event is critical to the subsequent activation of IFN-β mRNA synthesis[14].

Once released from the infected cell, IFN-β binds to the IFNαβ receptor and sets in motion a cascade of tyrosine phosphorylation reactions that result in the transcriptional upregulation of IFN stimulated genes (ISG). This second pathway is termed the IFN signaling (or JAK/STAT) pathway and involves activation of the latent transcription factors, STAT1 and STAT2, which once phosphorylated, heterodimerize and translocate to the nucleus where they interact with IRF9 to form the ISGF3 complex[6]. This transcription factor complex binds to specific sites within the promoters of ISGs, termed IFN sensitive response elements (ISRE). One such ISG encodes IRF7, which is required for the transcription of most IFN-α genes and like, IRF3, is activated in response to virus infection[15]. This allows for the formation of an amplification loop, whereby low levels of IFN produced at early stages post infection, leads to the induced expression of a specific transcription factor (IRF7) that activates a second, much greater wave of IFN production.

The third pathway, which is not so much one as a multitude of pathways, represents the various activities of the proteins encoded by the ISGs. These proteins are responsible for establishing the antiviral state within the cell, and they function through a variety of mechanisms. Well characterized examples of these effector proteins include protein kinase R (PKR), the family of 2'-5'oligoadenylate synthetases, RNaseL and the P56 protein, all of which function as translation inhibitors, as detailed in chapter 3. Others include the Mx GTPases[16], promyelocytic leukemia protein (PML)[17] and ADAR, a double-stranded RNA-specific adenosine deaminase[18]. The exact functions of many IFN-inducible proteins remain unknown but it is clear that a major role of these proteins is to dramatically reduce the activity of the host enzymatic machinery which viruses parasitize in order to replicate. Thus, this IFN-induced antiviral state halts virus replication in infected cells, prevents infection of neighboring, uninfected cells and buys time for the host before the adaptive arm of the immune response is activated.

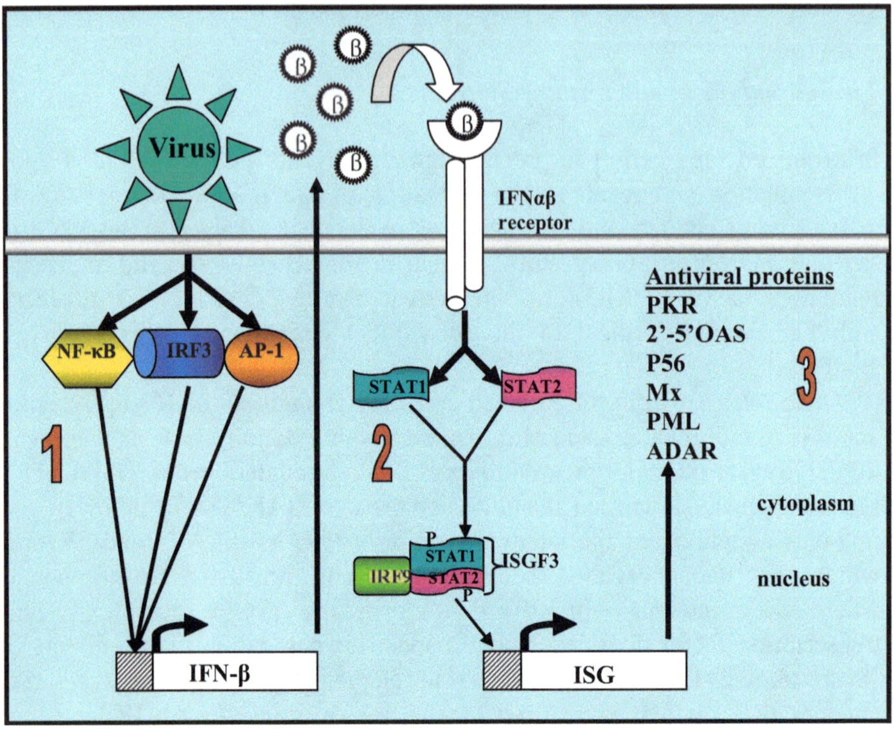

Figure 1. Schematic representation of the IFN response triggered by virus infection. 1) IFN production. Virus infection activates the transcription factors, NF-κB, IRF3 and AP-1 which translocate to the nucleus and bind to the IFN-β promoter to activate synthesis of IFN-β mRNA. 2) IFN signaling. IFN-β binds to the IFNαβ receptor which results in the activation of the transcription factors, STAT1 and STAT2. The phosphorylated STATs heterodimerize and translocate to the nucleus where they interact with IRF9 to form the ISGF3 complex. This binds to the promoters of IFN stimulated genes (ISGs) and activates transcription. 3) Activities of antiviral proteins. The protein products of the ISGs set up the antiviral state within the cell.

3.2 Viral anti-interferon activities

If the host IFN response described above functioned at maximum efficiency, virus infection would be of relatively minor consequence to the host. Of course, as the title of this book implies, this is not the case. In order to gain a foothold during the early stages of infection, viruses have devised ways of inhibiting the IFN response, thereby preventing the induction of an antiviral state and allowing replication to proceed. The viral proteins that encode this function are termed IFN antagonist proteins and examples of these have been described for an ever increasing number of viruses, covering both DNA and RNA virus families (Table 1). The mechanisms by which these antagonists act are diverse, and a select number are described in chapters 4-8. All three of the IFN signaling pathways described above have been shown to be targeted by viral IFN antagonists and some viruses encode multiple antagonists that enable the virus to inhibit more than one IFN pathway, while others encode a single multifunctional antagonist that can target multiple pathways (see Table 1). Because transcriptional upregulation of genes plays such a key role in the IFN response, it is not surprising that many viral IFN antagonists specifically target transcription factors, such as IRFs and STATs, either directly or indirectly preventing their activation. Other viruses take a less specialized approach and cause a general shut off of host transcription or translation.

Most viral IFN antagonists are accessory proteins in that they are not required for viral replication *in vitro*, however deletion or mutation of the antagonist gene often results in an attenuated phenotype *in vivo*[10,19-23]. Therefore these proteins are also functioning as virulence factors. The association with the IFN response can clearly be demonstrated by comparing infection in IFN competent and incompetent hosts. For example, an influenza virus lacking its IFN antagonist, NS1, is avirulent in wild-type mice but in $STAT1^{-/-}$ mice it is pathogenic[10]. This illustrates the importance of an intact IFN system for the host and also the requirement of a functional IFN antagonist protein for the virus. Recent studies also indicate that IFN antagonists may determine the host range of a virus[24-28]. This is most likely due to species-specific differences in the cellular targets of these proteins and therefore restricts a virus to the particular host whose IFN response can be overcome.

The ongoing battle of virus versus host IFN response should not only be viewed as the virus outwitting the host but also in terms of host adaptation to the anti-IFN strategies of the virus. The multifaceted nature of the human IFN system no doubt reflects our evolution in response to the barrage of diverse viral IFN antagonists that we have been exposed to. Therefore, elucidation of the cellular IFN signaling pathways and the mechanisms that viruses use to inhibit them, work hand in hand to guide us toward a clearer picture of the intricate nature of the innate immune response.

Table 1. Examples of viral IFN antagonists and the IFN pathways that they target.

VIRUS	IFN ANTAGONIST	TARGET PATHWAY/PROTEIN	REFERENCE
DNA viruses			
Vaccinia virus	E3L	IFN production, PKR, OAS (dsRNA binding)	7,29,30
	K3L	PKR	31,32
	B18R	IFN signaling	33,34
Herpes simplex virus	$\gamma_1$34.5	PKR	35
	US11	PKR	36,37
	ICP0	PML, IFN production	38,39
	unknown	IFN signaling	40
Human herpes virus 8	vIRF proteins	IFN production, PKR	41-44
Adenovirus	E1A	IFN production, IFN signaling	45-47
	VA RNAs	PKR	48
Human papilloma virus	E6	IFN production, IFN signaling	49,50
	E7	IFN signaling	51
Retroviruses			
HIV	Tat	PKR	52
	TAR RNA	PKR	53
	unknown	OAS/RNaseL	54
Double-strand RNA viruses			
Reovirus	σ3	PKR	55
Positive-strand RNA viruses			
Hepatitis C virus	NS3/4A	IFN production	56
	NS5A	PKR	57
	E2	PKR	58
	core	IFN signaling	59
Dengue virus	NS4B	IFN signaling	60
Poliovirus	$3C^{Pro}$	Transcriptional shut off	61,62
	unknown	PKR	63
Negative-strand RNA viruses			
Influenza A virus	NS1	IFN production, PKR, OAS (dsRNA binding), mRNA processing and transport	10,64-70

	unknown	PKR	[71]
Thogoto virus	ML	IFN production	[72]
Bunyamwera virus	NSs	Transcriptional shut off	[73,74]
Rift Valley fever virus	NSs	Transcriptional shut off	[75,76]
Ebola virus	VP35	IFN production	[77,78]
Sendai virus	C proteins	IFN signaling, IFN production	[79-86]
	V	IFN production	[86]
Simian virus 5	V	IFN signaling, IFN production	[87-92]
Human parainfluenza virus 2	V	IFN signaling, IFN production	[90,92-95]
Mumps virus	V	IFN signaling	[96,97]
Measles virus	V,C	IFN signaling	[98,99]
	unknown	IFN production	[100]
Nipah virus	P,V,W	IFN signaling	[101-104]
	C	unknown	[101]
Hendra virus	V	IFN signaling	[105]
Respiratory syncytial virus	NS1, NS2	IFN production	[106-108]

4. EXPLORATION OF NEW FRONTIERS

4.1 The influence of host genetics on viral disease outcome

In the post-genomic era, it is now possible to look at how mutations in the hosts' genes influence the antiviral response and how this affects the outcome of infection (see chapter 11). Mutations in IFN-related genes may have an impact on the ability of an individual to mount an immune response against a viral infection, which may have a direct effect on the course of the disease. In fact a lethal mutation in the *STAT1* gene, which obliterates IFN signaling has been described in two children, both of whom died as a result of viral infection[109].

The influence of mutations in IFN-related genes with respect to specific viruses has been addressed more extensively for hepatitis C virus (HCV) than any other virus, mainly because interferon is the current therapy for HCV-infected individuals but not all patients respond to this treatment. Polymorphism studies have identified single nucleotide polymorphisms

(SNPs) in the *MxA*, *IL-10*, *OAS1* and *PKR* genes that affect either the outcome of HCV infection or the response to interferon therapy[110-113]. Amongst respiratory viruses, SNPs in both the *IL-4* and *IL-4Ralpha* genes have been found to be associated with increased risk of severe respiratory syncytial virus infection in children[114,115]. The results of such studies may have a direct bearing on the design of future vaccines as well as on the choice of target population for vaccination.

4.2 The role of Toll-like receptors in the antiviral response

The members of the Toll-like receptor (TLR) family are responsible for the recognition of a broad range of pathogens and subsequent activation of the innate immune response. The growing interest in the relationship between TLRs and viruses is two-fold. Firstly, the signaling pathways activated by viruses and TLRs overlap to a large extent and stimulation of certain TLRs (TLR3, 4, 7 and 9) leads to IFN production and thus the establishment of an antiviral state. Secondly, the molecular components of some viruses (envelope proteins, nucleic acids) have been found to specifically activate select TLRs (TLR2, 3, 4, 7 and 9) (see Chapter 9). This has prompted speculation that these TLRs may serve as the sensors of virus infection and that recognition of virus-specific molecules triggers the induction of the IFN response. However this idea has been challenged for various reasons. These include cell-type specificity (e.g. most cells, apart from the immune cells, express only a subset of TLRs yet all cells can produce IFN) and subcellular distribution (e.g. TLRs are either found on the cell surface or in intracellular compartments, neither of which allows for sensing of the cytoplasm; the replication compartment of many viruses). The most compelling evidence that virus- and TLR-activated signaling pathways are not one and the same comes from gene knockout studies, where it has been shown that deletion of TLR-associated signaling components (e.g. TLR3, TRIF, TBK-1) does not prevent activation of the IFN response by virus infection[116-118]. A caveat to this is that these studies mostly use prototype viruses (such as Sendai virus) and while it may be said that TLR-signaling is not essential for the induction of an immune response to all viruses, it is possible that some TLRs do play an important role in the response to select viruses. Another point to consider is that the high levels of TLR expression in immune cells may reflect a more specialized role for TLRs. For example, the IFN response plays a key role in promoting maturation of dendritic cells (DC)[119], which express a wide range of TLRs and are major antigen presenting cells that are capable of priming naïve $CD4^+$ T cells. Thus DCs are considered to form the link between the innate and adaptive immune responses and TLR signaling may prove to play a crucial role in virus-induced DC maturation and hence influence subsequent T-cell activation.

4.3 Function of ISG products and the specificity of interferons

As much as the IFN field has advanced in the past years, we remain relatively in the dark when it comes to the functions of IFN-stimulated gene (ISG) products or IFN-induced proteins. Gene expression analyses on IFN-treated cells have identified a vast number of ISGs and based on these data, it is estimated that there may be anywhere from 600 to 2000 IFN-inducible human genes[120,121]. de Veer *et al*[120] have undertaken to group the identified ISGs into functional categories based on sequence homologies, which has revealed a diverse array of functions covering almost all cellular activities. The question of how each individual protein contributes to the establishment of an antiviral state and the relative importance of each protein with regard to inhibiting replication of specific viruses, will be key lines of inquiry in the coming years.

The need for so many IFN species poses another unanswered question: Are they all equal? Most likely not. The single IFN-β species has a defined role as the first IFN to be made in response to virus infection and acts to prime cells to make more IFN. The specific roles of all the IFN-α species remain unknown, although comparison of purified IFN-α proteins has shown that they have distinct activities[122]. Do they display cell-type specific expression patterns? Are different subsets of IFN-α species made in response to different viruses? Do some have more antiviral activity against specific viruses than others? Answers to these questions have remained elusive so far but further technological advances will hopefully allow them to be addressed in the near future.

4.4 Applications – antivirals and vaccines

IFN-α was first approved for therapy in 1986 and has been used to treat both viral diseases and cancers. A pegylated form of IFN, which has been modified by the addition of polyethylene glycol, is now being used for treating patients with hepatitis C virus with the benefit that larger doses can be given infrequently. Another way to improve on IFN therapy would be to use genetic engineering to create a hybrid IFN species that has higher activity than those currently available. This technique of DNA shuffling has been used successfully to make a hybrid IL-12 from several mammalian *IL-12* genes, resulting in a molecule with improved efficacy[123]. Rather than administering IFN directly, it is also possible to use drugs that induce the production of IFN. For example, a synthetic form of double-stranded RNA (poly I:C) is a potent inducer of IFN, and a modified form has been used to treat HIV patients[124] and is effective against several other viruses in animal models[125,126]. An alternative antiviral approach may be the targeting of the viral IFN antagonists with small molecular weight drugs. Reducing or blocking the anti-interferon activity of the invading virus should allow the host innate immune system to control virus replication.

Vaccines have proven to be one of the most effective ways of preventing viral infection. Successful live human virus vaccines have been selected by repeated passaging in tissue culture or animal hosts giving rise to strains that have lower virulence in humans than the wild type virus. These host range mutants have been the basis for vaccines against measles, mumps, rubella and chicken pox. Another strategy for the design of live attenuated vaccines that is being considered is based on modification of the viral IFN antagonist. As discussed in section 3.2, many IFN antagonists act as virulence factors and viruses lacking these factors are attenuated *in vivo*. The idea behind the development of a vaccine candidate is that a virus expressing an antagonist with intermediate anti-IFN capabilities is likely to replicate sufficiently well to induce an immune response, but ultimately the host response will win, and therefore, the virus remains attenuated. In practice this has been demonstrated for the NS1 protein of influenza virus using viruses expressing C-terminal truncated forms of NS1. When compared to wild type virus and a virus lacking NS1 (delNS1), these viruses displayed intermediate phenotypes with respect to growth properties and the ability to induce IFN but were still attenuated in mice[127,128]. More importantly, immunization with the truncated NS1 virus protected mice against lethal challenge, whereas immunization with the delNS1 virus was less effective, indicating that the deletion virus is too attenuated to induce a protective immune response[127]. These data establish proof of concept that a virus that expresses a modified IFN antagonist retains the balance between optimal levels of attenuation and immunogenic properties, two important features of a live virus vaccine.

The advantage of this strategy is that the molecular basis for attenuation is understood, which paves the way for the development of next generation candidates where additional mutations can be inserted into the virus to adjust the level of attenuation appropriately and to exploit what is known about the innate immune response. Prior functional characterization of the IFN antagonist protein is therefore crucial to the development of such vaccine candidates and of new antivirals and should fuel further research into the molecular mechanisms by which other viral IFN antagonists exert their action. In addition, these studies will lead to further elucidation of the complex mechanisms that determine species and tissue specificity of viruses and provide us with a more in depth understanding of viral pathogenicity in general.

REFERENCES

1. Isaacs, A. and Lindenmann, J., 1957, Virus interference. I. The interferon. *Proc R Soc Lond B Biol Sci* **147:**258-67
2. Lindenmann, J., 1960, [Interferon and inverse interference]. *Z Haut Geschlechtskr* **146:**287-309
3. Havell, E.A. and Vilcek, J., 1974, Mass production and some characteristics of human interferon from diploid cells. *In Vitro Monogr* 47
4. Strander, H. and Cantell, K., 1967, Further studies on the production of interferon by human leukocyte in vitro. *Ann Med Exp Biol Fenn* **45:**20-9
5. Nagata, S., Taira, H., Hall, A., Johnsrud, L., Streuli, M., Ecsodi, J., Boll, W., Cantell, K. and Weissmann, C., 1980, Synthesis in E. coli of a polypeptide with human leukocyte interferon activity. *Nature* **284:**316-20
6. Darnell, J.E., Jr., Kerr, I.M. and Stark, G.R., 1994, Jak-STAT pathways and transcriptional activation in response to IFNs and other extracellular signaling proteins. *Science* **264:**1415-21
7. Chang, H.W., Watson, J.C. and Jacobs, B.L., 1992, The E3L gene of vaccinia virus encodes an inhibitor of the interferon-induced, double-stranded RNA-dependent protein kinase. *Proc Natl Acad Sci USA* **89:**4825-9
8. Kitajewski, J., Schneider, R.J., Safer, B., Munemitsu, S.M., Samuel, C.E., Thimmappaya, B. and Shenk, T., 1986, Adenovirus VAI RNA antagonizes the antiviral action of interferon by preventing activation of the interferon-induced eIF-2 alpha kinase. *Cell* **45:**195-200
9. Roizman, B. and Knipe, D.M., 2001, Herpes Simplex Viruses and Their Replication. In *Fields Virology*, (eds. Knipe, D.M. and Howley, P.M.), Lippincott Williams & Wilkins, Philadelphia, Vol. 2, 2399-2460
10. Garcia-Sastre, A., Egorov, A., Matassov, D., Brandt, S., Levy, D.E., Durbin, J.E., Palese, P. and Muster, T., 1998, Influenza A virus lacking the NS1 gene replicates in interferon-deficient systems. *Virology* **252:**324-30
11. Muller, U., Steinhoff, U., Reis, L.F., Hemmi, S., Pavlovic, J., Zinkernagel, R.M. and Aguet, M., 1994, Functional role of type I and type II interferons in antiviral defense. *Science* **264:**1918-21
12. Durbin, J.E., Hackenmiller, R., Simon, M.C. and Levy, D.E., 1996, Targeted disruption of the mouse Stat1 gene results in compromised innate immunity to viral disease. *Cell* **84:**443-50
13. Deonarain, R., Alcami, A., Alexiou, M., Dallman, M.J., Gewert, D.R. and Porter, A.C., 2000, Impaired antiviral response and alpha/beta interferon induction in mice lacking beta interferon. *J Virol* **74:**3404-9
14. Wathelet, M.G., Lin, C.H., Parekh, B.S., Ronco, L.V., Howley, P.M. and Maniatis, T., 1998, Virus infection induces the assembly of coordinately activated transcription factors on the IFN-beta enhancer in vivo. *Mol Cell* **1:**507-18
15. Marie, I., Durbin, J.E. and Levy, D.E., 1998, Differential viral induction of distinct interferon-alpha genes by positive feedback through interferon regulatory factor-7. *Embo J* **17:**6660-9
16. Haller, O. and Kochs, G., 2002, Interferon-induced mx proteins: dynamin-like GTPases with antiviral activity. *Traffic* **3:**710-7

17. Regad, T. and Chelbi-Alix, M.K., 2001, Role and fate of PML nuclear bodies in response to interferon and viral infections. *Oncogene* **20:**7274-86
18. Liu, Y. and Samuel, C.E., 1996, Mechanism of interferon action: functionally distinct RNA-binding and catalytic domains in the interferon-inducible, double-stranded RNA-specific adenosine deaminase. *J Virol* **70:**1961-8
19. Bridgen, A., Weber, F., Fazakerley, J.K. and Elliott, R.M., 2001, Bunyamwera bunyavirus nonstructural protein NSs is a nonessential gene product that contributes to viral pathogenesis. *Proc Natl Acad Sci USA* **98:**664-9
20. Garcin, D., Itoh, M. and Kolakofsky, D., 1997, A point mutation in the Sendai virus accessory C proteins attenuates virulence for mice, but not virus growth in cell culture. *Virology* **238:**424-31
21. Patterson, J.B., Thomas, D., Lewicki, H., Billeter, M.A. and Oldstone, M.B., 2000, V and C proteins of measles virus function as virulence factors in vivo. *Virology* **267:**80-9
22. Chou, J., Kern, E.R., Whitley, R.J. and Roizman, B., 1990, Mapping of herpes simplex virus-1 neurovirulence to gamma 134.5, a gene nonessential for growth in culture. *Science* **250:**1262-6
23. Brandt, T.A. and Jacobs, B.L., 2001, Both carboxy- and amino-terminal domains of the vaccinia virus interferon resistance gene, E3L, are required for pathogenesis in a mouse model. *J Virol* **75:**850-6
24. Bossert, B. and Conzelmann, K.K., 2002, Respiratory syncytial virus (RSV) nonstructural (NS) proteins as host range determinants: a chimeric bovine RSV with NS genes from human RSV is attenuated in interferon-competent bovine cells. *J Virol* **76:**4287-93
25. Parisien, J.P., Lau, J.F. and Horvath, C.M., 2002, STAT2 acts as a host range determinant for species-specific paramyxovirus interferon antagonism and simian virus 5 replication. *J Virol* **76:**6435-41
26. Park, M.S., Garcia-Sastre, A., Cros, J.F., Basler, C.F. and Palese, P., 2003, Newcastle disease virus V protein is a determinant of host range restriction. *J Virol* **77:**9522-32
27. Young, D.F., Chatziandreou, N., He, B., Goodbourn, S., Lamb, R.A. and Randall, R.E., 2001, Single amino acid substitution in the V protein of simian virus 5 differentiates its ability to block interferon signaling in human and murine cells. *J Virol* **75:**3363-70
28. Langland, J.O. and Jacobs, B.L., 2002, The role of the PKR-inhibitory genes, E3L and K3L, in determining vaccinia virus host range. *Virology* **299:**133-41
29. Smith, E.J., Marie, I., Prakash, A., Garcia-Sastre, A. and Levy, D.E., 2001, IRF3 and IRF7 phosphorylation in virus-infected cells does not require double-stranded RNA-dependent protein kinase R or Ikappa B kinase but is blocked by Vaccinia virus E3L protein. *J Biol Chem* **276:**8951-7
30. Rivas, C., Gil, J., Melkova, Z., Esteban, M. and Diaz-Guerra, M., 1998, Vaccinia virus E3L protein is an inhibitor of the interferon (i.f.n.)-induced 2-5A synthetase enzyme. *Virology* **243:**406-14
31. Davies, M.V., Elroy-Stein, O., Jagus, R., Moss, B. and Kaufman, R.J., 1992, The vaccinia virus K3L gene product potentiates translation by inhibiting double-stranded-RNA-activated protein kinase and phosphorylation of the alpha subunit of eukaryotic initiation factor 2. *J Virol* **66:**1943-50
32. Carroll, K., Elroy-Stein, O., Moss, B. and Jagus, R., 1993, Recombinant vaccinia virus K3L gene product prevents activation of double-stranded RNA-dependent, initiation factor 2 alpha-specific protein kinase. *J Biol Chem* **268:**12837-42
33. Alcami, A., Symons, J.A. and Smith, G.L., 2000, The vaccinia virus soluble alpha/beta interferon (IFN) receptor binds to the cell surface and protects cells from the antiviral effects of IFN. *J Virol* **74:**11230-9
34. Symons, J.A., Alcami, A. and Smith, G.L., 1995, Vaccinia virus encodes a soluble type I interferon receptor of novel structure and broad species specificity. *Cell* **81:**551-60
35. He, B., Gross, M. and Roizman, B., 1997, The gamma(1)34.5 protein of herpes simplex virus 1 complexes with protein phosphatase 1alpha to dephosphorylate the alpha subunit

of the eukaryotic translation initiation factor 2 and preclude the shutoff of protein synthesis by double-stranded RNA-activated protein kinase. *Proc Natl Acad Sci USA* **94:**843-8

36. Poppers, J., Mulvey, M., Khoo, D. and Mohr, I., 2000, Inhibition of PKR activation by the proline-rich RNA binding domain of the herpes simplex virus type 1 Us11 protein. *J Virol* **74:**11215-21
37. Peters, G.A., Khoo, D., Mohr, I. and Sen, G.C., 2002, Inhibition of PACT-mediated activation of PKR by the herpes simplex virus type 1 Us11 protein. *J Virol* **76:**11054-64
38. Everett, R.D., Freemont, P., Saitoh, H., Dasso, M., Orr, A., Kathoria, M. and Parkinson, J., 1998, The disruption of ND10 during herpes simplex virus infection correlates with the Vmw110- and proteasome-dependent loss of several PML isoforms. *J Virol* **72:**6581-91
39. Melroe, G.T., DeLuca, N.A. and Knipe, D.M., 2004, Herpes simplex virus 1 has multiple mechanisms for blocking virus-induced interferon production. *J Virol* **78:**8411-20
40. Chee, A.V. and Roizman, B., 2004, Herpes simplex virus 1 gene products occlude the interferon signaling pathway at multiple sites. *J Virol* **78:**4185-96
41. Burysek, L., Yeow, W.S., Lubyova, B., Kellum, M., Schafer, S.L., Huang, Y.Q. and Pitha, P.M., 1999, Functional analysis of human herpesvirus 8-encoded viral interferon regulatory factor 1 and its association with cellular interferon regulatory factors and p300. *J Virol* **73:**7334-42
42. Li, M., Lee, H., Guo, J., Neipel, F., Fleckenstein, B., Ozato, K. and Jung, J.U., 1998, Kaposi's sarcoma-associated herpesvirus viral interferon regulatory factor. *J Virol* **72:**5433-40
43. Lubyova, B. and Pitha, P.M., 2000, Characterization of a novel human herpesvirus 8-encoded protein, vIRF-3, that shows homology to viral and cellular interferon regulatory factors. *J Virol* **74:**8194-201
44. Burysek, L. and Pitha, P.M., 2001, Latently expressed human herpesvirus 8-encoded interferon regulatory factor 2 inhibits double-stranded RNA-activated protein kinase. *J Virol* **75:**2345-52
45. Juang, Y.T., Lowther, W., Kellum, M., Au, W.C., Lin, R., Hiscott, J. and Pitha, P.M., 1998, Primary activation of interferon A and interferon B gene transcription by interferon regulatory factor 3. *Proc Natl Acad Sci USA* **95:**9837-42
46. Leonard, G.T. and Sen, G.C., 1996, Effects of adenovirus E1A protein on interferon-signaling. *Virology* **224:**25-33
47. Look, D.C., Roswit, W.T., Frick, A.G., Gris-Alevy, Y., Dickhaus, D.M., Walter, M.J. and Holtzman, M.J., 1998, Direct suppression of Stat1 function during adenoviral infection. *Immunity* **9:**871-80
48. Mathews, M.B. and Shenk, T., 1991, Adenovirus virus-associated RNA and translation control. *J Virol* **65:**5657-62
49. Ronco, L.V., Karpova, A.Y., Vidal, M. and Howley, P.M., 1998, Human papillomavirus 16 E6 oncoprotein binds to interferon regulatory factor-3 and inhibits its transcriptional activity. *Genes Dev* **12:**2061-72
50. Li, S., Labrecque, S., Gauzzi, M.C., Cuddihy, A.R., Wong, A.H., Pellegrini, S., Matlashewski, G.J. and Koromilas, A.E., 1999, The human papilloma virus (HPV)-18 E6 oncoprotein physically associates with Tyk2 and impairs Jak-STAT activation by interferon-alpha. *Oncogene* **18:**5727-37
51. Barnard, P. and McMillan, N.A., 1999, The human papillomavirus E7 oncoprotein abrogates signaling mediated by interferon-alpha. *Virology* **259:**305-13
52. Roy, S., Katze, M.G., Parkin, N.T., Edery, I., Hovanessian, A.G. and Sonenberg, N., 1990, Control of the interferon-induced 68-kilodalton protein kinase by the HIV-1 tat gene product. *Science* **247:**1216-9

53. Gunnery, S., Rice, A.P., Robertson, H.D. and Mathews, M.B., 1990, Tat-responsive region RNA of human immunodeficiency virus 1 can prevent activation of the double-stranded-RNA-activated protein kinase. *Proc Natl Acad Sci USA* **87:**8687-91
54. Martinand, C., Montavon, C., Salehzada, T., Silhol, M., Lebleu, B. and Bisbal, C., 1999, RNase L inhibitor is induced during human immunodeficiency virus type 1 infection and down regulates the 2-5A/RNase L pathway in human T cells. *J Virol* **73:**290-6
55. Imani, F. and Jacobs, B.L., 1988, Inhibitory activity for the interferon-induced protein kinase is associated with the reovirus serotype 1 sigma 3 protein. *Proc Natl Acad Sci USA* **85:**7887-91
56. Foy, E., Li, K., Wang, C., Sumpter, R., Jr., Ikeda, M., Lemon, S.M. and Gale, M., Jr., 2003, Regulation of interferon regulatory factor-3 by the hepatitis C virus serine protease. *Science* **300:**1145-8
57. Gale, M.J., Jr., Korth, M.J. and Katze, M.G., 1998, Repression of the PKR protein kinase by the hepatitis C virus NS5A protein: a potential mechanism of interferon resistance. *Clin Diagn Virol* **10:**157-62
58. Taylor, D.R., Shi, S.T., Romano, P.R., Barber, G.N. and Lai, M.M., 1999, Inhibition of the interferon-inducible protein kinase PKR by HCV E2 protein. *Science* **285:**107-10
59. Melen, K., Fagerlund, R., Nyqvist, M., Keskinen, P. and Julkunen, I., 2004, Expression of hepatitis C virus core protein inhibits interferon-induced nuclear import of STATs. *J Med Virol* **73:**536-47
60. Munoz-Jordan, J.L., Sanchez-Burgos, G.G., Laurent-Rolle, M. and Garcia-Sastre, A., 2003, Inhibition of interferon signaling by dengue virus. *Proc Natl Acad Sci USA* **100:**14333-8
61. Yalamanchili, P., Datta, U. and Dasgupta, A., 1997, Inhibition of host cell transcription by poliovirus: cleavage of transcription factor CREB by poliovirus-encoded protease 3Cpro. *J Virol* **71:**1220-6
62. Yalamanchili, P., Harris, K., Wimmer, E. and Dasgupta, A., 1996, Inhibition of basal transcription by poliovirus: a virus- encoded protease (3Cpro) inhibits formation of TBP-TATA box complex in vitro. *J Virol* **70:**2922-9
63. Black, T.L., Barber, G.N. and Katze, M.G., 1993, Degradation of the interferon-induced 68,000-M(r) protein kinase by poliovirus requires RNA. *J Virol* **67:**791-800
64. Ludwig, S., Wang, X., Ehrhardt, C., Zheng, H., Donelan, N., Planz, O., Pleschka, S., Garcia-Sastre, A., Heins, G. and Wolff, T., 2002, The influenza A virus NS1 protein inhibits activation of Jun N-terminal kinase and AP-1 transcription factors. *J Virol* **76:**11166-71
65. Wang, X., Li, M., Zheng, H., Muster, T., Palese, P., Beg, A.A. and Garcia-Sastre, A., 2000, Influenza A virus NS1 protein prevents activation of NF-kappaB and induction of alpha/beta interferon. *J Virol* **74:**11566-73
66. Lu, Y., Wambach, M., Katze, M.G. and Krug, R.M., 1995, Binding of the influenza virus NS1 protein to double-stranded RNA inhibits the activation of the protein kinase that phosphorylates the elF-2 translation initiation factor. *Virology* **214:**222-8
67. Bergmann, M., Garcia-Sastre, A., Carnero, E., Pehamberger, H., Wolff, K., Palese, P. and Muster, T., 2000, Influenza virus NS1 protein counteracts PKR-mediated inhibition of replication. *J Virol* **74:**6203-6
68. Chen, Z., Li, Y. and Krug, R.M., 1999, Influenza A virus NS1 protein targets poly(A)-binding protein II of the cellular 3'-end processing machinery. *Embo J* **18:**2273-83
69. Fortes, P., Beloso, A. and Ortin, J., 1994, Influenza virus NS1 protein inhibits pre-mRNA splicing and blocks mRNA nucleocytoplasmic transport. *Embo J* **13:**704-12
70. Noah, D.L., Twu, K.Y. and Krug, R.M., 2003, Cellular antiviral responses against influenza A virus are countered at the posttranscriptional level by the viral NS1A protein via its binding to a cellular protein required for the 3' end processing of cellular pre-mRNAS. *Virology* **307:**386-95

71. Lee, T.G., Tomita, J., Hovanessian, A.G. and Katze, M.G., 1990, Purification and partial characterization of a cellular inhibitor of the interferon-induced protein kinase of Mr 68,000 from influenza virus-infected cells. *Proc Natl Acad Sci USA* **87:**6208-12
72. Hagmaier, K., Jennings, S., Buse, J., Weber, F. and Kochs, G., 2003, Novel gene product of Thogoto virus segment 6 codes for an interferon antagonist. *J Virol* **77:**2747-52
73. Thomas, D., Blakqori, G., Wagner, V., Banholzer, M., Kessler, N., Elliott, R.M., Haller, O. and Weber, F., 2004, Inhibition of RNA polymerase II phosphorylation by a viral interferon antagonist. *J Biol Chem* **279:**31471-7
74. Weber, F., Bridgen, A., Fazakerley, J.K., Streitenfeld, H., Kessler, N., Randall, R.E. and Elliott, R.M., 2002, Bunyamwera bunyavirus nonstructural protein NSs counteracts the induction of alpha/beta interferon. *J Virol* **76:**7949-55
75. Billecocq, A., Spiegel, M., Vialat, P., Kohl, A., Weber, F., Bouloy, M. and Haller, O., 2004, NSs protein of rift valley fever virus blocks interferon production by inhibiting host gene transcription. *J Virol* **78:**9798-806
76. Le May, N., Dubaele, S., De Santis, L.P., Billecocq, A., Bouloy, M. and Egly, J.M., 2004, TFIIH transcription factor, a target for the Rift Valley hemorrhagic fever virus. *Cell* **116:**541-50
77. Basler, C.F., Mikulasova, A., Martinez-Sobrido, L., Paragas, J., Muhlberger, E., Bray, M., Klenk, H.D., Palese, P. and Garcia-Sastre, A., 2003, The Ebola virus VP35 protein inhibits activation of interferon regulatory factor 3. *J Virol* **77:**7945-56
78. Basler, C.F., Wang, X., Muhlberger, E., Volchkov, V., Paragas, J., Klenk, H.D., Garcia-Sastre, A. and Palese, P., 2000, The Ebola virus VP35 protein functions as a type I IFN antagonist. *Proc Natl Acad Sci USA* **97:**12289-94
79. Garcin, D., Latorre, P. and Kolakofsky, D., 1999, Sendai virus C proteins counteract the interferon-mediated induction of an antiviral state. *J Virol* **73:**6559-65
80. Garcin, D., Marq, J.B., Goodbourn, S. and Kolakofsky, D., 2003, The amino-terminal extensions of the longer Sendai virus C proteins modulate pY701-Stat1 and bulk Stat1 levels independently of interferon signaling. *J Virol* **77:**2321-9
81. Garcin, D., Marq, J.B., Iseni, F., Martin, S. and Kolakofsky, D., 2004, A short peptide at the amino terminus of the Sendai virus C protein acts as an independent element that induces STAT1 instability. *J Virol* **78:**8799-811
82. Garcin, D., Marq, J.B., Strahle, L., le Mercier, P. and Kolakofsky, D., 2002, All four Sendai Virus C proteins bind Stat1, but only the larger forms also induce its mono-ubiquitination and degradation. *Virology* **295:**256-65
83. Gotoh, B., Takeuchi, K., Komatsu, T. and Yokoo, J., 2003, The STAT2 activation process is a crucial target of Sendai virus C protein for the blockade of alpha interferon signaling. *J Virol* **77:**3360-70
84. Takeuchi, K., Komatsu, T., Yokoo, J., Kato, A., Shioda, T., Nagai, Y. and Gotoh, B., 2001, Sendai virus C protein physically associates with Stat1. *Genes Cells* **6:**545-57
85. Saito, S., Ogino, T., Miyajima, N., Kato, A. and Kohase, M., 2002, Dephosphorylation failure of tyrosine-phosphorylated STAT1 in IFN-stimulated Sendai virus C protein-expressing cells. *Virology* **293:**205-9
86. Komatsu, T., Takeuchi, K., Yokoo, J. and Gotoh, B., 2004, C and V proteins of Sendai virus target signaling pathways leading to IRF-3 activation for the negative regulation of interferon-beta production. *Virology* **325:**137-48
87. Didcock, L., Young, D.F., Goodbourn, S. and Randall, R.E., 1999, The V protein of simian virus 5 inhibits interferon signalling by targeting STAT1 for proteasome-mediated degradation. *J Virol* **73:**9928-33
88. Andrejeva, J., Poole, E., Young, D.F., Goodbourn, S. and Randall, R.E., 2002, The p127 subunit (DDB1) of the UV-DNA damage repair binding protein is essential for the targeted degradation of STAT1 by the V protein of the paramyxovirus simian virus 5. *J Virol* **76:**11379-86

89. Parisien, J.P., Lau, J.F., Rodriguez, J.J., Ulane, C.M. and Horvath, C.M., 2002, Selective STAT protein degradation induced by paramyxoviruses requires both STAT1 and STAT2 but is independent of alpha/beta interferon signal transduction. *J Virol* **76:**4190-8
90. Ulane, C.M. and Horvath, C.M., 2002, Paramyxoviruses SV5 and HPIV2 assemble STAT protein ubiquitin ligase complexes from cellular components. *Virology* **304:**160-6
91. He, B., Paterson, R.G., Stock, N., Durbin, J.E., Durbin, R.K., Goodbourn, S., Randall, R.E. and Lamb, R.A., 2002, Recovery of paramyxovirus simian virus 5 with a V protein lacking the conserved cysteine-rich domain: the multifunctional V protein blocks both interferon-beta induction and interferon signaling. *Virology* **303:**15-32
92. Poole, E., He, B., Lamb, R.A., Randall, R.E. and Goodbourn, S., 2002, The V proteins of simian virus 5 and other paramyxoviruses inhibit induction of interferon-beta. *Virology* **303:**33-46
93. Young, D.F., Didcock, L., Goodbourn, S. and Randall, R.E., 2000, Paramyxoviridae use distinct virus-specific mechanisms to circumvent the interferon response. *Virology* **269:**383-90
94. Nishio, M., Tsurudome, M., Ito, M., Kawano, M., Komada, H. and Ito, Y., 2001, High resistance of human parainfluenza type 2 virus protein-expressing cells to the antiviral and anti-cell proliferative activities of alpha/beta interferons: cysteine-rich V-specific domain is required for high resistance to the interferons. *J Virol* **75:**9165-76
95. Parisien, J.P., Lau, J.F., Rodriguez, J.J., Sullivan, B.M., Moscona, A., Parks, G.D., Lamb, R.A. and Horvath, C.M., 2001, The V protein of human parainfluenza virus 2 antagonizes type I interferon responses by destabilizing signal transducer and activator of transcription 2. *Virology* **283:**230-9
96. Kubota, T., Yokosawa, N., Yokota, S. and Fujii, N., 2001, C terminal CYS-RICH region of mumps virus structural V protein correlates with block of interferon alpha and gamma signal transduction pathway through decrease of STAT 1-alpha. *Biochem Biophys Res Commun* **283:**255-9
97. Yokosawa, N., Yokota, S., Kubota, T. and Fujii, N., 2002, C-terminal region of STAT-1alpha is not necessary for its ubiquitination and degradation caused by mumps virus V protein. *J Virol* **76:**12683-90
98. Palosaari, H., Parisien, J.P., Rodriguez, J.J., Ulane, C.M. and Horvath, C.M., 2003, STAT protein interference and suppression of cytokine signal transduction by measles virus V protein. *J Virol* **77:**7635-44
99. Shaffer, J.A., Bellini, W.J. and Rota, P.A., 2003, The C protein of measles virus inhibits the type I interferon response. *Virology* **315:**389-97
100. Naniche, D., Yeh, A., Eto, D., Manchester, M., Friedman, R.M. and Oldstone, M.B., 2000, Evasion of host defenses by measles virus: wild-type measles virus infection interferes with induction of Alpha/Beta interferon production. *J Virol* **74:**7478-84
101. Park, M.S., Shaw, M.L., Munoz-Jordan, J., Cros, J.F., Nakaya, T., Bouvier, N., Palese, P., Garcia-Sastre, A. and Basler, C.F., 2003, Newcastle disease virus (NDV)-based assay demonstrates interferon antagonist activity for the NDV V protein and the Nipah virus V, W, and C proteins. *J Virol* **77:**1501-11
102. Shaw, M.L., Garcia-Sastre, A., Palese, P. and Basler, C.F., 2004, Nipah virus V and W proteins have a common STAT1-binding domain yet inhibit STAT1 activation from the cytoplasmic and nuclear compartments, respectively. *J Virol* **78:**5633-41
103. Rodriguez, J.J., Cruz, C.D. and Horvath, C.M., 2004, Identification of the nuclear export signal and STAT-binding domains of the Nipah virus V protein reveals mechanisms underlying interferon evasion. *J Virol* **78:**5358-67
104. Rodriguez, J.J., Parisien, J.P. and Horvath, C.M., 2002, Nipah virus V protein evades alpha and gamma interferons by preventing STAT1 and STAT2 activation and nuclear accumulation. *J Virol* **76:**11476-83

105. Rodriguez, J.J., Wang, L.F. and Horvath, C.M., 2003, Hendra virus V protein inhibits interferon signaling by preventing STAT1 and STAT2 nuclear accumulation. *J Virol* **77:**11842-5
106. Bossert, B., Marozin, S. and Conzelmann, K.K., 2003, Nonstructural proteins NS1 and NS2 of bovine respiratory syncytial virus block activation of interferon regulatory factor 3. *J Virol* **77:**8661-8
107. Spann, K.M., Tran, K.C., Chi, B., Rabin, R.L. and Collins, P.L., 2004, Suppression of the induction of alpha, beta, and lambda interferons by the NS1 and NS2 proteins of human respiratory syncytial virus in human epithelial cells and macrophages [corrected]. *J Virol* **78:**4363-9
108. Valarcher, J.F., Furze, J., Wyld, S., Cook, R., Conzelmann, K.K. and Taylor, G., 2003, Role of alpha/beta interferons in the attenuation and immunogenicity of recombinant bovine respiratory syncytial viruses lacking NS proteins. *J Virol* **77:**8426-39
109. Dupuis, S., Jouanguy, E., Al-Hajjar, S., Fieschi, C., Al-Mohsen, I.Z., Al-Jumaah, S., Yang, K., Chapgier, A., Eidenschenk, C., Eid, P., Al Ghonaium, A., Tufenkeji, H., Frayha, H., Al-Gazlan, S., Al-Rayes, H., Schreiber, R.D., Gresser, I. and Casanova, J.L., 2003, Impaired response to interferon-alpha/beta and lethal viral disease in human STAT1 deficiency. *Nat Genet* **33:**388-91
110. Hijikata, M., Mishiro, S., Miyamoto, C., Furuichi, Y., Hashimoto, M. and Ohta, Y., 2001, Genetic polymorphism of the MxA gene promoter and interferon responsiveness of hepatitis C patients: revisited by analyzing two SNP sites (-123 and -88) in vivo and in vitro. *Intervirology* **44:**379-82
111. Yee, L.J., Tang, J., Gibson, A.W., Kimberly, R., Van Leeuwen, D.J. and Kaslow, R.A., 2001, Interleukin 10 polymorphisms as predictors of sustained response in antiviral therapy for chronic hepatitis C infection. *Hepatology* **33:**708-12
112. Knapp, S., Hennig, B.J., Frodsham, A.J., Zhang, L., Hellier, S., Wright, M., Goldin, R., Hill, A.V., Thomas, H.C. and Thursz, M.R., 2003, Interleukin-10 promoter polymorphisms and the outcome of hepatitis C virus infection. *Immunogenetics* **55:**362-9
113. Knapp, S., Yee, L.J., Frodsham, A.J., Hennig, B.J., Hellier, S., Zhang, L., Wright, M., Chiaramonte, M., Graves, M., Thomas, H.C., Hill, A.V. and Thursz, M.R., 2003, Polymorphisms in interferon-induced genes and the outcome of hepatitis C virus infection: roles of MxA, OAS-1 and PKR. *Genes Immun* **4:**411-9
114. Choi, E.H., Lee, H.J., Yoo, T. and Chanock, S.J., 2002, A common haplotype of interleukin-4 gene IL4 is associated with severe respiratory syncytial virus disease in Korean children. *J Infect Dis* **186:**1207-11
115. Hoebee, B., Rietveld, E., Bont, L., Oosten, M., Hodemaekers, H.M., Nagelkerke, N.J., Neijens, H.J., Kimpen, J.L. and Kimman, T.G., 2003, Association of severe respiratory syncytial virus bronchiolitis with interleukin-4 and interleukin-4 receptor alpha polymorphisms. *J Infect Dis* **187:**2-11
116. Honda, K., Sakaguchi, S., Nakajima, C., Watanabe, A., Yanai, H., Matsumoto, M., Ohteki, T., Kaisho, T., Takaoka, A., Akira, S., Seya, T. and Taniguchi, T., 2003, Selective contribution of IFN-alpha/beta signaling to the maturation of dendritic cells induced by double-stranded RNA or viral infection. *Proc Natl Acad Sci USA* **100:**10872-7
117. Edelmann, K.H., Richardson-Burns, S., Alexopoulou, L., Tyler, K.L., Flavell, R.A. and Oldstone, M.B., 2004, Does Toll-like receptor 3 play a biological role in virus infections? *Virology* **322:**231-8
118. Perry, A.K., Chow, E.K., Goodnough, J.B., Yeh, W.C. and Cheng, G., 2004, Differential requirement for TANK-binding kinase-1 in type I interferon responses to toll-like receptor activation and viral infection. *J Exp Med* **199:**1651-8
119. Santini, S.M., Di Pucchio, T., Lapenta, C., Parlato, S., Logozzi, M. and Belardelli, F., 2002, The natural alliance between type I interferon and dendritic cells and its role in linking innate and adaptive immunity. *J Interferon Cytokine Res* **22:**1071-80

120. de Veer, M.J., Holko, M., Frevel, M., Walker, E., Der, S., Paranjape, J.M., Silverman, R.H. and Williams, B.R., 2001, Functional classification of interferon-stimulated genes identified using microarrays. *J Leukoc Biol* **69:**912-20
121. Geiss, G.K., Carter, V.S., He, Y., Kwieciszewski, B.K., Holzman, T., Korth, M.J., Lazaro, C.A., Fausto, N., Bumgarner, R.E. and Katze, M.G., 2003, Gene expression profiling of the cellular transcriptional network regulated by alpha/beta interferon and its partial attenuation by the hepatitis C virus nonstructural 5A protein. *J Virol* **77:**6367-75
122. Pestka, S., 2000, The human interferon alpha species and receptors. *Biopolymers* **55:**254-87
123. Leong, S.R., Chang, J.C., Ong, R., Dawes, G., Stemmer, W.P. and Punnonen, J., 2003, Optimized expression and specific activity of IL-12 by directed molecular evolution. *Proc Natl Acad Sci USA* **100:**1163-8
124. Thompson, K.A., Strayer, D.R., Salvato, P.D., Thompson, C.E., Klimas, N., Molavi, A., Hamill, A.K., Zheng, Z., Ventura, D. and Carter, W.A., 1996, Results of a double-blind placebo-controlled study of the double-stranded RNA drug polyI:polyC12U in the treatment of HIV infection. *Eur J Clin Microbiol Infect Dis* **15:**580-7
125. Morrey, J.D., Day, C.W., Julander, J.G., Blatt, L.M., Smee, D.F. and Sidwell, R.W., 2004, Effect of interferon-alpha and interferon-inducers on West Nile virus in mouse and hamster animal models. *Antivir Chem Chemother* **15:**101-9
126. Padalko, E., Nuyens, D., De Palma, A., Verbeken, E., Aerts, J.L., De Clercq, E., Carmeliet, P. and Neyts, J., 2004, The interferon inducer ampligen [poly(I)-poly(C12U)] markedly protects mice against coxsackie B3 virus-induced myocarditis. *Antimicrob Agents Chemother* **48:**267-74
127. Talon, J., Salvatore, M., O'Neill, R.E., Nakaya, Y., Zheng, H., Muster, T., Garcia-Sastre, A. and Palese, P., 2000, Influenza A and B viruses expressing altered NS1 proteins: A vaccine approach. *Proc Natl Acad Sci USA* **97:**4309-14
128. Wang, X., Basler, C.F., Williams, B.R., Silverman, R.H., Palese, P. and Garcia-Sastre, A., 2002, Functional replacement of the carboxy-terminal two-thirds of the influenza A virus NS1 protein with short heterologous dimerization domains. *J Virol* **76:**12951-62

Chapter 2

HOW VIRUSES ELICIT INTERFERON PRODUCTION

Triggering the innate immune response to viral infection

DAVID E. LEVY and ISABELLE J. MARIÉ
Molecular Oncology and Immunology Program, Departments of Pathology and Microbiology and the NYU Cancer Institute, New York University School of Medicine, 550 First Avenue, New York, NY, USA

1. INTRODUCTION

The type I interferons (IFNα and IFNβ) were first characterized as cytokines capable of inducing an antiviral state in sensitive target cells[1]. They were discovered as a substance produced by virus-infected cells that was capable of conferring protection of uninfected naïve cells from subsequent infection. Since their discovery and characterization as founding members of the type II cytokine family, investigations into IFN biology have served not only to elucidate their potent antiviral properties, but also as a system for understanding molecular mechanisms of gene expression control. IFN genes are stringently and acutely regulated, being expressed only in infected cells and only transiently following infection. Their immediate biological actions are mediated by a set of acutely regulated cellular target genes that are also stringently regulated. The aim of this review is to discuss some of the recent discoveries and controversies surrounding the signaling pathways and gene expression control mechanisms that regulate this important innate immune system.

IFNα and IFNβ were originally classified as leukocyte IFN and fibroblast IFN, respectively, to designate their distinct presumptive cellular origins. This designation has been replaced by a more precise nomenclature,

P. Palese (ed.), Modulation of Host Gene Expression and Innate Immunity by Viruses, 19-34.

based on molecular characterization following the isolation, cloning, and sequencing of the IFN multigene family. The originally detected IFNα activity is encoded by a multigene family of closely related and clustered genes, while IFNβ is encoded by a single, somewhat more distantly related gene, but still retained in the IFN gene cluster on chromosome 9 in humans and the syntenic region on chromosome 4 in the mouse. It has also become clear that despite their original designations, both the IFNα family and IFNβ can be synthesized by many, if not all, nucleated cells following viral infection, just as virtually all nucleated cells have the capacity to respond to secreted IFN to induce an antiviral state. This ability of most cells to secrete and respond to IFN makes the IFN system a powerful first line of defense against pathogens and an essential component of innate antiviral immunity[2,3].

2. TRANSCRIPTIONAL CONTROL OF IFN GENE EXPRESSION

2.1 Regulation of the IFNβ gene

One of the molecular hallmarks of the IFN system is the precision with which the expression of these genes is regulated[4]. Basal IFN is produced by most cells at extremely low or virtually undetectable levels, probably reflecting a biological need to keep this powerful compound in check. However, viral infection rapidly leads to extremely high levels of expression, producing relatively abundant levels of IFN mRNA and secreted protein. Induction of IFNβ was the first to be examined at a molecular level and remains the most intensely studied[5]. Interestingly, its transcriptional induction relies on a relatively short *cis* regulatory element in its promoter that serves as a binding site for three distinct transcription factor complexes, NF-κB, ATF2/c-jun, and IFN regulatory factors (IRFs). Each of these factors is activated by serine phosphorylation in virus-infected cells, either directly or through phosphorylation of associated inhibitory proteins, as in the case of NF-κB.

NF-κB is activated by phosphorylation-dependent degradation of its inhibitor, I-κB, releasing the active NF-κB protein for complex formation and nuclear translocation. In addition, the activity of NF-κB can be stimulated by phosphorylation of its transactivation domain[6]. The transactivation potency of the ATF2/c-jun complex is increased through phosphorylation by c-Jun kinase, leading to increased activity of this preformed heterodimeric transcription factor. Finally, at least some members of the IRF family are activated by direct phosphorylation on a regulatory domain, leading to dimerization, nuclear accumulation, DNA binding, and increased transcriptional potency. Due to this phosphorylation-dependent process of transcription factor activation, kinase activation is the key initial biochemical event that is triggered by viral infection.

The IFNβ promoter is controlled by an enhancer element that contains binding sites for all three of these transcription factors, and their cooperative interaction and concerted recruitment of coactivator proteins is necessary for efficient transcriptional induction of IFNβ gene expression. The IFNβ enhancer is located immediately upstream of the promoter and contains an ATF2/c-jun binding site, two tandem binding sites for IRF proteins and a NFκB binding site. There are also binding sites for negative-acting proteins that are presumably displaced by the binding of activators, including the IRF2 protein that competes for binding at the IRF sites with activating IRF family members. The simultaneous binding of activating protein complexes at the individual sites of the IFNβ enhancer creates a multimeric structure, which together with recruited coactivator proteins has been named an enhanceosome[5]. This structure is additionally stabilized by induced DNA bending, at least in part through the action of non-histone HMG chromatin proteins[7].

While it has been well established that an AP-1 complex composed of ATF2/c-jun dimers and a NFκB complex composed of p50/p65 dimers contribute to the IFNβ enhanceosome, the identity of the IRF protein(s) contributing to this activity has been less clear. The first IRF protein to be implicated in IFNβ gene regulation was IRF1[8]. IRF1 is capable of binding the IFNβ enhancer and will induce IFN gene transcription when over-expressed[9] or when assayed in vitro[10]. However, IRF1 is only minimally expressed in untreated cells due to its detrimental effect on cell growth[11], and it can be readily detected only following induction in response to a variety of stimuli[12]. IFNβ gene expression on the other hand, can be readily induced in the absence of cellular protein synthesis, suggesting that all necessary factors pre-exist in the cell. IFNβ gene expression in response to viral infection remains intact in mice deleted for the IRF1 gene[13], providing formal genetic proof that IRF1 is not essential. On the other hand, several other IRF family members have been implicated in IFN gene transcription, and gene ablation studies in mice have clearly demonstrated the importance of IRF3 and IRF7 in this process[14].

The structure of the enhanceosome appears to contribute several unique attributes. First, simultaneous interaction of multiple independent transcription factors enhances the subsequent recruitment of coactivator proteins, since several coactivators interact with more than one enhancer-binding protein complex. For instance, the transactivation domains of c-jun, NFκB, and IRF3 are all capable of recruiting the coactivator CBP, allowing more efficient coactivator recruitment through the concerted effort of the linked transcription factors. Moreover, the assembly of the three transcription factors into a unit structure appears to create a novel composite protein interaction surface that more efficiently recruits CBP than the combined effect of the individual transactivation domains[10]. Thus, the activity of the assembled complex is greater than the sum of its parts.

Another attribute of the enhanceosome structure is derived from DNA bending. The physical structure of the DNA at the IFNβ promoter becomes altered following binding of the transcription factors. This bent structure favors cooperative binding, thereby increasing the efficiency of enhanceosome assembly[15]. It is likely that DNA bending also facilitates recruitment of the basal transcription machinery by juxtaposing the enhanceosome and its associated coactivator proteins with transcriptional initiator elements. Finally, enhanceosome assembly occurs in a largely nucleosome-free region of DNA, causing a shift in phased nucleosomes. This change in nucleosome positioning results in movement of a fixed nucleosome that normally occludes the TATA box of the promoter and therefore inhibiting transcription. After enhanceosome-dependent nucleosome sliding, which involves recruitment of chromatin remodeling complexes, the TATA box becomes available for binding by TBP and nucleation of the transcriptional preinitiation complex[16].

2.2 Complex regulation of IFNα genes

The IFNα genes are also transcriptionally induced in response to viral infection[17], and similar to IFNβ, they require serine-phosphorylated IRF proteins for their expression[14]. In some respects, IFNα gene regulation appears less complex than activation of IFNβ gene expression. The only well-characterized enhancer elements controlling IFNα genes contain binding sites for IRF proteins, although roles for negative-acting factors have also been described[18]. Regulation of these genes appears to be entirely dependent on IRF proteins, with no contribution from AP1 or NFκB complexes. Negative regulation of these genes also appears to operate largely through inhibition of IRF proteins. Binding competition between the repressive IRF2 and activating members of the family maintains the very low levels of basal IFN observed in the absence of viral infection[8]. The homeobox repressor Pitx1 also appears to contribute to IFNα gene silencing through its ability to interact with and inhibit IRF3 and IRF7 proteins[19]. It is not currently known whether IFNα gene transcription involves formation of an enhanceosome structure or requires DNA bending or nucleosome repositioning.

In spite of the superficial simplicity of a gene regulatory scheme involving binding of a single transcriptional activator type at a positive regulatory element, there is an aspect of complexity within the IFNα family that is not seen with IFNβ. Recently, it was found that IFNβ and the multigene family of IFNα proteins are not uniformly regulated during viral infection[20], and that their differential expression is at least partially regulated through a positive-feedback loop involving induction of IRF proteins[20, 21]. IFNβ and the IFNα4 isotype of mouse IFNα are induced with immediate-early kinetics through the action of the constitutively expressed IRF3 protein. However, the enhancers of other members of the IFNα gene family cannot bind IRF3, and are instead activated only by IRF7[22]. IRF7, unlike IRF3, is not constitutively expressed in most cell types, but rather its expression is induced by IFN signaling through the Jak-Stat pathway. Thus, in response to early secretion of IFNβ and IFNα4 through the action of IRF3, induction of IRF7 makes cells sensitized for induction of additional IFNα subtypes (referred to as the non-IFNα4 subset), leading to robust production of multiple IFNα species and potent antiviral activity. Significantly, the robust expression of the complete complement of IFNα genes occurs only in IFN responsive cells, due to the requirement for IRF7 induction in response to IFN signaling[23].

A similar pattern of differential regulation of IFNα genes through the positive feedback induction of IRF7 occurs for human IFN genes[24]. In addition, recent studies have suggested a further layer of complexity for IFNα gene regulation. Just as the presence and activation of IRF3 or IRF7 program induction of distinct subtypes of IFNα genes, other IRF family members may also target specific isotypes of IFNα. In particular, it was found that IRF5 can participate in IFNα gene induction by certain viruses, with NDV leading to the preferential induction of the human IFNA8 gene, when IRF5 was present[25]. One might speculate whether virus-specific induction of particular IFNα isotypes is related to unique biological functions for individual members of this multigene family.

3. THE VIRUS-ACTIVATED SWITCH

A unique attribute of IFN gene regulation is regulation in response to virus infection. Whether gene induction occurs through the complex assembly of an enhanceosome at the IFNβ promoter with subsequent chromatin alterations or through the sequential action of IRF3 and feedback-induced IRF7, gene expression is strongly induced in virus-infected cells. The controlling event that confines IFN gene expression to virus-infected cells is the regulated activity of the transcription factors required for gene induction, and serine phosphorylation is the key regulatory event for transcription factor activation. Therefore, the problem of understanding gene regulation is reduced to the question of how phosphorylation is regulated.

3.1 Virus-activated kinases

IFNβ gene induction requires phosphorylation-dependent activation of AP-1, NFκB, and IRF proteins while IFNα gene induction appears to require only activation of IRF proteins. Since AP-1 and NFκB activation have been recently reviewed[26, 27] and IRF activation is the common event for IFNα and IFNβ gene induction, we will largely confine our discussion to the mechanisms of IRF activation by virus infection.

Since the discovery of phosphorylated IRF3 in virus-infected cells[28], it has been clear that the key to understanding virus-induced gene expression was identification of virus-activated IRF kinases and understanding their mechanism of activation. Recently, it was found that two serine kinases of the IKK family, TBK1 (also known as T2K and NAP) and IKKε (also known as IKKi) were able to phosphorylate IRF3 and IRF7 in virus-infected cells[29,30]. Since their original discovery based on similarity to classical IKKα and β and their potential involvement in LPS or TNF signaling[31,32], it has become clear that they are critically involved in virus-dependent signaling. Interestingly, however, their involvement as activators of NFκB or in response to LPS or TNF remains unclear. For instance, TBK1 and IKKε will phosphorylate IRF3 and IRF7 on appropriate residues leading to activation, but they appear unable to phosphorylate IκB in a manner to cause its degradation by the proteosome, since they directly phosphorylate only one of the two critical serine residues[32]. However, in association with other proteins, they may also have an important role in NFκB activation. Significantly, gene-targeted deletion of TBK1 results in an embryonic lethal phenotype closely resembling loss of NFκB p65 or IKKβ[33], suggesting that loss of TBK1 results in impaired NFκB activation. Moreover, this embryonic lethality can be suppressed by deletion of the gene for TNFα or TNFR1[34, 35], strongly suggesting that TBK1 is essential for NFκB, at least under some circumstances.

In contrast to the somewhat ambiguous role of TBK1 in activation of NFκB, its role and that of IKKε in activation of IRF3 is well documented. These kinases cause the phosphorylation of IRF3 when over-expressed in mammalian cells, and RNAi-mediated knockdown impairs the phosphorylation of IRF in virus-infected cells[29,30]. Moreover, cells from TBK1 knockout mice are impaired for IRF3 phosphorylation[36]. Due to the partial redundancy between TBK1 and IKKε, complete loss of IRF3 phosphorylation in response to virus infection is only seen when both kinases are eliminated[37].

3.2 Virus-induced signals

The identification of the role of TBK1 and IKKε in IRF3 phosphorylation explained some of the proximal events in IFN gene induction, but it left open the question of what virus signal caused kinase activation. The viral signals important for kinase activation remain incompletely defined, although it is clear that in many cases viral replication is critical, especially in the case of negative-sense RNA viruses[23]. Existing evidence suggests that virus-encoded double-stranded RNA (dsRNA) provides one of the inducing signals for IFN induction, but additional virus-encoded components also appear to be necessary[38-41]. Current evidence favors a ribonucleoprotein (RNP) complex as a likely kinase inducer formed in virus-infected cells[42]. For viruses capable of inducing IFN in the absence of replication, the presumption is that during infection they deliver a payload of pre-formed inducing molecules to the cell upon infection.

One mechanism by which viral RNP might lead to kinase activation is through its RNA component. Considerable evidence suggests that RNA, in particular double-stranded RNA (dsRNA) which is a common viral replication intermediate, is capable of inducing IFN gene expression. For instance, treatment of mammalian cells in culture or injecting mice with synthetic dsRNA causes robust IFN induction and has led to the development of a variety of artificial IFN inducers of potential therapeutic utility. The first potential cellular sensor of dsRNA to be identified was the dsRNA-dependent protein kinase, PKR, whose catalytic activity is stimulated by binding dsRNA[43]. PKR phosphorylates the translation initiation factor eIF2α, leading to inhibition of cellular protein synthesis, an essential component of innate antiviral resistance. It has also been suggested that PKR stimulated by dsRNA leads to the activation of NFκB, although the mechanism of this activation is obscure[44]. While PKR is capable of phosphorylating IκB in vitro[45], it is not clear that this phosphorylation occurs on residues not required for proteosomal degradation. It is also unclear whether the ability of PKR to augment NFκB activation requires its catalytic activity[44]. Current evidence would favor some kind of non-catalytic adapter role for PKR in activation of downstream signaling events important in host defense[46]. While it is clear that PKR is required for IFN production in response to purified dsRNA, gene targeting experiments in mice showed that it was superfluous for IFN responses to viral infection[47-49]. These genetic experiments distinguish dsRNA-dependent activation of IFN through a PKR pathway from virus-induced signaling that is independent of the catalytic function of PKR.

TLR3 is another cellular sensor that recognizes dsRNA, and like PKR, it is capable of stimulating IFN production in response to purified ligand[50]. Again, however, gene targeting experiments showed that TLR3 was also not required for IFN production in virally-infected cells[51], although it is important for responses to extracellular dsRNA. It is a reasonable hypothesis that TLR3 plays a critical role in the cellular response to extracellular dsRNA, possibly generated and released during a lytic viral infection following death of initially infected cells.

TLR7 is another cellular sensor capable of inducing IFN production. It was originally found to be stimulated by artificial IFN inducers[52], but has recently been found to be activated by single-stranded RNA[53], leading to the suggestion that it plays a role in IFN production in response to negative-strand RNA viral infections[54,55]. However, TLR7 is unlikely to be a universal mediator of IFN production in virus-infected cells. TLR7 is expressed mainly by dendritic cells, and many cells capable of producing IFN following viral infection lack TLR7 expression. Moreover, TLR7, like all TLR proteins, is expressed in cells with its amino-terminal ligand-binding domain exposed to the extracellular or luminal environment. RNA viruses, in contrast, replicate in the cell cytoplasm and do not expose single-stranded or double-stranded RNA intermediates to the cell exterior prior to cell lysis. Therefore, like TLR3, TLR7 may play its major role in detecting virus at post-replication times following cell lysis.

Another inconsistency in the idea of RNA sensors such as PKR or TLR proteins as primary viral sensors for IFN production is the question of whether dsRNA *per se* ever exists inside cells during virus infection, even if TLR proteins were there to detect it. Viruses wrap their nucleic acid into protein particles; even during replication, the template for viral RNA-dependent RNA polymerases is probably an RNP complex that severely restricts the accumulation of naked dsRNA. These viral nucleocapsid structures, while probably being poor ligands for PKR or TLR proteins, are potent inducers of IFN production[56]. It is likely that PKR is an essential mediator of the antiviral action of IFN by inhibiting viral protein synthesis and possibly inducing cellular apoptosis[57], and TLR7 and TLR9 are important for the systemic response to viral infection. However, these proteins do not appear to explain the initial, cell-autonomous recognition of viral infection that induces the first wave of IFN production.

3.3 IFN induction through an RNA helicase switch

Recently, Fujita and colleagues[58] reported that the RNA helicase RIG-I is required for IFN production in response to the parainfluenza virus, Newcastle disease virus (NDV). NDV enters cells by membrane fusion at the cell surface, injecting its nucleocapsids into the cell cytoplasm, and inducing IFN within hours of entering cells, prior to the appearance of significant viral progeny or cellular cytopathic effect. IFN production in response to NDV infection is dependent on viral RNA and protein synthesis[38,39], presumably due to a requirement for viral replication intermediates for signaling. RIG-I is an unusual helicase in that, in addition to a catalytic domain, it possesses protein interaction CARD modules. Both the ATP-dependent helicase activity as well as the CARD domains are required to activate the transcription factor IRF3 during viral infection. In fact, the CARD domains when expressed alone activate IRF3 phosphorylation, NF-κB release, and IFN production, suggesting that they have a direct role in downstream signaling. CARD domains, like other structurally related protein interaction modules found in innate immune adaptor molecules, such as death domains, TIR domains, and death effector domains, promote the creation of platforms for nucleating signaling events. CARD domains mediate homotypic interactions and were first characterized in the network of interactions that recruit caspases to receptor complexes that signal apoptosis[59]. More recently, they have been implicated in signaling from both TLRs and the intracellular bacterial sensors, Nod1 and Nod2[60]. The CARD-domain-containing kinase RIP2 has also been implicated in cytokine production, particularly downstream of TLR2, 3, and 4. However, RIG-I is the first CARD-domain signaling molecule directly implicated in IFN production in virus-infected cells.

A model for how RIG-I acts as a sensor for cytoplasmic virus replication posits that it acts as an RNA-activated switch. Silent in the absence of viral nucleic acid, presumably due to some kind of closed configuration that hides the CARD domains from further interactions, a conformational shift caused by the activity of the helicase domain would promote additional protein-protein interactions. This switch requires both RNA binding as well as helicase catalytic function, suggesting that active unwinding of a viral RNA is key to signaling. Expression of the isolated CARD modules, on the other hand, signal constitutively, suggesting that release of the CARD domain from inhibitory intramolecular interactions makes it available for activating interactions with downstream signaling components.

Given the common action of CARD domains as homotypic interaction surfaces, it is likely that the direct downstream target of RIG-I is another CARD-containing adaptor protein. A likely target of such a postulated adaptor is a kinase complex capable of phosphorylating IRF3, although whether this would be a direct interaction or indirectly through one or more intermediates has not been determined. Given the current evidence that TBK1 and IKK-ε are the direct IRF3 kinases[29,30], it is assumed that additional intermediates are required to transmit the RIG-I since neither TBK1 or IKKε has an identifiable CARD domain. Therefore, the catalytic process of unwinding a viral RNP converts RIG-I into a signaling intermediate capable of interacting with and activating yet to be defined downstream components that serve as a platform for recruitment and activation of the IRF3 kinase complex.

An adaptor protein that has been implicated in activation of TBK1 and IKK-ε is the TIR domain protein, TRIF[61,62]. TRIF acts downstream of TLR3 and 4, both of which can induce IFN synthesis in response to their respective ligands, and it has been implicated in some responses to viral infection as well[63]. Although TRIF activates IFN production through TBK1[36], it does not appear to be an intermediary for RIG-I function. TRIF is not required for RIG-I-mediated responses to viral infection, even though it is required for responses to extracellular dsRNA[58]. Therefore, RIG-I responses can be clearly distinguished from IRF3 activation through a TRIF-dependent pathway involving TLR3 stimulated by dsRNA.

3.4 Additional mechanisms of IFN induction

IFN induction has been most intensely studied in response to negative-strand RNA viruses, which are excellent inducers of type I IFN. However, other signals induce the production of IFN, particularly in specialized dendritic cells known as plasmacytoid dendritic cells or IFN producing cells[64]. Dendritic cells and macrophages express abundant TLR proteins, making them responsive to a variety of pathogen-encoded signals, many of which induce IFN. For example, bacterial infection or stimulation of cells with LPS leads to activation of TLR4, which activates gene expression through a MyD88-dependent pathway involving the adapters MAL and TRAM, leading to activation of IRAK, TRAF6, TAK1, and IKKβ-mediated release of NFκB. A MyD88-independent pathway involving the adapter TRIF leads to the activation of TBK1 and IKKε and phosphorylation of IRF3 and IRF7. TLR7, 8, and 9, however, cannot activate the MyD88-independent signaling pathway, since they do not associate with TRIF, even though they are efficient inducers of IFN production[65]. Since herpes simplex virus induces IFN production, at least in dendritic cells, through a TLR9-dependent pathway[66-68], it is critical to understand how TLR9 links to IFN signaling.

Recently, Kawai et al. provided a mechanism for IFN induction by TLR9 ligands[34]. They found that MyD88 and TRAF6, known mediators of TLR9 signaling, can directly associate with IRF7 and stimulate its activity, leading to efficient IFNα gene induction. Interestingly, IFNβ was induced inefficiently by this mechanism, consistent with the preference of IRF7 for IFNα promoters. An interesting aspect of this novel mechanism is that the ubiquitinylation activity of TRAF6 is required, leading to the direct ubiquitinylation of IRF7 in a region that normally inhibits transcriptional activity. Whereas activation of IRF7 during negative-strand virus infection involves derepression of this inhibitory domain through carboxyl-terminal phosphorylation[69], it appears that TLR9-induced activation involves a different mechanism. However, IRF7 appears to be phosphorylated as well as ubiquitinylated in response to TLR9, so the two mechanisms may not be as fundamentally distinct as they initially appear. Although Kawai et al. suggest that neither TBK1 or IKKε is the IRF7 kinase in response to MyD88 and TRAF6, they draw this conclusion from the analysis of single mutants of each kinase. It will obviously be important to examine the consequences of simultaneous inactivation of both kinases since, unlike fibroblasts, TLR9-expressing dendritic cells express abundant amounts of both proteins[35].

4. CONCLUSIONS

Studies of IFN gene regulation in virus-infected cells have contributed significant information to our overall understanding of transcriptional control mechanisms in vertebrate cells. Enhanceosome theory was largely established by the analysis of IFNβ gene expression, and similarly important concepts for differential regulation of closely related genes will likely emerge from future studies of the IFNα locus. This system has also contributed to the understanding of mammalian signaling systems, with the identification of a variety of pathways that impact on gene induction through novel mechanisms. Signal transduction dependent on the action of RNA helicases and on ubiquitin-directed transcription factor phosphorylation are just some of the novel concepts emerging from the study of IFN gene induction.

Analysis of IFN gene expression has also contributed to the understanding of the structure of the innate arm of the immune system and its interface with adaptive immunity. IFN-producing dendritic cells represent an important component of the early response to pathogens. A more precise molecular understanding of the unique mechanisms governing gene transcription in this cell type will undoubtedly contribute additional novel concepts to our picture of mammalian gene expression.

ACKNOWLEDGEMENTS

Helpful discussion with members of the laboratory and with colleagues is gratefully acknowledged. Work in the authors' laboratory was funded by the National Institutes of Health and the American Heart Association.

REFERENCES

1. Pestka S, Langer JA, Zoon KC, and Samuel CE, Interferons and their actions. *Ann Rev Biochem* (1987) **56:**727-777
2. Stark GR, Kerr IM, Williams BR, Silverman RH, and Schreiber RD, How cells respond to interferons. *Annu. Rev. Biochem.* (1998) **67:**227-264
3. Levy DE and García-Sastre A, The virus battles: IFN induction of the antiviral state and mechanisms of viral evasion. *Cytokine Growth Factor Rev.* (2001) **12:**143-156
4. Maniatis T, Goodbourn S, and Fischer JA, Regulation of inducible and tissue-specific gene expression. *Science* (1987) **236:**1237-1245
5. Merika M and Thanos D, Enhanceosomes. Curr. Opin. *Genet. Dev.* (2001) **11:**205-208
6. Hayden MS and Ghosh S, Signaling to NF-κB. *Genes Dev.* (2004) **18:**2195-2224
7. Thanos D and Maniatis T, Virus induction of human IFN beta gene expression requires the assembly of an enhanceosome. *Cell* (1995) **83:**1091-1100
8. Harada H, Fujita T, Miyamoto M, Kimura Y, Maruyama M, Furia A, Miyata T, and Taniguchi T, Structurally similar but functionally distinct factors, IRF-1 and IRF-2, bind to the same regulatory elements of IFN and IFN-inducible genes. *Cell* (1989) **58:**729-739
9. Fujita T, Kimura Y, Miyamoto M, Barsoumian EL, and Taniguchi T, Induction of endogenous IFN-alpha and IFN-beta genes by a regulatory transcription factor, IRF-1. *Nature* (1989) **337:**270-272
10. Merika M, Williams AJ, Chen G, Collins T, and Thanos D, Recruitment of CBP/p300 by the IFN beta enhanceosome is required for synergistic activation of transcription. *Mol. Cell* (1998) **1:**277-287
11. Pine R, Constitutive expression of an ISGF2/IRF1 transgene leads to interferon-independent activation of interferon-inducible genes and resistance to virus infection. *J. Virol.* (1992) **66:**4470-4478
12. Taniguchi T, Lamphier MS, and Tanaka N, IRF-1: the transcription factor linking the interferon response and oncogenesis. *Biochim. Biophys. Acta* (1997) **1333:**M9-17
13. Ruffner H, Reis LF, Naf D, and Weissmann C, Induction of type I interferon genes and interferon-inducible genes in embryonal stem cells devoid of interferon regulatory factor 1. *Proc. Natl. Acad. Sci. USA* (1993) **90:**11503-11507
14. Sato M, Suemori H, Hata N, Asagiri M, Ogasawara K, Nakao K, Nakaya T, Katsuki M, Noguchi S, Tanaka N, and Taniguchi T, Distinct and essential roles of transcription factors IRF-3 and IRF-7 in response to viruses for IFN-α/β gene induction. *Immunity* (2000) **13:**539-548
15. Falvo JV, Thanos D, and Maniatis T, Reversal of intrinsic DNA bends in the IFN beta gene enhancer by transcription factors and the architectural protein HMG I(Y). *Cell* (1995) **83:**1101-1111
16. Lomvardas S and Thanos D, Nucleosome sliding via TBP DNA binding in vivo. *Cell* (2001) **106:**685-696
17. Ryals J, Dierks P, Ragg H, and Weissmann C, A 46-nucleotide promoter segment from an IFN-alpha gene renders an unrelated promoter inducible by virus. *Cell* (1985) **41:**497-507

18. Lopez S, Reeves R, Island ML, Bandu MT, Christeff N, Doly J, and Navarro S, Silencer activity in the interferon-A gene promoters. *J. Biol. Chem.* (1997) **272:**22788-22799
19. Island ML, Mesplede T, Darracq N, Bandu MT, Christeff N, Djian P, Drouin J, and Navarro S, Repression by homeoprotein pitx1 of virus-induced interferon a promoters is mediated by physical interaction and trans repression of IRF3 and IRF7. *Mol. Cell. Biol.* (2002) **22:**7120-7133
20. Marié I, Durbin JE, and Levy DE, Differential viral induction of distinct interferon-alpha genes by positive feedback through interferon regulatory factor-7. *EMBO J.* (1998) **17:**6660-6669
21. Sato M, Hata N, Asagiri M, Nakaya T, Taniguchi T, and Tanaka N, Positive feedback regulation of type I IFN genes by the IFN-inducible transcription factor IRF-7. *FEBS Lett* (1998) **441:**106-110
22. Lin R, Genin P, Mamane Y, and Hiscott J, Selective DNA binding and association with the CREB binding protein coactivator contribute to differential activation of Alpha/Beta interferon genes by interferon regulatory factors 3 and 7. *Mol. Cell. Biol.* (2000) **20:**6342-6353
23. Levy DE, Marié I, Smith E, and Prakash A, Enhancement and Diversification of IFN Induction by IRF7-Mediated Positive Feedback. *J. Interferon Cytokine Res.* (2002) **22:**87-93
24. Yeow WS, Au WC, Juang YT, Fields CD, Dent CL, Gewert DR, and Pitha PM, Reconstitution of virus-mediated expression of interferon alpha genes in human fibroblast cells by ectopic interferon regulatory factor-7. *J. Biol. Chem.* (2000) **275:**6313-6320
25. Barnes BJ, Moore PA, and Pitha PM, Virus-specific activation of a novel interferon regulatory factor, IRF-5, results in the induction of distinct interferon alpha genes. *J. Biol. Chem.* (2001) **276:**23382-23390
26. Weston CR and Davis RJ, The JNK signal transduction pathway. *Curr. Opin. Genet. Dev.* (2002) **12:**14-21
27. Zhang D, Zhang G, Hayden MS, Greenblatt MB, Bussey C, Flavell RA, and Ghosh S, A toll-like receptor that prevents infection by uropathogenic bacteria. *Science* (2004) **303:**1522-1526
28. Yoneyama M, Suhara W, Fukuhara Y, Fukuda M, Nishida E, and Fujita T, Direct triggering of the type I interferon system by virus infection: activation of a transcription factor complex containing IRF-3 and CBP/p300. *EMBO J.* (1998) **17:**1087-1095
29. Fitzgerald KA, McWhirter SM, Faia KL, Rowe DC, Latz E, Golenbock DT, Coyle AJ, Liao SM, and Maniatis T, IKKepsilon and TBK1 are essential components of the IRF3 signaling pathway. *Nat. Immunol.* (2003) **4:**491-496
30. Sharma S, tenOever BR, Grandvaux N, Zhou GP, Lin R, and Hiscott J, Triggering the interferon antiviral response through an IKK-related pathway. *Science* (2003) **300:**1148-1151
31. Pomerantz JL and Baltimore D, NF-kappaB activation by a signaling complex containing TRAF2, TANK and TBK1, a novel IKK-related kinase. *EMBO J.* (1999) **18:**6694-6704.
32. Peters RT, Liao SM, and Maniatis T, IKKepsilon is part of a novel PMA-inducible IkappaB kinase complex. *Mol. Cell* (2000) **5:**513-522
33. Bonnard M, Mirtsos C, Suzuki S, Graham K, Huang J, Ng M, Itie A, Wakeham A, Shahinian A, Henzel WJ, Elia AJ, Shillinglaw W, Mak TW, Cao Z, and Yeh WC, Deficiency of T2K leads to apoptotic liver degeneration and impaired NF-kappaB-dependent gene transcription. *EMBO J.* (2000) **19:**4976-4985
34. Kawai T, Sato S, Ishii KJ, Coban C, Hemmi H, Yamamoto M, Terai K, Matsuda M, Inoue JI, Uematsu S, Takeuchi O, and Akira S, Interferon-alpha induction through Toll-like receptors involves a direct interaction of IRF7 with MyD88 and TRAF6. *Nat. Immunol.* (2004) Advance Online Publication: DOI:10.1038/ni1118

35. Perry AK, Chow EK, Goodnough JB, Yeh WC, and Cheng G, Differential requirement for TANK-binding kinase-1 in type I interferon responses to toll-like receptor activation and viral infection. *J. Exp. Med.* (2004) **199:**1651-1658
36. McWhirter SM, Fitzgerald KA, Rosains J, Rowe DC, Golenbock DT, and Maniatis T, IFN-regulatory factor 3-dependent gene expression is defective in Tbk1-deficient mouse embryonic fibroblasts. *Proc. Natl. Acad. Sci. USA* (2004) **101:**233-238
37. Hemmi H, Takeuchi O, Sato S, Yamamoto M, Kaisho T, Sanjo H, Kawai T, Hoshino K, Takeda K, and Akira S, The roles of two IkappaB kinase-related kinases in lipopolysaccharide and double stranded RNA signaling and viral infection. *J. Exp. Med.* (2004) **199:**1641-1650
38. Servant MJ, ten Oever B, LePage C, Conti L, Gessani S, Julkunen I, Lin R, and Hiscott J, Identification of distinct signaling pathways leading to the phosphorylation of interferon regulatory factor 3. *J. Biol. Chem.* (2001) **276:**355-363
39. Smith E, Marié I, Prakash A, García-Sastre A, and Levy DE, IRF3 and IRF7 phosphorylation in virus-infected cells does not require double-stranded RNA-dependent protein kinase R or IkB kinase but is blocked by vaccinia virus E3L protein. *J. Biol. Chem.* (2001) **276:**8951-8957
40. Iwamura T, Yoneyama M, Yamaguchi K, Suhara W, Mori W, Shiota K, Okabe Y, Namiki H, and Fujita T, Induction of IRF-3/-7 kinase and NF-kappaB in response to double-stranded RNA and virus infection: common and unique pathways. *Genes Cells* (2001) **6:**375-388
41. Kumar A, Yang YL, Flati V, Der S, Kadereit S, Deb A, Haque J, Reis L, Weissmann C, and Williams BR, Deficient cytokine signaling in mouse embryo fibroblasts with a targeted deletion in the PKR gene: role of IRF-1 and NF-kappaB. *EMBO J.* (1997) **16:**406-416
42. TenOever BR, Sharma S, Zou W, Sun Q, Grandvaux N, Julkunen I, Hemmi H, Yamamoto M, Akira S, Yeh WC, Lin R, and Hiscott J, Activation of TBK1 and IKK{varepsilon} Kinases by Vesicular Stomatitis Virus Infection and the Role of Viral Ribonucleoprotein in the Development of Interferon Antiviral Immunity. *J. Virol.* (2004) **78:**10636-10649
43. Roberts WK, Hovanessian A, Brown RE, Clemens MJ, and Kerr IM, Interferon-mediated protein kinase and low-molecular-weight inhibitor of protein synthesis. *Nature* (1976) **264:**477-480
44. Chu WM, Ostertag D, Li ZW, Chang L, Chen Y, Hu Y, Williams B, Perrault J, and Karin M, JNK2 and IKKbeta are required for activating the innate response to viral infection. *Immunity* (1999) **11:**721-731
45. Kumar A, Haque J, Lacoste J, Hiscott J, and Williams BR, Double-stranded RNA-dependent protein kinase activates transcription factor NF-kappa B by phosphorylating I kappa B. *Proc. Natl. Acad. Sci. USA* (1994) **91:**6288-6292
46. Jiang Z, Zamanian-Daryoush M, Nie H, Silva AM, Williams BR, and Li X, Poly(I-C)-induced Toll-like receptor 3 (TLR3)-mediated activation of NFkappa B and MAP kinase is through an interleukin-1 receptor-associated kinase (IRAK)-independent pathway employing the signaling components TLR3-TRAF6-TAK1-TAB2-PKR. *J. Biol. Chem.* (2003) **278:**16713-16719
47. Yang YL, Reis LF, Pavlovic J, Aguzzi A, Schafer R, Kumar A, Williams BR, Aguet M, and Weissmann C, Deficient signaling in mice devoid of double-stranded RNA-dependent protein kinase. *EMBO J.* (1995) **14:**6095-6106
48. Iordanov MS, Wong J, Bell JC, and Magun BE, Activation of NF-B by double-stranded RNA (dsRNA) in the absence of protein kinase R and RNase L demonstrates the existence of two separate dsRNA-triggered antiviral programs. *Mol. Cell. Biol.* (2001) **21:**61-72
49. Abraham N, Stojdl DF, Duncan PI, Methot N, Ishii T, Dube M, Vanderhyden BC, Atkins HL, Gray DA, McBurney MW, Koromilas AE, Brown EG, Sonenberg N, and Bell JC,

Characterization of transgenic mice with targeted disruption of the catalytic domain of the double-stranded RNA-dependent protein kinase, PKR. *J. Biol. Chem.* (1999) **274:**5953-5962

50. Alexopoulou L, Holt AC, Medzhitov R, and Flavell RA, Recognition of double-stranded RNA and activation of NF-kappaB by Toll-like receptor 3. *Nature* (2001) **413:**732-738
51. Honda K, Sakaguchi S, Sakaguchi S, Nakajima C, Watanabe A, Yanai H, Matsumoto M, Ohteki T, Kaisho T, Takaoka A, Akira S, Seya T, and Taniguchi T, Selective contribution of IFN-α/β signaling to the maturation of dendritic cells induced by double-stranded RNA or viral infection. *Proc. Natl. Acad. Sci. USA* (2003) **100:**10872-10877
52. Hemmi H, Kaisho T, Takeuchi O, Sato S, Sanjo H, Hoshino K, Horiuchi T, Tomizawa H, Takeda K, and Akira S, Small antiviral compounds activate immune cells via the TLR7 MyD88-dependent signaling pathway. *Nat. Immunol.* (2002) **3:**196-200
53. Heil F, Hemmi H, Hochrein H, Ampenberger F, Kirschning C, Akira S, Lipford G, Wagner H, and Bauer S, Species-specific recognition of single-stranded RNA via toll-like receptor 7 and 8. *Science* (2004) **303:**1526-1529
54. Lund JM, Alexopoulou L, Sato A, Karow M, Adams NC, Gale NW, Iwasaki A, and Flavell RA, Recognition of single-stranded RNA viruses by Toll-like receptor 7. *Proc. Natl. Acad. Sci. USA* (2004) **101:**5598-5603
55. Diebold SS, Kaisho T, Hemmi H, Akira S, and Reis e Sousa C, Innate antiviral responses by means of TLR7-mediated recognition of single-stranded RNA. *Science* (2004) **303:**1529-1531
56. tenOever BR, Servant MJ, Grandvaux N, Lin R, and Hiscott J, Recognition of the measles virus nucleocapsid as a mechanism of IRF-3 activation. *J. Virol.* (2002) **76:**3659-3669
57. Hsu LC, Park JM, Zhang K, Luo JL, Maeda S, Kaufman RJ, Eckmann L, Guiney DG, and Karin M, The protein kinase PKR is required for macrophage apoptosis after activation of Toll-like receptor 4. *Nature* (2004) **428:**341-345
58. Yoneyama M, Kikuchi M, Natsukawa T, Shinobu N, Imaizumi T, Miyagishi M, Taira K, Akira S, and Fujita T, CARD-containing DExD/H box RNA helicase, RIG-I, plays an essential role for doublestranded RNA-induced innate antiviral responses. *Nat. Immunol.* (2004) **5:**730-737
59. Hofmann K, Bucher P, and Tschopp J, The CARD domain: a new apoptotic signalling motif. *Trends Biochem Sci* (1997) **22:**155-156
60. Kobayashi K, Inohara N, Hernandez LD, Galan JE, Nunez G, Janeway CA, Medzhitov R, and Flavell RA, RICK/Rip2/CARDIAK mediates signalling for receptors of the innate and adaptive immune systems. *Nature* (2002) **416:**194-199
61. Yamamoto M, Sato S, Mori K, Hoshino K, Takeuchi O, Takeda K, and Akira S, Cutting edge: a novel Toll/IL-1 receptor domain-containing adapter that preferentially activates the IFN-beta promoter in the Toll-like receptor signaling. *J. Immunol.* (2002) **169:**6668-6672
62. Oshiumi H, Matsumoto M, Funami K, Akazawa T, and Seya T, TICAM-1, an adaptor molecule that participates in Toll-like receptor 3-mediated interferon-beta induction. *Nat. Immunol.* (2003) **4:**161-167
63. Hoebe K, Du X, Georgel P, Janssen E, Tabeta K, Kim SO, Goode J, Lin P, Mann N, Mudd S, Crozat K, Sovath S, Han J, and Beutler B, Identification of Lps2 as a key transducer of MyD88-independent TIR signalling. *Nature* (2003) **424:**743-748
64. Rothenfusser S, Tuma E, Endres S, and Hartmann G, Plasmacytoid dendritic cells: the key to CpG. *Hum Immunol* (2002) **63:**1111-1119
65. Yamamoto M, Takeda K, and Akira S, TIR domain-containing adaptors define the specificity of TLR signaling. *Mol Immunol* (2004) **40:**861-868
66. Lund J, Sato A, Akira S, Medzhitov R, and Iwasaki A, Toll-like receptor 9-mediated recognition of Herpes simplex virus-2 by plasmacytoid dendritic cells. *J. Exp. Med.* (2003) **198:**513-520

67. Krug A, Luker GD, Barchet W, Leib DA, Akira S, and Colonna M, Herpes simplex virus type 1 activates murine natural interferon-producing cells through toll-like receptor 9. *Blood* (2004) **103:**1433-1437
68. Hochrein H, Schlatter B, O'Keeffe M, Wagner C, Schmitz F, Schiemann M, Bauer S, Suter M, and Wagner H, Herpes simplex virus type-1 induces IFN-alpha production via Toll-like receptor 9-dependent and -independent pathways. *Proc. Natl. Acad. Sci. USA* (2004) **101:**11416-11421
69. Marié I, Smith E, Prakash A, and Levy DE, Phosphorylation-induced dimerization of interferon regulatory factor 7 unmasks DNA binding and a bipartite transactivation domain. *Mol. Cell. Biol.* (2000) **20:**8803-8814

Chapter 3

GENES MODULATED BY INTERFERONS AND DOUBLE-STRANDED RNA

SAUMENDRA N. SARKAR, GREGORY A. PETERS, and GANES C. SEN
Department of Molecular Biology, Lerner Research Institute, The Cleveland Clinic Foundation, 9500 Euclid Avenue, Cleveland, OH, USA

1. INTRODUCTION

Specific host responses to virus infection play major roles in determining not only the fate of the infected cells but also the efficacy of virus replication. Many of these responses are mediated by the products of cellular genes that are transcriptionally induced upon virus infection. In this induction process, interferons (IFN) and double-stranded (ds) RNA are often used as the proximal mediators. Consequently, in experimental systems, many of the same cellular genes can be independently induced by virus infection or treatments with IFN or dsRNA. This article focuses on this group of genes, and we discuss their induction characteristics and the functions of proteins encoded by selected members.

P. Palese (ed.), Modulation of Host Gene Expression and Innate Immunity by Viruses, 35-63.

2. GENE INDUCTION BY VIRUSES, INTERFERONS AND DOUBLE-STRANDED RNA

2.1 Characteristics of gene induction

Virus infection of host cells induces the transcription of multiple cellular genes, including cytokines and chemokines involved in the establishment of an antiviral state. IFNs are the first known members of the cytokine family of proteins that inhibit virus replication. They are induced in mammalian cells upon virus infection, secreted, and act on uninfected cells to activate a global antiviral state[1]. Like many other hormones, IFNs have pleiotropic effects on cell physiology affecting cell growth, cell motility and other cell functions as well. In the context of host-virus interaction, the regulation of the components of the interferon system is multifaceted and complex. Viruses infect cells, and viral gene products, most notably dsRNA, induce the synthesis of IFNs that are secreted. They bind to specific cell surface receptors and transcriptionally activate the IFN-stimulated genes whose products inhibit various stages of virus replication[2,3]. Some of the same genes can be directly induced by dsRNA as well, so that the encoded proteins are induced in virus-infected cells without the involvement of IFNs (Figure 1). Viral infection also causes induction of many cellular genes including some that are induced by IFNs and dsRNA. The common set of genes induced by virus infection, IFNs and dsRNA are called viral stress-inducible genes (VSIGs) (Table 1). These genes are usually not expressed in uninfected cells, but are strongly and transiently induced upon exposure to viral stresses.

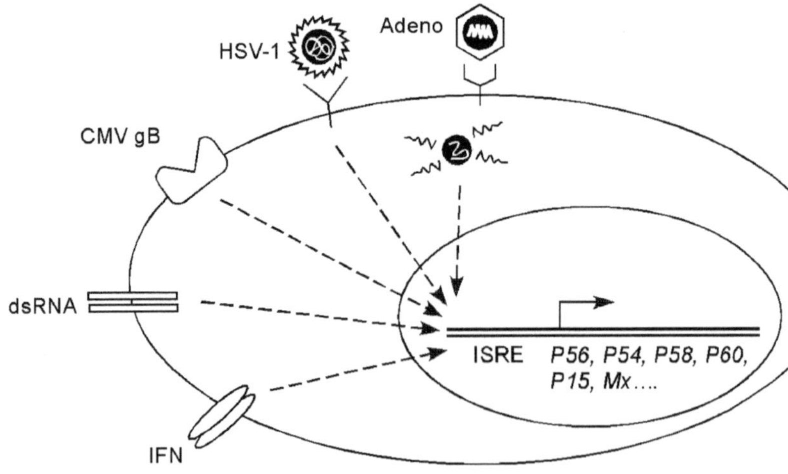

Figure 1. Different modes of induction of IFN-stimulated genes.

Table 1. Induction of VSIGs by IFN, dsRNA and Viruses. Microarray gene profiling data for selected genes from: Der et al. 1998[4] (IFN α, β, γ); Geiss et al. 2001[5] (dsRNA); Mossman et al. 2001[6] (HSV-1); Geiss et al. 2002[7] (Influenza A); Zhu et al. 1998[8] (hCMV) are summarized in a qualitative manner. Fold inductions in the range of 0-5 are represented by '+', 5-25 by '++', 25-100 by '+++' and >100 by '++++'.

	Fold Induction By						
Genes	**HSV-1**	**Influenza A**	**hCMV**	**IFN α**	**IFN β**	**IFN γ**	**dsRNA**
9-27	+	+	++	++	++	++	
6-16			+	++	++		
ADAR	+			+	+	+	+
GBP I		+	++	+	++	++	++
IL-11			+				
IL-6			++				
IP-10	++		+++				+++
IRF1		+	++	+	++	++	+
IRF9		+		++	++	++	
ISG15	++	+	++++	++	++	+	++
ISG54	++		++	+	+	+	
ISG56	++	+		+++	++++		+++
ISG58					+		
MxA	++	++	+++	++	+++		
OAS1			++	++	++		
OAS2	+		++	++	++		
OAS3	+						
PKR					++		
RING4				+++	++	+++	+
STAT1	+	+		+	++	+	

Investigation of cellular genes whose transcription is induced upon infection with different viruses or treatments with IFN or dsRNA has revealed that there is much overlap amongst these sets of genes[4,5]. Because many of them were originally discovered as IFN-inducible genes, analyses of their functions have been biased toward their putative roles in the IFN system. For a long time it was thought that induction of these genes by virus infection or dsRNA is a secondary consequence of IFN production. But, by using various IFN-response defective mutant cell lines, it has been unequivocally established that these genes can be induced independent of IFNs[9-11]. In fact, the different inducers use different signaling pathways to induce the same genes (Figure 1). IFN-signaling is initiated from the cognate cell-surface receptors, whereas dsRNA signals are initiated through Toll-like receptor 3 or other intracellular dsRNA-binding proteins. Different viruses can trigger their induction differently: some, such as adenoviruses, use newly synthesized viral gene products as the inducer whereas others, such as HSV-1, use the incoming virions as the triggering agent. In the case of CMV, not only virus infection, but the viral glycoprotein gB itself, can induce these genes. Recently, use of microarray analysis for gene expression profiling of virus-infected cells has helped the cataloging of virus induced genes[4-8,12-14]. However, because appropriate mutant lines have not been used in most studies, it is difficult to determine whether viruses can directly induce some of these genes. Nonetheless, even from incomplete information, consistent patterns have emerged (Table 1). For example, it is clear that the ISG56 family of genes (ISG56, ISG54 and ISG58) and ISG15 are induced by IFN α/β, dsRNA and infections with a number of viruses.

2.2 Signaling pathways

IFNs, dsRNA and viruses use distinct but partially overlapping signaling pathways to induce transcription of the same genes. It is the same cis-acting sequence, the IFN-stimulated response element (ISRE), present in the promoters of these genes, which receives the signals generated by these diverse pathways (Figure 2). Although, in all cases the interferon regulatory factors (IRF) are the critical components of the transcription complexes that recognize the ISRE, the specific members of the IRF family, their activation mechanisms, and their partners in action are distinct for the different signaling pathways. Among these, the IFN signaling pathways are the best understood[15].

IFNs are divided into two types, type I and type II, both of which have antiviral activity, but act through different cell surface receptors and are structurally unrelated. There are many members of type I family: IFN α and its subfamilies, IFN β, IFN ω and IFN τ, but there is only one human type II interferon, IFN γ. The signaling pathway for type I IFNs is mediated by a homologous receptor complex IFNAR (Figure 2)[16]. The ligand-induced stimulation of IFNAR results in cross activation of the two receptor-associated Janus protein tyrosine kinases (Jaks). Activated Jaks phosphorylate themselves, the receptor subunits, and specific members of the family of proteins called signal transducers and activators of transcription (STAT). Tyrosine phosphorylated STAT1 and STAT2 form a trimeric complex with IRF9 (P48) and the complex, ISGF3, translocates to the nucleus, binds to the ISRE, the specific DNA sequences in promoters, and activates transcription[2,17]. On the other hand, IFN-γ induced transcription of genes use multiple mechanisms. The most well understood mechanism involves STAT1 containing transcription factor GAF (Gamma Activated Factor), which is activated by phosphorylation by Jak1 and Jak2 and binds to the GAS sequence element present in the promoter regions of many genes. Besides GAF mediated transcriptional activation, IFN γ can also signal to the ISRE through a transcription factor distinct from ISGF3, or by using the CIITA transcription factor to class II MHC genes[3].

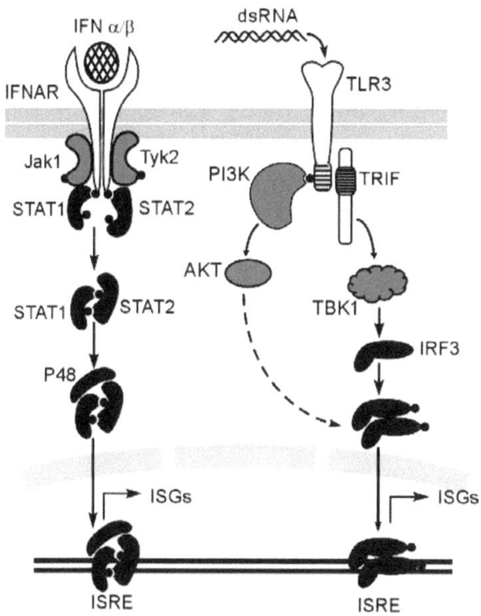

Figure 1. Signaling pathways leading to the induction of IFN-stimulated genes. Schematic representation of the signal transduction pathways involved in the transcriptional induction of VSIGs. On the left, is the canonical JAK-STAT pathway, where sequential tyrosine phosphorylations triggered by IFN α/β interaction with IFN receptor (IFNAR), leads to transcriptional activation of ISRE containing ISG promoters. On the right, is the newly discovered pathway of TLR3 dependent activation of IRF3 transcription factor.

Toll like receptor 3 (TLR3) is needed in one pathway for dsRNA-mediated gene induction. A TLR3 knock out mouse showed markedly reduced induction of inflammatory cytokines in response to dsRNA. Also, when TLR3, but not other TLRs is expressed in dsRNA unresponsive 293 cells, they respond to dsRNA[18]. dsRNA triggers at least three signaling pathways leading to the activation of the transcription factors NFκB, IRF-3/IRF-7 and ATF-2[9,19,20]. All of these factors together drive transcription from complex promoters present in genes such as the IFN-β gene. In contrast, the promoters of IFN-inducible genes, such as P56, contain only ISREs to which activated IRF-3 binds. Activation of neither NFκB nor ATF-2 is necessary for ISG56 gene induction[21]. Among the kinases activated in dsRNA-treated cells are PKR, IKK, P38 and JNK. For the dsRNA induced

NFκB activation via TLR3, both the adaptor proteins MyD88 and TRIF/TICAM1 are important[22]. The adaptor in turn interacts with the IRAK family of kinases, through their death domains. The signal is transmitted downstream by the TRAF6 signalosome complex, finally leading to the activation of IKK and degradation of IκB, releasing the active transcription factor NFκB[23-25]. The other arm of the dsRNA signaling pathway involves transcription factor IRF-3[21,26]. The adaptor protein TRIF/TICAM1 seems to be the sole adaptor for this pathway (Figure 2). Thus, TRIF/TICAM1 is the adaptor for TLR3, from which the IRF-3 and NFκB pathways diverge[27]. For the activation of the IRF-3 pathway, two Ser/Thr kinases are necessary, TBK-1 and PI3K. Upon activation, IRF-3 is phosphorylated leading to its dimerization and translocation to the nucleus to induce transcription[28,29]. This activation is a two-step process achieved by sequential phosphorylation of the protein: TBK-1 is responsible for phosphorylation of one set of residues whereas a PI3K-initiated pathway phosphorylates another set of residues[30]. Double stranded RNA-elicited phosphorylation of specific tyrosine residues present in the cytoplasmic domain of TLR3 is necessary for all of the above process[26]. Two specific phosphotyrosines of TLR3 initiate the two pathways mediated by TBK-1 and PI3K (Figure 2). Surprisingly, viruses can induce these genes in the absence of many of the above-described signaling components. For example, Sendai virus can induce P56 in cells lacking TLR3 or STAT1 (Elco, C and Sen, G. unpublished observation) indicating that additional signaling pathways exist.

3. FUNCTIONS OF IFN-INDUCED PROTEINS

3.1 Proteins induced by IFN and dsRNA

The proteins encoded by IFN-inducible genes have diverse functions. We discuss below the functions of selected members of this group, all of which inhibit cellular and viral protein synthesis using different strategies (Figure 3). The P56 proteins block initiation of translation by interfering with the functions of eIF-3, a protein synthesis initiation factor. The protein kinase, PKR, interferes with the functions of a different translation initiation factor, eIF-2, by phosphorylating its alpha subunit. The third family of proteins, the 2-5 (A) synthetases, synthesize the oligoadenylates, 2-5(A), which, in turn, activate the latent ribonuclease, RNaseL, and cause degradation of mRNA. P56 is distinguished from PKR and 2-5(A) synthetases by its non-enzymatic functions. Moreover, it does not require any co-factor for its actions whereas both the IFN-induced enzymes need to be activated by co-factors, such as dsRNA.

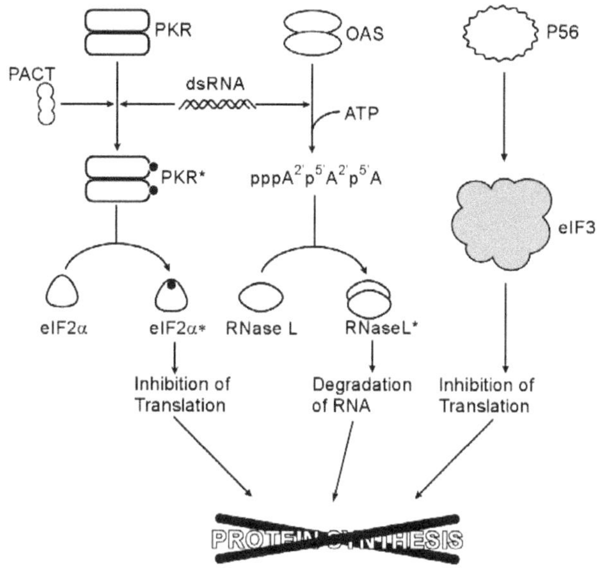

Figure 2. Translational inhibition pathways induced by IFNs. Two well known IFN-induced proteins, PKR and OAS, are activated by co-factor dsRNA and inhibit protein synthesis, whereas another, P56, binds to eIF3 to cause translational inhibition without dsRNA.

3.2 2-5 (A) Synthetases

2'-5' Oligoadenylate Synthetases (OAS), an important class of IFN inducible proteins, was discovered as one of the "factors" responsible for the inhibition of cell free protein synthesis by dsRNA[31,32]. It was purified from cells treated with IFN and found to be capable of synthesizing small 2'-5' linked oligomers of adenosine (2-5 (A)) in the presence of dsRNA and ATP. The only known function of 2-5 (A) is to bind and activate the latent ribonuclease, RNase L, through its dimerization (Figure 4). The activated RNase L degrades cellular and viral RNA to inhibit protein synthesis and viral replication[3,33].

The OAS family proteins are unique and highly conserved with no significant sequence homology with other proteins. Humans have four OAS genes: *OAS1, OAS2, OAS3* and *OASL*[34], whereas there are 8 *OAS* genes present in mouse. There are three major forms of enzymatically active OAS proteins found in human cells: 40-46 kDa small isoforms (OAS1), 69-71 kDa medium isoforms (OAS2) and 100 kDa large isoform (OAS3). Other OAS like proteins (OASL) are inactive. In the case of OAS1 and OAS2, alternative splicing produces multiple isozymes with different carboxyl terminal ends (Figure 4). The small isozymes function as tetramers, the medium isozymes as dimers and the large isozymes as monomers. Although all three active isozymes are capable of producing 2-5 (A), the length of 2-5

(A) varies. The OAS1 isozymes, synthesize up to hexamers of 2-5 (A), and tend to be processive. The OAS2, P69 isozyme, synthesizes longer, up to 30mers of 2-5 (A), and the reaction is non-processive[35]. OAS3 makes mostly dimers of 2-5 (A) that is incapable of activating RNase L[36].

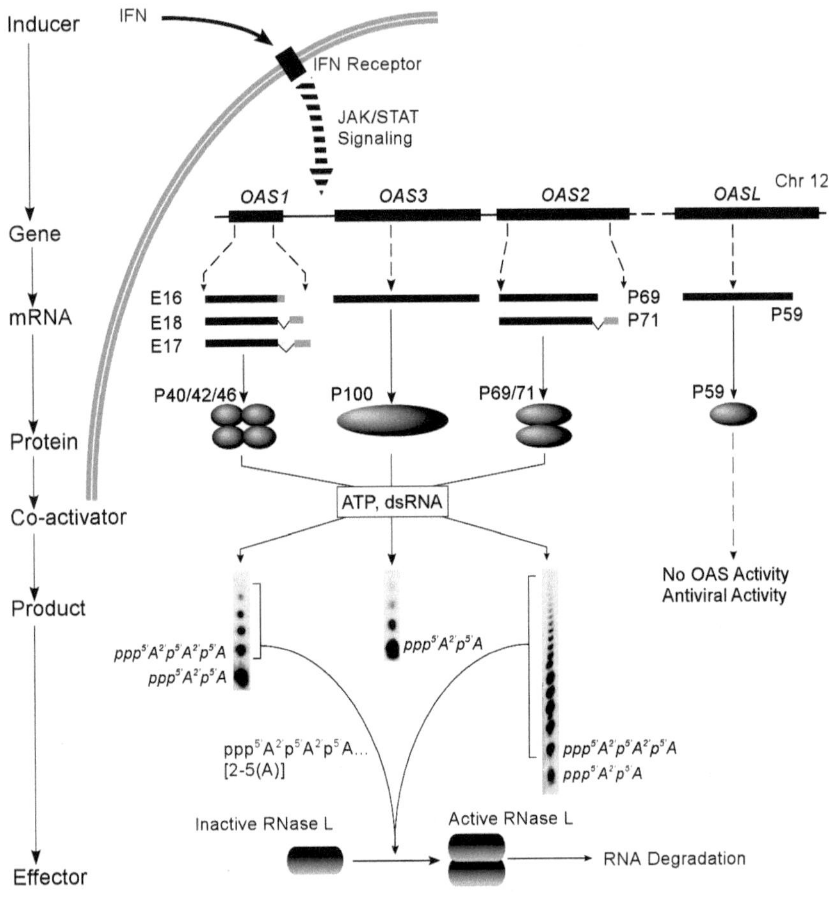

Figure 3. The 2-5 (A) System. Type I Interferons through IFN receptors and JAK-STAT signaling molecules induce mRNA for all 4 OAS isozymes. The mRNA for two of these isozymes, OAS1 and OAS2, undergo alternative splicing to generate further variants. Proteins encoded by these mRNAs (except P59) catalyze the polymerization of ATP to 2-5 (A) in the presence of dsRNA. 2-5 (A), in turn, binds to latent RNase L, activates it to an active ribonuclease, which degrades RNA.

The primary structure of OAS proteins does not show the presence of any significant recognizable motif. Structure-function studies using recombinant proteins have identified several functional domains important for dsRNA-dependent 2-5 (A) synthesis. All members of the OAS gene family share the same general organization consisting of a basic 2-5 (A) synthetase domain.

This domain is found once in the OAS1 and OASL proteins; it is repeated twice in the OAS2 proteins and three times in the OAS3 protein. Although in OAS2 and OAS3 multiple 2-5 (A) synthetase domains are present, only the C-terminal domain seems to actively contribute to 2-5 (A) synthesis. The unique primer and template independent 2-5 (A) synthesis activity indicates that the enzyme should have three distinct sites involved in catalysis: an acceptor site (A) to which the first ATP or elongating 2-5 (A) molecules should bind, a donor site (D) to which the donor ATP should bind and a catalytic site (C) that catalyzes the covalent bond formation between the two substrates. Substrate-crosslinking, protein modeling and mutagenesis studies have identified these three sites within a short linear sequence of OAS2 P69 C-terminal region[37,38]. Three aspartic acid residues D408, D410 and D481 form the catalytic site (C), which overlaps with the acceptor-binding site (A) consisting of Y421. The donor-binding site (D) is composed of positively charged residues R544 and K547 (Figure 5). This isozyme is a dimer and its dimerization is essential for enzyme activity. The amino acids needed for dimerization of P69 or tetramerization of P46 were identified (C668, F669 and K670) and their mutations destroy the enzyme activity[35,37,39]. The reason for the need of dimerization of P69 to maintain its enzyme activity was studied by co-expressing differentially tagged wt and mutant proteins in insect cells, and purifying the heterodimers and examining their activities. Since the mutant/mutant heterodimers, A*/D* and C*/D*, are both enzymatically active, although none of the parent protein is active, a criss-cross model of P69 action was proposed[40]. In this model, the donor bound to the 'D' site of one subunit is covalently linked to the acceptor bound to the 'A' site of the other subunit, which also contains the catalytic site (C). It remains to be seen whether the small isozymes also catalyze in a similar manner.

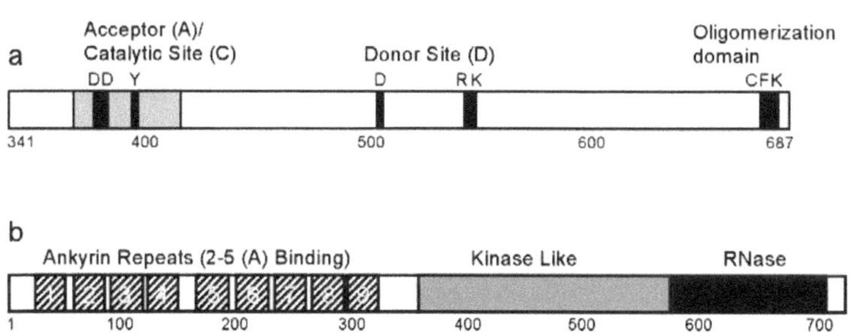

Figure 4. Functional domains in P69 OAS and RNase L. (A) Functionally distinct domains and important amino acid residues in the C-terminal half of P69 OAS2. The Catalytic Site (C) is composed of D408, D410, D481; Y421 contributes to Acceptor Binding (A) and R544 is responsible for Donor binding (D) Three consecutive residues, C668, F669, K670 are responsible for dimerization. (B) Domain Structure of RNase L showing 9 ankyrin repeats, kinase like domain and the RNase domain.

The crystal structure of a porcine OAS1 isozyme has recently been determined at 1.74 Å resolution[41] which shows a U-shaped structure with two major domains. The N-terminal domain consists of a five-stranded anti-parallel β sheets and two α helices. This is followed by a 35 amino acid stretch containing two α-helices bridging the N-terminal and the C-terminal domains. The second domain consists of a four-helix bundle. The first 20 residues of the N-terminus of the protein pack tightly against the C-terminal domain. The three Asp residues constituting the catalytic site are found in β strands 2 and 5 of the amino-terminal lobe. This structure has strong similarities with those of poly (A) polymerases that are confined to their N-terminal and central domains. The Asp residues in the catalytic centers are similarly located in the two enzymes. Thus, it seems plausible that the reason for the two enzymes forming different bonds, one 2'-5' and the other 3'-5', is because of the way the substrates become accessible to the active centers of the proteins. Another interesting aspect of the OAS family of enzymes is their dsRNA activation mechanism. Unlike PKR family of dsRNA binding proteins, OAS proteins do not have any defined dsRNA-binding motif. The crystal structure, mutagenesis and enzyme kinetic studies suggest, that the activation is a two-step process. First, dsRNA bind to a positively charged groove in the OAS, followed by a structural rearrangement that widens the active site cleft.

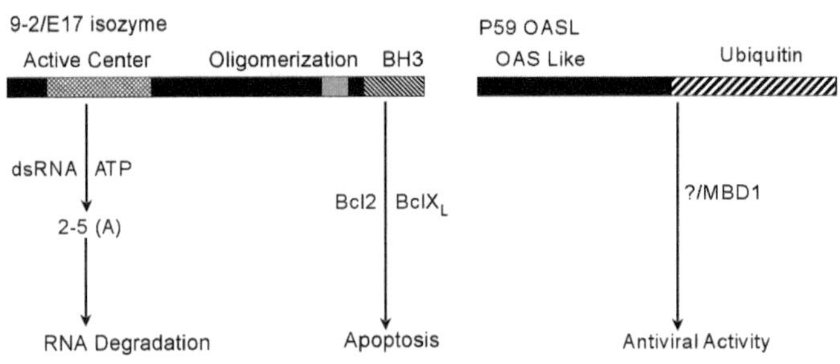

Figure 5. Non-classical functions of OAS. The 9-2/E17 isozyme, like all other isozymes, polymerizes ATP to 2'-5' oligoadenylates and activates RNaseL. In addition, it also mediates apoptosis by interacting with anti-apoptotic proteins Bcl2 and BclXL. The P59 OASL has been shown to cause inhibition of viral replication, through its ubiquitine like domain, which also interacts with methyl CpG-binding protein 1[42].

Constitutive expression of OAS1 and OAS2 inhibits the replication of picornaviruses, but not rhabdoviruses[43,44]. OAS enzymes can be activated by many viral RNAs: HIV TAR RNA, the adenoviral VA RNA and Epstein-Barr virus EBER-1 RNA[45-47]. Recently, the non-enzymatic functions of OAS family of proteins have become apparent from several studies. Resistance to West Nile virus infection was mapped to one of the OAS genes, *OAS1b,* in mouse[48,49]. This allele encodes an enzymatically inactive protein. Mouse strains that are susceptible to West Nile virus were found to have a C to T transition in the fourth exon resulting in a truncated *OAS1b* gene product. One of the alternatively spliced OAS1 isozyme, 9-2/E17, has been shown to cause apoptosis through its C-terminal BH3 domain (Figure 6)[50]. This activity is independent of its enzymatic activity and RNase L. The inactive OAS like protein, P59 OAS L, has also been shown to confer antiviral activity against EMCV, when expressed in cell line[51]. This activity was attributed to the C-terminal ubiquitin-like domain.

Because the 2-5 (A) Synthetase/RNase L system is an important part of interferon's antiviral machinery, viruses have evolved mechanisms to evade this pathway. In the case of hepatitis C virus (HCV), interferon-resistant strains have significantly fewer RNase L cleavage sites (UA and UU), thus preventing the genomic RNA from RNase L mediated degradation[52]. The HCV NS5A protein specifically binds to OAS1 and inhibits the antiviral activity of IFN[53].

3.3 RNase L

RNase L is a regulated endoribonuclease that is the second major component of the 2-5 (A) system. It is activated by 2'-5' oligoadenylates with at least three adenyl residues and one 5' phosphoryl group[54,55]. Human RNase L is a 741 amino acid protein with several identifiable structural motifs with functional implications (Figure 5). The N-terminal ankyrin repeats are involved in 2-5 (A) binding and the ribonuclease activity resides in the C-terminal RNase domain[56,57]. There are several protein kinase like domains in the C-terminal half of RNase L, whose precise functional importance is still unclear.

2-5 (A) binds to RNase L with very high affinity ($K_d = 4 \times 10^{-11}$ M), converting it from an inactive monomeric state to a potent dimeric RNase[58]. Binding of 2-5 (A) to the N-terminal ankyrin repeats, presumably causes a conformational change of the protein unmasking its dimerization and ribonuclease domains. RNase L cleaves after UpNp dinucleotide sequences (primarily after UU and UA sequence) in single-stranded RNA. The importance of N-terminal inhibitory domain is demonstrated by a truncated, recombinant RNase L that is constitutively active without 2-5 (A)[57]. These findings suggest another mechanism of RNase L activation, independent of 2-5 (A), by proteolysis. In fact, in the case of chronic fatigue syndrome, cleaved RNase L fragments have been detected in the extracts of peripheral blood mononuclear cells[59,60]. However, these fragments are still dependent on 2-5 (A). An inhibitor of RNase L (RLI) has also been described, which is present in the mitochondria, and presumably involved in down-regulation of mitochondrial mRNA after IFN α treatment[61].

The physiological roles of RNase L have been explored by generating a mouse with targeted disruption of the RNase L gene[62]. Slightly elevated levels of EMCV replication were found in RNase L -/- mice and their fibroblasts after IFN treatment. More significantly, these mice have enlarged thymuses, containing more thymocytes, due to diminished apoptosis. Though the signaling mechanism by which RNase L causes apoptosis, is not clear, involvement of JNK and Caspase 9 have been indicated[63]. RNase L has also been implicated in the onset of familial and sporadic prostate cancer[64-66]. Several loci associated with prostate cancer have been mapped to different mutations in RNase L, mostly in the N-terminal and the kinase like domain. Though some of these mutations have been shown to have moderate effects on the ribonuclease activity, a direct link is still missing.

3.4 PKR

PKR (Protein Kinase RNA regulated) is a ubiquitously expressed, interferon-inducible, serine/theonine kinase whose enzymatic activity is latent, and needs to be activated by autophosphorylation. PKR was discovered as a key mediator of protein synthesis inhibition when dsRNA was added to IFN-treated cell-lysate in cell-free translation systems. The most well known activators of PKR are dsRNA (often produced as an intermediate product or by-product of viral replication) and the cellular protein PACT (Figure 7). Once activated, PKR can phosphorylate a limited set of cellular proteins, the most well characterized of which is the α subunit of the translation initiation factor eIF2. eIF2-α phosphorylation leads to an inhibition of translation, thus contributing to PKR-mediated control of certain viral infections and cell growth. PKR has also been shown to be an important element in the transcriptional signaling pathways activated by specific cytokines, growth factors, dsRNA, and extracellular stresses (reviewed in 20). In addition, PKR has been implicated in a broad array of cellular processes such as differentiation, cell growth, apoptosis, and oncogenic transformation.

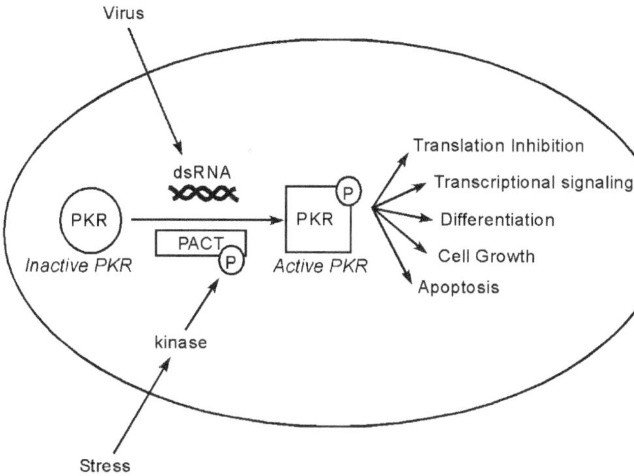

Figure 6. PKR activation by dsRNA or PACT and its biological effects. PKR becomes active as a result of its own autophosphorylation. PKR can be activated by dsRNA, produced as a result of viral replication, or the cellular protein PACT, which itself is phosphorylated by a stress-activated protein kinase. Once active, PKR has a myriad of biological effects.

While viral dsRNA is the most likely activator of PKR in virus-infected cells, PACT is probably the more physiologically relevant activator in an uninfected cell. PACT-mediated PKR activation requires an additional stress signal for efficient PKR activation *in vivo*. A variety of cellular stresses, including withdrawal of growth factors, or treatment of cells with a low dose of actinomycin D, arsenite, thapsigargin, or peroxide, can promote PKR activation by PACT[67-69]. Upon stressing cells, PACT becomes phosphorylated at specific serine residues that allow it to become a better activator of PKR *in vivo* (Peters, G., Li, S. and Sen, G. unpublished observation).

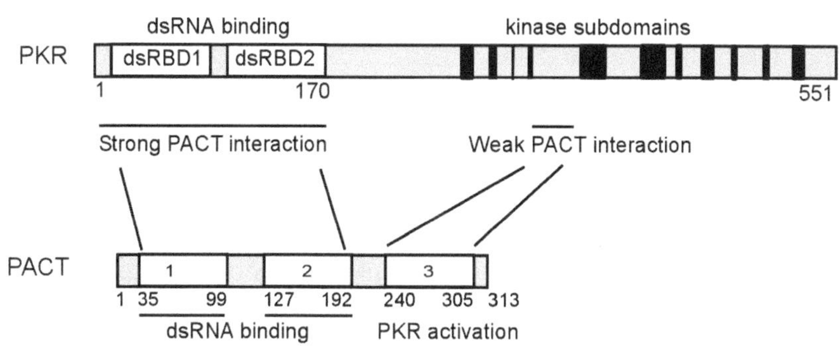

Figure 7. Domain structure of PKR and PACT. The PKR regulatory domain includes the N-terminal 170 amino acids, which contain 2 dsRBDs that bind dsRNA. This same region also interacts strongly with PACT. The PKR catalytic domain spans the C-terminal region and

contains 11 kinase subdomains, and a weak PACT interaction domain. PACT contains two dsRBDs, domain 1 and 2, which strongly bind to PKR. PACT domain 3 interacts weakly with the PKR catalytic domain and is responsible for PKR activation. PACT can also bind dsRNA through domains 1 and 2. Amino acid numbers are indicated.

The 68 kDa PKR protein contains two functionally distinct domains: an N-terminal dsRNA binding regulatory domain containing two dsRNA binding motifs, and a C terminal catalytic domain containing eleven kinase subdomains (Figure 8). The two dsRNA-binding motifs of PKR, dsRBD1 and dsRBD2, also participate in dsRNA-independent protein-protein interactions with proteins containing similar domains. Many structure-function studies have identified critical residues in the regulatory domain responsible for dsRNA binding and activation of PKR (reviewed in 70). Upon binding dsRNA, PKR activation is achieved by a conformational change that exposes the ATP-binding site, causing its autophosphorylation and permanent activation. Based on NMR structural studies with the PKR dsRNA-binding domain, along with many previous mutational and biochemical studies, Nanduri et al have described a model for PKR activation by dsRNA[71]. According to their model, the dsRBD2 of PKR acts as a negative regulator of the kinase domain. In the absence of dsRNA, the intramolecular interaction of dsRBD2 and the kinase domain keeps the protein in a closed/inactive conformation, preventing the binding of the substrate, ATP. The dsRBD1, with its higher dsRNA binding affinity and motional flexibility, binds to dsRNA, inducing cooperative binding of dsRNA to dsRBD2. This co-operative dsRNA binding to the dsRBDs releases the intramolecular interaction of dsRBD2 with the C-terminal kinase domain and opens the protein to an active conformation (Figure 9). Once in the open/active conformation, PKR becomes autophosphorylated and stays as an active kinase independent of dsRNA.

Similar structures of PACT-PKR complexes have not been solved; however, it has been suggested that PACT activates PKR by a similar, but distinct mechanism, as dsRNA. PACT can directly bind to PKR and activate it independent of dsRNA[72]. PACT contains an N-terminal PKR binding domain, which is composed of two consecutive dsRBDs (domain 1 and 2), and a C-terminal domain (domain 3) for PKR activation (Figure 8)[69]. Although PACT strongly interacts with PKR through domains 1 and 2, the critical event is the interaction of PACT domain 3 with a distinct region in the C-terminal half of PKR to initiate its autophosphorylation and activation. This model is supported by the fact that only domain 3 of PACT is necessary and sufficient to activate PKR *in vitro*. Moreover, mutants of PKR that fail to bind dsRNA can be activated by PACT. *In vivo*, PACT causes apoptosis by a PKR dependent mechanism, but only under cellular stress conditions. PACT, like PKR, is a dsRNA binding protein expressed in most cell types in small amounts. It has been proposed that stress-induced phosphorylation of PACT, *in vivo*, makes it a better PKR activator[67]. We have observed that phosphorylated PACT has higher affinity for PKR and is a better PKR activator (Peters, G., Li, S. and Sen, G. unpublished observation).

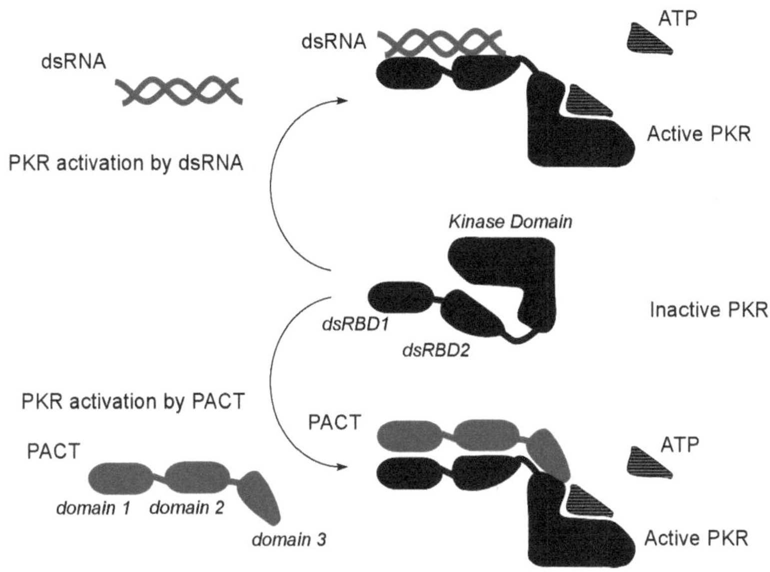

Figure 8. Models of PKR activation by dsRNA and PACT. Inactive PKR exists in a closed form in which the ATP binding site in the kinase domain is blocked by dsRBD2. Upon dsRNA binding, a structural rearrangement of PKR opens the ATP binding site, and allows its autophosphorylation and activation. On the other hand, although PACT binds to the dsRBDs of PKR with its two dsRBDs, activation requires the interaction of PACT domain 3 with the kinase domain of PKR.

As mentioned above, the dsRNA-binding domain of PKR also mediates dsRNA-independent protein-protein interactions with protein containing similar domains. This property has been exploited to identify several additional PKR interacting proteins. One such protein is a substrate of PKR, DRBP76[73]. The functional significance of this modification of DRBP76 by PKR is presently not known. However, a homolog of DRBP76, NF90, plays a role in mRNA stabilization and is controlled by JNK MAP kinase activation[74]. Other PKR-interacting cellular proteins are subunits of protein phosphatase1 and 2[75,76].

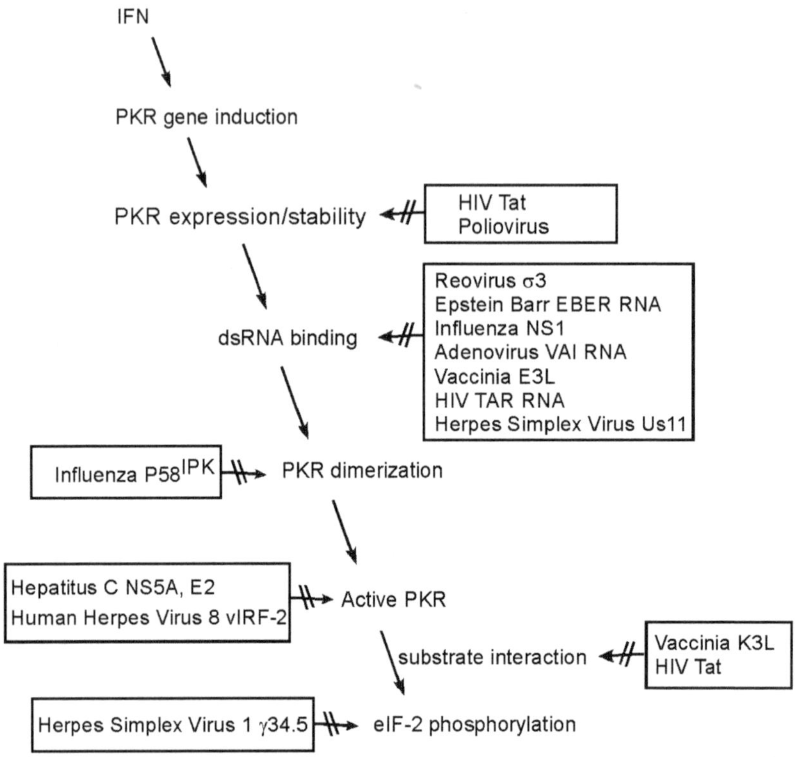

Figure 9. The sites of viral-directed inhibitors along the PKR activation pathway. Transcription of the PKR gene in response to interferons results in PKR expression. The PKR protein becomes activated by binding to dsRNA and dimerizing. Active PKR phosphorylates its substrate, eIF-2a, resulting in protein synthesis inhibition, and inhibition of viral replication. Arrows containing hatch marks indicate specific areas of the PKR pathway targeted by the viruses indicated.

Viruses encode many of the additionally known PKR-interacting proteins. To evade the host antiviral response, viruses have developed effective measures to repress nearly all aspects of the IFN system. Among these mechanisms, inhibition of PKR function by viral proteins is the most studied one (reviewed in 77). To inhibit PKR, viruses produce inhibitors that 1) modulate PKR levels; 2) interfere with dsRNA binding to PKR or that sequester dsRNA; 3) interfere with PKR dimerization; 4) block PKR catalysis site/substrate interaction and 5) regulate eIF-2 phosphorylation (Figure 10). In the case of poliovirus, PKR levels are dramatically decreased in infected cells, likely due to the activation of a cellular pathway to degrade PKR. HIV-1 Tat protein also has been reported to down regulate PKR levels, as well as to act as a pseudosubstrate to inhibit PKR. Another virus, Reovirus, encodes the σ3 protein, a dsRNA binding protein that likely sequesters dsRNA to prevent PKR activation. Epstein Barr virus EBER and

Adenovirus VAI RNAs are virally encoded RNAs that bind to the dsRBDs of PKR to inhibit its activation. Influenza virus directs at least two approaches to target PKR for inhibition, one involving the viral-encoded NS1 protein, and the other using the cellular protein $p58^{IPK}$. Viral NS1 is an RNA binding protein that blocks dsRNA-mediated PKR activation by binding to PKR and sequestering it in the nucleus[78]. Cellular $p58^{IPK}$, activated by influenza infection, binds to PKR and prevents its dimerization. HIV1 TAR RNA and cellular TAR RNA binding protein (TRBP) are also negative regulators of PKR activation that target dsRNA binding. Two inhibitory proteins expressed by Hepatitus C virus are the nonstructural protein NS5A and the envelope protein E2. NS5A binds to the PKR catalytic domain to prevent PKR activation and remove the IFN-induced block on viral replication[79]. On the other hand, HCV envelope protein E2 contains a sequence identical with phosphorylation sites of PKR and eIF2-α, thereby inhibiting the kinase activity and effects of active PKR[80]. Vaccinia virus, with its large protein coding capacity, encodes at least two proteins, E3L and K3L, to interfere with PKR mediated translational inhibition. E3L targets the dsRNA-PKR interaction, whereas K3L behaves as a pseudosubstrate inhibitor. Human Herpes Virus 8 produces the protein vIRF-2 that physically interacts with PKR, consequently inhibiting autophosphorylation of the kinase[81]. Herpes Simplex Virus-1 also targets multiple levels of the PKR activation pathway by encoding Us11 and γ34.5. Us11 is a RNA binding protein that, by binding to the N-terminal half of PKR, can inhibit PKR activation by dsRNA[82]. Us11 is the first example of a viral protein that can also inhibit PACT-mediated activation of PKR[83]. In this case, although Us11 can bind to both PKR and PACT, it is the binding of Us11 to PKR, and not to PACT, that is critical for blocking PACT activation of PKR. The other HSV-1 protein, γ34.5, targets an additional step of the PKR activation pathway by recruiting a cellular phosphatase to eIF-2, which restores eIF-2α to its unphosphorylated state[84]. Undoubtedly there are other novel mechanisms of viral PKR regulation that remain to be discovered.

In addition to its pivotal role in innate immunity, in recent years the role of PKR in cellular signal transduction has been of focus. In several reports, dsRNA mediated gene induction has been shown to be dependent on PKR[85-89]. It was proposed that PKR mediated this effect by associating with IκB complex and modulating the release of NFκB. However, the presence of dsRNA signaling in PKR-/- fibroblast suggested that dsRNA signaling functions independently of PKR[90]. With the discovery of TLR3, the PKR independent dsRNA signaling pathway has become more clear. However, recent reports suggest TLR3 independent dsRNA signaling is mediated by PKR[91]. PKR has also been suggested to contribute to activation of signaling pathways by proinflammatory stimuli, including TNF-α, LPS, and IL-1, as well as mediating other stress activated signaling pathways that trigger apoptosis.

3.5 P56 family of proteins

P56 is the product of mRNA 561 encoded by the *ISG56* gene (also known as IFIT1), a gene that is highly induced by IFN, dsRNA and many viruses[4,5,92]. Untreated cells do not express P56, but viral infection and other stresses induce transcription of the ISG56 gene rapidly and strongly. The induction is transient and both the mRNA and the protein turn over quickly[10,93]. All members of the P56 family are induced by viral stresses. In humans, there are three other members, HuP60, HuP58 and HuP54. In mouse, MuP56, MuP54 and MuP49 are the three members of this family (Table 2). The P56 proteins are structurally and evolutionally related and among the human and the murine P56-related proteins, the cognate members of the two species are more closely related than two members of the same species. For example, at the level of protein sequence, HuP54 and MuP54 have 73% sequence identity. In contrast, there is only 42% sequence conservation between HuP56 and HuP54. The hallmark of the structures of the P56 family members is that they all contain multiple tetratricopeptides (TPR) motifs. The TPR motif is a degenerate 34 amino acid protein-protein interaction module found in multiple copies in a number of functionally different proteins that facilitates specific interactions with partner proteins[94-96]. The consensus sequence of a TPR is defined by a pattern of small and large hydrophobic amino acids, although no position is completely invariant. Residues are conserved at only a few positions, e.g. Gly or Ala at position 8 and Ala at 20. Crystal structures of proteins containing TPR motifs revealed that TPRs adopt helix-turn-helix arrangements. Adjacent TPR motifs pack in parallel fashion, resulting in a spiral of repeating anti-parallel helices. In several TPR containing proteins, an additional capping or solubility helix is present at the C-terminus. Most TPR containing proteins associate with multiprotein complexes involved in diverse function, and as discussed below, the P56 family of proteins also associates with large protein complexes and modifies their functions.

Table 2. Interferon inducible P56 family of proteins.

Species	Name	Calculated M.W. (kDa)
Human	IFIT-1, ISG56, P56, IFI-56K, **HuP56**	55.4
Human	IFIT-2, ISG54, P54, IFI-54K, **HuP54**	54.6

Species	**Name**	**Calculated M.W. (kDa)**
Human	IFIT-4, ISG60, IFI-60K, Retinoic acid-induced gene G protein (RIG-G), **HuP60**	55.9
Human	IFIT-5, ISG58, **HuP58**	55.8
Mouse	IFIT-1, ISG56, Glucocorticoid-attenuated response gene 16 protein (GARG-16), **MuP56**	53.7
Mouse	IFIT-2, ISG54, Glucocorticoid-attenuated response gene 39 protein (GARG-39), **MuP54**	55.0
Mouse	IFIT-3, ISG49, Glucocorticoid-attenuated response gene 49 protein (GARG-49), **MuP49**	47.2

Several cellular proteins interact with HuP56. The most well characterized HuP56-interacting protein is the Int-6/p48 protein, which is identical in human and mouse. It is encoded by the Int-6 gene whose disruption by the integration of the mouse mammary tumor virus genome causes breast cancer in mice[97]. P48 is identical to the eIF-3e subunit of the eukaryotic translation initiation factor 3[98]. Through its interaction with P48, HuP56 can bind to eIF-3, a large complex of 11 protein subunits. This interaction causes an impairment of eIF-3 function and resultant inhibition of protein synthesis[99]. In the case of Hepatitis C virus, it has been shown that HuP56 can suppress the HCV IRES function and translation *in vivo* and *in vitro* by the same mechanism. This effect was also shown to be independent of PKR mediated translational inhibition[100]. eIF-3 catalyzes many steps of translation initiation, such as promoting dissociation of the 40S ribosomal subunit from the 60S subunit, promoting mRNA loading to 40S subunits, and promoting the formation of the eIF2.GTP.tRNA Met_i ternary complex[101]. Among the steps of translation initiation catalyzed by eIF-3, only the enhancement of eIF2.GTP.tRNA Met_i ternary complex formation is impaired by HuP56 binding (Figure 11)[102]. The HuP56/P48 interaction is mediated by specific TPR motifs in HuP56 and a motif in P48 called the PCI motif (Proteasome, COP9-signalosome, Initiation factor). The PCI motif, a long α-helix, is present in specific subunits of three multi-protein complexes: the regulatory subunit of proteasome, subunits of the COP9/signalosome complex which routes target proteins in the nucleus for proteasomal degradation, and the eukaryotic translation initiation factor 3[103]. Three subunits of eIF3 contain the PCI motif: eIF3e/P48, eIF3c/P110 and eIF3a/P170. Surprisingly, MuP56 interacts with eIF3c, not eIF3e. Thus, although the TPR motifs in the P56-related proteins probably recognize the PCI motifs in other proteins, there is a high degree of specificity among cognate pairs. Binding of MuP56 to eIF3 also leads to translation inhibition. But the specific function of eIF3 impaired by MuP56 is different from that affected by HuP56. Unlike HuP56, MuP56 does not appreciably inhibit the stabilization of the eIF2.GTP.tRNA Met_i ternary complex by eIF3. It also does not inhibit eIF3-mediated ribosomal subunit dissociation, eIF3-interaction with the 40S subunit, eIF3-interaction with eIF4F or the interaction of the ternary complex, eIF3 and 40S ribosomal subunit. The ternary complex, eIF3 and eIF4F, can form a 20S complex which is also unaffected by MuP56. However, the interaction of the above 20S complex with the 40S ribosomal subunit is completely blocked by the MuP56 protein (Hui, D., Merrick, W. and Sen, G. unpublished observations). Thus, it appears that different subunits of eIF3 interface with different components of the translation initiation pathway, and two members of the P56 family can block different steps of this process by interacting with relevant eIF3 subunits. It remains to be seen if the same general principle applies to other members of the P56 family as well. Interactions of the P56 family of proteins with other cellular proteins containing PCI motifs remain to be carefully examined. Our preliminary studies indicate that there are such interactions, and the interacting pairs are highly specific (Terenzi, F. and Sen G. unpublished observation). The structural basis of such specificity remains unclear, as do the effects of these interactions on cellular physiology. We can anticipate much new information related to the functional consequences of interactions among members of these two families of proteins, the constitutively expressed PCI proteins and the P56 proteins expressed in response to viral stresses.

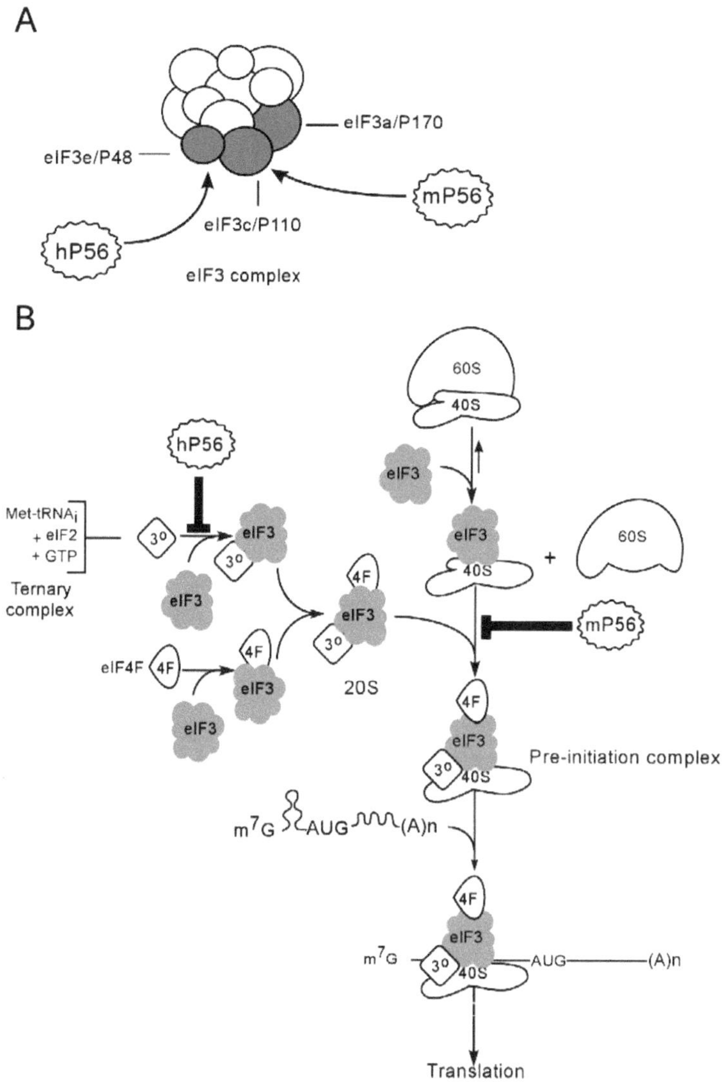

Figure 10. Different steps of eIF3 functions in eukaryotic translational initiation process and blockades by HuP56 and MuP56. (A) Schematic representation of the eIF3 complex, which contains 11 subunits. The PCI domain containing subunits are shown with dark shade. HuP56 interacts with the eIF3e/P48/Int6 subunit, whereas, MuP56 interacts with the eIF3c/P110 subunit. (B) Steps of transnational initiation carried out with the help of eIF3. There are three major steps that are carried out with the help of eIF3: the stabilization of the ternary complex (3°) containing Met-tRNAi, eIF2 and GTP; the association of eIF3 with eIF4F followed by the association with the above ternary complex to form the 20S complex and the release of the 40S subunit of ribosomes to form eIF3-40S complex. Finally, the end products of the above three steps combine together to form the pre-initiation complex. HuP56 inhibits stabilization of the ternary complex by eIF3, whereas MuP56 blocks formation of the pre-initiation complex.

4. CONCLUSIONS

The IFN-inducible genes, the topic of this article, remain to be more intensively studied. Although in this article we have focused on dsRNA and IFNs as the inducers of these genes, other cytokines or TLR-ligands may also fit this bill. Many cytokines, in addition to IFNs, can be induced by virus infection, and many of them can activate some of the signaling pathways that lead to induction of these genes. Similarly, engagement of not only TLR3, but several other TLRs as well, causes activation of NFκB, IRFs or AP-1 signaling pathways. The relevant ligands for these additional TLRs could be viral DNA, single stranded RNAs or proteins[104-107]. More focused studies will be required in the future to examine the above possibilities. A similar question that remains relatively unexplored is "how overlapping are the sets of genes induced by a different family of viruses?"

With regard to the converging signaling pathways used by different viral stress inducers, the type I IFN-signaling pathway is pretty well defined. The corresponding dsRNA signaling pathways are being delineated by several laboratories now, and it is reasonable to expect that all of their major factors will be described in the near future. In contrast, pathways activated by virus infection itself, without IFN or dsRNA participation, are not well understood yet. Even the specific viral products that trigger the signaling pathways remain elusive. For every host-virus combination, determination of the relative contributions of different viral stress-inducers and the corresponding signaling pathways to gene induction, requires systematic studies using genetically deficient mutant cells lines. Unfortunately, such studies have been few and far between, and the concept of IFNs and dsRNA being the major, if not the sole, proximal inducers of genes in virus-infected cells has remained unchallenged.

Understanding the biochemical and cellular functions of proteins encoded by these genes remains a daunting task. The relevant studies have focused primarily on only a few proteins of this large family. But even for the relatively well studied proteins, such as the OAS enzymes, there are large gaps in our knowledge: for example, we do not know why there are so many OAS isozymes, what are their cell-type specificities, what substrates, other than ATP, can they use, what are their roles in uninfected cells and what are their non-enzymatic and non-RNase L-dependent cellular functions? With respect to their functions in virus-infected cells, most of the information is anecdotal and incomplete. The traditional paradigm is that these proteins serve as weapons to protect the cells or the organism against the deleterious effects of virus infection. Given the fact that viruses allow their induction suggests that at least some of them may be beneficial to viruses. Thus, these proteins may have been designed to maintain "viral homeostasis" in infected cells, an equilibrium reached by the host and the virus during a long co-evolution.

Among the specific IFN-induced proteins discussed in this article, HuP56 was the first to be cloned[108]. However, any clues to its functions have come only recently[99]. It appears that all functions of the P56 family of proteins are mediated by their interactions with other cellular proteins. Their TPR motifs indicate that such interactions, although highly specific, could be large in numbers; consequently, they may affect many cellular functions in addition to their effects on translation. Because these proteins are induced for a short time and are short-lived, the cells can probably tolerate their presence only transiently. These proteins have the potential to interact with specific viral proteins and directly inhibit their functions as well, a hypothesis yet to be tested. On the other hand, viruses may have developed defense systems to evade the effects of the P56 family members.

Solution of the first crystal structure of an OAS isozyme has opened the way to understand, at the structural level, how it gets activated and why it catalyzes 2'-5', not 3'-5', bond formation. The recent observations regarding non-enzymatic functions of specific OAS members have broadened our expectations of the plethora of cellular functions that these enzymes may have. Although PKR has been implicated in many regulatory systems beyond its possible role in antiviral defense, what could activate PKR in uninfected cells had remained a mystery until the discovery of PACT, which itself can be activated in response to a cellular stress. Thus, it is quite reasonable to speculate that PACT is the critical mediator between many extracellular stimuli and PKR activation.

ACKNOWLEDGEMENTS

Helpful discussions with other members of the Sen Laboratory are gratefully acknowledged. Our research is supported by National Institutes of Health grants CA 68782 and CA62220.

REFERENCES

1. Biron, C. & Sen, G.C., 2001 Interferons and other cytokines. In Fields Virology, Vol. 1 (Knipe, D. et al. eds.) Lippincott, Williams & Wilkins, Philadelphia, pp.321-351.
2. Stark, G.R., Kerr, I.M., Williams, B.R., Silverman, R.H. & Schreiber, R.D., 1998, How cells respond to interferons. *Annu. Rev. Biochem.* **67**:227-264
3. Sen, G.C., 2001, Viruses and interferons. *Annu. Rev. Microbiol.* **55**:255-281

4. Der, S.D., Zhou, A., Williams, B.R. & Silverman, R.H., 1998, Identification of genes differentially regulated by interferon alpha, beta, or gamma using oligonucleotide arrays. *Proc. Natl. Acad. Sci. USA* **95:**15623-15628
5. Geiss, G. et al., 2001, A comprehensive view of regulation of gene expression by double-stranded RNA-mediated cell signaling. *J. Biol. Chem.* **276:**30178-30182
6. Mossman, K.L. et al., 2001, Herpes simplex virus triggers and then disarms a host antiviral response. *J. Virol.* **75:**750-758
7. Geiss, G.K. et al., 2002, Cellular transcriptional profiling in influenza A virus-infected lung epithelial cells: the role of the nonstructural NS1 protein in the evasion of the host innate defense and its potential contribution to pandemic influenza. *Proc. Natl. Acad. Sci. USA* **99:**10736-10741
8. Zhu, H., Cong, J.P., Mamtora, G., Gingeras, T. & Shenk, T., 1998, Cellular gene expression altered by human cytomegalovirus: global monitoring with oligonucleotide arrays. *Proc. Natl. Acad. Sci. USA* **95:**14470-14475
9. Weaver, B.K., Kumar, K.P. & Reich, N.C., 1998, Interferon Regulatory Factor 3 and CREB-Binding Protein/p300 Are Subunits of Double-Stranded RNA-Activated Transcription Factor DRAF1. *Mol. Cell. Biol.* **18:**1359-1368
10. Bandyopadhyay, S.K., Leonard, G.T., Jr., Bandyopadhyay, T., Stark, G.R. & Sen, G.C., 1995, Transcriptional induction by double-stranded RNA is mediated by interferon-stimulated response elements without activation of interferon-stimulated gene factor 3. *J. Biol. Chem.* **270:**19624-19629
11. Daly, C. & Reich, N., 1993, Double-stranded RNA activates novel factors that bind to the interferon- stimulated response element. *Mol. Cell. Biol.* **13:**3756-3764
12. Chang, Y.E. & Laimins, L.A., 2000, Microarray analysis identifies interferon-inducible genes and Stat-1 as major transcriptional targets of human papillomavirus type 31. *J. Virol.* **74:**4174-4182
13. Geiss, G.K. et al., 2003, Gene expression profiling of the cellular transcriptional network regulated by alpha/beta interferon and its partial attenuation by the hepatitis C virus nonstructural 5A protein. *J. Virol.* **77:**6367-6375
14. Johnston, C., Jiang, W., Chu, T. & Levine, B., 2001, Identification of genes involved in the host response to neurovirulent alphavirus infection. *J. Virol.* **75:**10431-10445
15. Levy, D.E. & Darnell, J.E., Jr., 2002, Stats: transcriptional control and biological impact. *Nat. Rev. Mol. Cell Biol.* **3:**651-662
16. Pestka, S., 1997, The interferon receptors. *Semin. Oncol.* **24:**S9-18-S19-40
17. Darnell, J.E., Jr., Kerr, I.M. & Stark, G.R., 1994, Jak-STAT pathways and transcriptional activation in response to IFNs and other extracellular signaling proteins. *Science.* **264:**1415-1421
18. Alexopoulou, L., Holt, A.C., Medzhitov, R. & Flavell, R.A., 2001, Recognition of double-stranded RNA and activation of NF-kappaB by Toll-like receptor 3. *Nature.* **413:**732-738
19. Chu, W.M. et al., 1999, JNK2 and IKKbeta are required for activating the innate response to viral infection. *Immunity.* **11:**721-731
20. Williams, B.R., 2001, Signal integration via PKR. *Sci STKE.* **2001:**RE2
21. Peters, K.L., Smith, H.L., Stark, G.R. & Sen, G.C., 2002, IRF-3-dependent, NFkappa B- and JNK-independent activation of the 561 and IFN-beta genes in response to double-stranded RNA. *Proc. Natl. Acad. Sci. USA* **99:**6322-6327
22. Oshiumi, H., Matsumoto, M., Funami, K., Akazawa, T. & Seya, T., 2003, TICAM-1, an adaptor molecule that participates in Toll-like receptor 3-mediated interferon-beta induction. *Nat. Immunol.* **4:**161-167
23. Aderem, A. & Ulevitch, R.J., 2000, Toll-like receptors in the induction of the innate immune response. *Nature.* **406:**782-787
24. Akira, S., 2003, Toll-like receptor signaling. *J. Biol. Chem.* **278:**38105-38108

25. Takeda, K., Kaisho, T. & Akira, S., 2003, Toll-like receptors. *Annu. Rev. Immunol.* **21:**335-376
26. Sarkar, S.N., Smith, H.L., Rowe, T.M. & Sen, G.C., 2003, Double-stranded RNA Signaling by Toll-like Receptor 3 Requires Specific Tyrosine Residues in Its Cytoplasmic Domain. *J. Biol. Chem.* **278:**4393-4396
27. Jiang, Z., Mak, T.W., Sen, G. & Li, X., 2004, Toll-like receptor 3-mediated activation of NF-kappaB and IRF3 diverges at Toll-IL-1 receptor domain-containing adapter inducing IFN-beta. *Proc. Natl. Acad. Sci. USA* **101:**3533-3538
28. Fitzgerald, K.A. et al., 2003, IKKepsilon and TBK1 are essential components of the IRF3 signaling pathway. *Nat. Immunol.* **4:**491-496
29. Sharma, S. et al., 2003, Triggering the interferon antiviral response through an IKK-related pathway. *Science.* **300:**1148-1151
30. Sarkar, S.N. et al., 2004, Novel Roles of TLR3 tyrosine phosphorylation and PI3 kinase in dsRNA signaling. *Nat. Sturct. Mol. Biol.* in press
31. Roberts, W.K., Hovanessian, A., Brown, R.E., Clemens, M.J. & Kerr, I.M., 1976, Interferon-mediated protein kinase and low-molecular-weight inhibitor of protein synthesis. *Nature.* **264:**477-480
32. Hovanessian, A.G., Brown, R.E. & Kerr, I.M., 1977, Synthesis of low molecular weight inhibitor of protein synthesis with enzyme from interferon-treated cells. *Nature.* **268:**537-540
33. Rebouillat, D. & Hovanessian, A.G., 1999, The human 2',5'-oligoadenylate synthetase family: interferon-induced proteins with unique enzymatic properties. *J. Interferon Cytokine Res.* **19:**295-308
34. Justesen, J., Hartmann, R. & Kjeldgaard, N.O., 2000, Gene structure and function of the 2'-5'-oligoadenylate synthetase family. *Cell. Mol. Life Sci.* **57:**1593-1612
35. Sarkar, S.N., Bandyopadhyay, S., Ghosh, A. & Sen, G.C., 1999, Enzymatic characteristics of recombinant medium isozyme of 2'-5' oligoadenylate synthetase. *J. Biol. Chem.* **274:**1848-1855
36. Rebouillat, D., Hovnanian, A., Marie, I. & Hovanessian, A.G., 1999, The 100-kDa 2',5'-oligoadenylate synthetase catalyzing preferentially the synthesis of dimeric pppA2'p5'A molecules is composed of three homologous domains [In Process Citation]. *J. Biol. Chem.* **274:**1557-1565
37. Sarkar, S.N., Ghosh, A., Wang, H.W., Sung, S.S. & Sen, G.C., 1999, The nature of the catalytic domain of 2'-5'-oligoadenylate synthetases. *J. Biol. Chem.* **274:**25535-25542
38. Sarkar, S.N., Miyagi, M., Crabb, J.W. & Sen, G.C., 2002, Identification of the substrate-binding sites of 2'-5'-oligoadenylate synthetase. *J. Biol. Chem.* **277:**24321-24330
39. Ghosh, A., Sarkar, S.N., Guo, W., Bandyopadhyay, S. & Sen, G.C., 1997, Enzymatic activity of 2'-5'-oligoadenylate synthetase is impaired by specific mutations that affect oligomerization of the protein. *J. Biol. Chem.* **272:**33220-33226
40. Sarkar, S.N., Pal, S. & Sen, G.C., 2002, Crisscross enzymatic reaction between the two molecules in the active dimeric P69 form of the 2'-5' oligodenylate synthetase. *J. Biol. Chem.* **277:**44760-44764
41. Hartmann, R., Justesen, J., Sarkar, S.N., Sen, G.C. & Yee, V.C., 2003, Crystal structure of the 2'-specific and double-stranded RNA-activated interferon-induced antiviral protein 2'-5'-oligoadenylate synthetase. *Mol. Cell.* **12:**1173-1185
42. Andersen, J.B., Strandbygard, D.J., Hartmann, R. & Justesen, J., 2004, Interaction between the 2'-5' oligoadenylate synthetase-like protein p59 OASL and the transcriptional repressor methyl CpG-binding protein 1. *Eur. J. Biochem.* **271:**628-636
43. Chebath, J., Benech, P., Revel, M. & Vigneron, M., 1987, Constitutive expression of (2'-5') oligo A synthetase confers resistance to picornavirus infection. *Nature.* **330:**587-588
44. Ghosh, A., Sarkar, S.N. & Sen, G.C., 2000, Cell growth regulatory and antiviral effects of the P69 isozyme of 2-5 (A) synthetase. *Virology.* **266:**319-328

45. Maitra, R.K. et al., 1994, HIV-1 TAR RNA has an intrinsic ability to activate interferon-inducible enzymes. *Virology.* **204:**823-827
46. Sharp, T.V. et al., 1999, Activation of the interferon-inducible (2'-5') oligoadenylate synthetase by the Epstein-Barr virus RNA, EBER-1. *Virology.* **257:**303-313
47. Desai, S.Y. et al., 1995, Activation of interferon-inducible 2'-5' oligoadenylate synthetase by adenoviral VAI RNA. *J. Biol. Chem.* **270:**3454-3461
48. Perelygin, A.A. et al., 2002, Positional cloning of the murine flavivirus resistance gene. *Proc. Natl. Acad. Sci. USA* **99:**9322-9327
49. Mashimo, T. et al., 2002, A nonsense mutation in the gene encoding 2'-5'-oligoadenylate synthetase/L1 isoform is associated with West Nile virus susceptibility in laboratory mice. *Proc. Natl. Acad. Sci. USA* **99:**11311-11316
50. Ghosh, A., Sarkar, S.N., Rowe, T.M. & Sen, G.C., 2001, A specific isozyme of 2'-5' oligoadenylate synthetase is a dual function proapoptotic protein of the bcl-2 family. *J. Biol. Chem.* **276:**25447-25455
51. Hartmann, R., Rebouillat, D., Justesen, J., Sen, G.C. & Williams, B.R., 2001, The P59 Oligoadenylate Synthetase like Protein (P59OASL) does not display oligoadenylate synthetase activity but posses antiviral properties conferred by an ubiquitin-like domain. *J. Interferon Cytokine Res.* **21:**S-69
52. Han, J.Q. & Barton, D.J., 2002, Activation and evasion of the antiviral 2'-5' oligoadenylate synthetase/ribonuclease L pathway by hepatitis C virus mRNA. *RNA.* **8:**512-525
53. Taguchi, T. et al., 2004, Hepatitis C virus NS5A protein interacts with 2',5'-oligoadenylate synthetase and inhibits antiviral activity of IFN in an IFN sensitivity-determining region-independent manner. *J. Gen. Virol.* **85:**959-969
54. Silverman, R.H., 2003, Implications for RNase L in prostate cancer biology. *Biochemistry.* **42:**1805-1812
55. Player, M.R. & Torrence, P.F., 1998, The 2-5A system: modulation of viral and cellular processes through acceleration of RNA degradation. *Pharmacol. Ther.* **78:**55-113
56. Dong, B., Niwa, M., Walter, P. & Silverman, R.H., 2001, Basis for regulated RNA cleavage by functional analysis of RNase L and Ire1p. *RNA.* **7:**361-373
57. Dong, B. & Silverman, R.H., 1997, A bipartite model of 2-5A-dependent RNase L. *J. Biol. Chem.* **272:**22236-22242
58. Dong, B. & Silverman, R.H., 1995, 2-5A-dependent RNase molecules dimerize during activation by 2-5A. *J. Biol. Chem.* **270:**4133-4137
59. Suhadolnik, R.J. et al., 1997, Biochemical evidence for a novel low molecular weight 2-5A-dependent RNase L in chronic fatigue syndrome. *J. Interferon Cytokine Res.* **17:**377-385
60. Demettre, E. et al., 2002, Ribonuclease L proteolysis in peripheral blood mononuclear cells of chronic fatigue syndrome patients. *J. Biol. Chem.* **277:**35746-35751
61. Le Roy, F. et al., 2001, The 2-5A/RNase L/RNase L inhibitor (RLI) [correction of (RNI)] pathway regulates mitochondrial mRNAs stability in interferon alpha-treated H9 cells. *J. Biol. Chem.* **276:**48473-48482
62. Zhou, A. et al., 1997, Interferon action and apoptosis are defective in mice devoid of 2',5'-oligoadenylate-dependent RNase L. *EMBO J.* **16:**6355-6363
63. Li, G., Xiang, Y., Sabapathy, K. & Silverman, R.H., 2004, An apoptotic signaling pathway in the interferon antiviral response mediated by RNase L and c-Jun NH2-terminal kinase. *J. Biol. Chem.* **279:**1123-1131
64. Carpten, J. et al., 2002, Germline mutations in the ribonuclease L gene in families showing linkage with HPC1. *Nat. Genet.* **30:**181-184
65. Wang, L. et al., 2002, Analysis of the RNASEL gene in familial and sporadic prostate cancer. *Am. J. Hum. Genet.* **71:**116-123
66. Casey, G. et al., 2002, RNASEL Arg462Gln variant is implicated in up to 13% of prostate cancer cases. *Nat. Genet.* **32:**581-583

67. Ito, T., Yang, M. & May, W.S., 1999, RAX, a cellular activator for double-stranded RNA-dependent protein kinase during stress signaling. *J. Biol. Chem.* **274:**15427-15432
68. Patel, C.V., Handy, I., Goldsmith, T. & Patel, R.C., 2000, PACT, a stress-modulated cellular activator of interferon-induced double-stranded RNA-activated protein kinase, PKR. *J. Biol. Chem.* **275:**37993-37998
69. Peters, G.A., Hartmann, R., Qin, J. & Sen, G.C., 2001, Modular structure of PACT: distinct domains for binding and activating PKR. *Mol. Cell. Biol.* **21:**1908-1920
70. Clemens, M.J. & Elia, A., 1997, The double-stranded RNA-dependent protein kinase PKR: structure and function. *J. Interferon Cytokine Res.* **17:**503-524
71. Nanduri, S., Rahman, F., Williams, B.R. & Qin, J., 2000, A dynamically tuned double-stranded RNA binding mechanism for the activation of antiviral kinase PKR. *EMBO J.* **19:**5567-5574
72. Patel, R.C. & Sen, G.C., 1998, PACT, a protein activator of the interferon-induced protein kinase, PKR. *EMBO J.* **17:**4379-4390
73. Patel, R.C. et al., 1999, DRBP76, a double-stranded RNA-binding nuclear protein, is phosphorylated by the interferon-induced protein kinase, PKR. *J. Biol. Chem.* **274:**20432-20437
74. Shim, J., Lim, H., J, R.Y. & Karin, M., 2002, Nuclear export of NF90 is required for interleukin-2 mRNA stabilization. *Mol. Cell.* **10:**1331-1344
75. Xu, Z. & Williams, B.R., 2000, The B56alpha regulatory subunit of protein phosphatase 2A is a target for regulation by double-stranded RNA-dependent protein kinase PKR. *Mol. Cell. Biol.* **20:**5285-5299
76. Tan, S.L., Tareen, S.U., Melville, M.W., Blakely, C.M. & Katze, M.G., 2002, The direct binding of the catalytic subunit of protein phosphatase 1 to the PKR protein kinase is necessary but not sufficient for inactivation and disruption of enzyme dimer formation. *J. Biol. Chem.* **277:**36109-36117
77. Gale, M., Jr. & Katze, M.G., 1998, Molecular mechanisms of interferon resistance mediated by viral-directed inhibition of PKR, the interferon-induced protein kinase. *Pharmacol. Ther.* **78:**29-46
78. Katze, M.G., He, Y. & Gale, M., Jr., 2002, Viruses and interferon: a fight for supremacy. *Nat. Rev. Immunol.* **2:**675-687
79. Gale, M., Jr. et al., 1998, Control of PKR protein kinase by hepatitis C virus nonstructural 5A protein: molecular mechanisms of kinase regulation. *Mol. Cell. Biol.* **18:**5208-5218
80. Taylor, D.R., Shi, S.T., Romano, P.R., Barber, G.N. & Lai, M.M., 1999, Inhibition of the interferon-inducible protein kinase PKR by HCV E2 protein. *Science.* **285:**107-110
81. Burysek, L. & Pitha, P.M., 2001, Latently expressed human herpesvirus 8-encoded interferon regulatory factor 2 inhibits double-stranded RNA-activated protein kinase. *J. Virol.* **75:**2345-2352
82. Poppers, J., Mulvey, M., Khoo, D. & Mohr, I., 2000, Inhibition of PKR activation by the proline-rich RNA binding domain of the herpes simplex virus type 1 Us11 protein. *J. Virol.* **74:**11215-11221
83. Peters, G.A., Khoo, D., Mohr, I. & Sen, G.C., 2002, Inhibition of PACT-mediated activation of PKR by the herpes simplex virus type 1 Us11 protein. *J. Virol.* **76:**11054-11064
84. He, B., Gross, M. & Roizman, B., 1997, The gamma(1)34.5 protein of herpes simplex virus 1 complexes with protein phosphatase 1alpha to dephosphorylate the alpha subunit of the eukaryotic translation initiation factor 2 and preclude the shutoff of protein synthesis by double-stranded RNA-activated protein kinase. *Proc. Natl. Acad. Sci. USA* **94:**843-848
85. Zinn, K., Keller, A., Whittemore, L.A. & Maniatis, T., 1988, 2-Aminopurine selectively inhibits the induction of beta-interferon, c-fos, and c-myc gene expression. *Science.* **240:**210-213

86. Tiwari, R.K., Kusari, J., Kumar, R. & Sen, G.C., 1988, Gene induction by interferons and double-stranded RNA: selective inhibition by 2-aminopurine. *Mol. Cell. Biol.* **8:**4289-4294
87. Kumar, A., Haque, J., Lacoste, J., Hiscott, J. & Williams, B.R., 1994, Double-stranded RNA-dependent protein kinase activates transcription factor NF-kappa B by phosphorylating I kappa B. *Proc. Natl. Acad. Sci. USA* **91:**6288-6292
88. Maran, A. et al., 1994, Blockage of NF-kappa B signaling by selective ablation of an mRNA target by 2-5A antisense chimeras. *Science.* **265:**789-792
89. Yang, Y.L. et al., 1995, Deficient signaling in mice devoid of double-stranded RNA-dependent protein kinase. *EMBO J.* **14:**6095-6106
90. Kumar, A. et al., 1997, Deficient cytokine signaling in mouse embryo fibroblasts with a targeted deletion in the PKR gene: role of IRF-1 and NF-kappaB. *EMBO J.* **16:**406-416
91. Diebold, S.S. et al., 2003, Viral infection switches non-plasmacytoid dendritic cells into high interferon producers. *Nature.* **424:**324-328
92. Grandvaux, N. et al., 2002, Transcriptional profiling of interferon regulatory factor 3 target genes: direct involvement in the regulation of interferon-stimulated genes. *J. Virol.* **76:**5532-5539
93. Guo, J., Peters, K.L. & Sen, G.C., 2000, Induction of the human protein P56 by interferon, double-stranded RNA, or virus infection. *Virology.* **267:**209-219
94. Lamb, J.R., Tugendreich, S. & Hieter, P., 1995, Tetratrico peptide repeat interactions: to TPR or not to TPR? *Trends Biochem. Sci.* **20:**257-259
95. Blatch, G.L. & Lassle, M., 1999, The tetratricopeptide repeat: a structural motif mediating protein-protein interactions. *Bioessays.* **21:**932-939
96. D'Andrea, L.D. & Regan, L., 2003, TPR proteins: the versatile helix. *Trends Biochem. Sci.* **28:**655-662
97. Marchetti, A. et al., 1995, Int-6, a highly conserved, widely expressed gene, is mutated by mouse mammary tumor virus in mammary preneoplasia. *J. Virol.* **69:**1932-1938
98. Asano, K., Merrick, W.C. & Hershey, J.W., 1997, The translation initiation factor eIF3-p48 subunit is encoded by int-6, a site of frequent integration by the mouse mammary tumor virus genome. *J. Biol. Chem.* **272:**23477-23480
99. Guo, J., Hui, D.J., Merrick, W.C. & Sen, G.C., 2000, A new pathway of translational regulation mediated by eukaryotic initiation factor 3. *EMBO J.* **19:**6891-6899
100. Gale, M., Jr. et al., 2002, Translational control of Hepatitis C Virus (HCV) RNA replication through PKR and P56. *J. Interferon Cytokine Res.* **22:**S-63
101. Hershey, J.W. & Merrick, W.C., 2000 Pathway and mechanism of initiation of protein synthesis. In *Translational Control of Gene Expression* (Sonenberg, N., Hershey, J.W. & Mathews, M.B. eds.) Cold Spring Harbor Laboratory Press., New York, pp185–243
102. Hui, D.J., Bhasker, C.R., Merrick, W.C. & Sen, G.C., 2003, Viral stress inducible protein P56 inhibits translation by blocking the interaction of eIF3 with the ternary complex eIF22GTP2Met-tRNAi. *J. Biol. Chem.* **278:**39477-39482
103. Hofmann, K. & Bucher, P., 1998, The PCI domain: a common theme in three multiprotein complexes. *Trends Biochem. Sci.* **23:**204-205
104. Heil, F. et al., 2004, Species-specific recognition of single-stranded RNA via toll-like receptor 7 and 8. *Science.* **303:**1526-1529
105. Chin, K.-C. & Cresswell, P., 2001, Inaugural Article: Viperin (cig5), an IFN-inducible antiviral protein directly induced by human cytomegalovirus. *PNAS.* **98:**15125-15130
106. Lund, J., Sato, A., Akira, S., Medzhitov, R. & Iwasaki, A., 2003, Toll-like receptor 9-mediated recognition of Herpes simplex virus-2 by plasmacytoid dendritic cells. *J. Exp. Med.* **198:**513-520
107. Akira, S. & Hemmi, H., 2003, Recognition of pathogen-associated molecular patterns by TLR family. *Immunol. Lett.* **85:**85-95

108. Chebath, J., Merlin, G., Metz, R., Benech, P. & Revel, M., 1983, Interferon-induced 56,000 Mr protein and its mRNA in human cells: molecular cloning and partial sequence of the cDNA. *Nucleic Acids Res.* **11:**1213-1226

Chapter 4

IMMUNOEVASIVE STRATEGIES: HOST AND VIRUS

MARKUS WAGNER, SHAHRAM MISAGHI, and HIDDE L. PLOEGH
Harvard Medical School, Department of Pathology, 77 Avenue Louis Pasteur, Boston, MA, USA

1. INTRODUCTION

Despite their heterogeneity, viruses from different families, including RNA and DNA viruses, have evolved similar gene functions that target many common cellular targets for immunoevasion. In most cases the goal is not a complete escape from recognition of the immune system, since this would destroy the host and limit virus replication and spread, but rather a balance between clearance and persistence that allows coexistence of the virus and its immunocompetent host. Nevertheless, there are fundamental differences between large DNA viruses (e.g. herpesviruses, poxviruses and adenoviruses) and small RNA viruses (e.g. retroviruses, picornaviruses and myxoviruses) where immunoevasion is concerned. The latter have genomes of limited size, a trait possibly linked to the low fidelity of the viral RNA polymerase. These RNA viruses therefore mainly carry multifunctional genes that are essential for virus replication and their genome does not have a lot of exon space for accessory immunomodulatory genes. Rather, they rely on a high mutation rate to alter the antigenicity of the virus, thus escaping B- and T-cell recognition and antiviral antibodies (see example of Influenza virus type A below). Another general strategy of the fast-replicating small RNA viruses is to overwhelm the host with an enormous number of infectious particles (high virus load) following the motto "mass versus cleverness." Before complete clearance of the infection from the host,

P. Palese (ed.), Modulation of Host Gene Expression and Innate Immunity by Viruses, 65-94.

infectious progeny can already infect new hosts. For instance, one single aerosolized droplet produced upon sneezing can contain up to 100 million rhinovirus particles.

In contrast, DNA viruses tend to have larger genomes. This allows maintenance of a large number of accessory genes that are non-essential for virus replication but important for immunomodulation. This results in more sophisticated strategies of co-existence in the host and virus spread over time. Murine and human cytomegaloviruses (CMVs) each contain more than 100 genes that are dispensable for replication in cell culture. Almost certainly their products are required for immunomodulation in the host (Wagner M., unpublished data and [1]). Several of these gene functions are homologous to those of cellular genes and were originally "stolen" from the host and are misused now by the virus to its own benefit (e.g. homologs of major histocompatibility complex (MHC) class I molecules, Fc receptors, GPCRs and chemokines). A second group of host- and immunomodulatory proteins, capable of affecting the humoral immune response, the cellular immune response and immune effector functions, does not show any sequence homology to cellular genes. In the latter case they might have been evolved by the viruses during co-evolution in the host or their cellular homolog remains to be identified yet.

First, we shall give a very brief overview on the different layers of the antiviral immune defense actively targeted by viral protein functions. Viruses are intracellular parasites that use the blood and lymph system only for spread within the body after initial infection. Virus replication always takes place within a host cell. For extracellular elimination of viruses, the innate immune system, comprised of the complement system and phagocytosing immune cells (neutrophils and macrophages) are important. When infection recurs, the adaptive immune response, in the form of B-cell derived antibodies is essential for extracellular elimination and may lead to sterilizing immunity. Once within a cell, the virus takes over the protein synthesis machinery and produces a large number of progeny viruses. Therefore, the host must try to eliminate infected cells before they can release new viruses. This may be achieved by triggering apoptosis (programmed cell death) in infected cells, or by killing of infected cells by natural killer (NK) cells, macrophages (innate immunity) or by cytotoxic T-lymphocytes (CTLs) (adaptive immunity). The immune cells recognize infected cells by antigens presented by class I and II MHC molecules and by alterations of surface expression of other molecules due to infection. Consequently, a prevalent strategy of viruses to avoid elimination is to interfere with the processing of antigenic peptides and the expression and trafficking of class I and II MHC molecules as well as those of other co-stimulatory molecules to avoid recognition and activation of immune cells, like CTLs and NK cells. Since cytokines can exert a direct antiviral effect (e.g. interferons) and are also important for activation and recruitment of cells involved in the immune response, viruses manipulate and even mimic cytokine signaling for their benefit. Finally, some viruses can directly infect cells of the immune system (e.g. T-, B-, NK-cells, macrophages and dendritic cells) and so impair their function.

Besides the interest of virologists in these host-modulatory viral proteins, their characterization and use of them as tools also allow immunologists and cell biologists to learn more about the cellular principles that are affected by these viral proteins. This review provides an overview of the basic concepts of active manipulation of the host by immunomodulatory viral functions, with a more detailed description on viral interference with antigen presentation by class I MHC and with NK cells. Those readers interested in the details of specific immunoevasive interaction besides class I MHC and NK cells, are referred to relevant review articles.

2. AVOIDANCE OF IMMUNE RECOGNITION BY PASSIVE STRATEGIES

Immune evasion of viruses from the host's immune system is not necessarily an active process in which the virus manipulates the host for its benefit. General aspects of virus replication in the host already contribute to efficient immune evasion. Unlike a cytopathic infection, a non-cytopathic infection that does not lead to cell death and subsequent activation of neutrophils, macrophages and dendritic cells is already an immune evasion strategy, since it avoids or delays an immune response. The same is true for cell-to-cell spread of viruses, which circumvents extracellular exposure of infectious virus particles to antibodies or complement.

Several cell types and organs of the body show less stringent immune surveillance and reduced accessibility for components of the immune system ("privileged" tissues). For instance, neurons, muscle cells, glandular epithelial cells or keratocytes in the skin do not express sufficient levels of class I MHC molecules for activation of CD8+ T cells. Restriction of virus replication to such cells or organs (tropism) may allow establishment of a persistent infection and can therefore be considered an immuoevasive strategy. An example of this is the persistent infection of salivary glands with cytomegalovirus (CMV), a beta-herpesvirus, which allows long-term secretion of virus in the saliva. The replication of papillomavirus in the terminally differentiated outer skin layer, causing skin warts, is beyond the reach of immune control and can be seen as an extreme example of such a strategy. Furthermore, viruses that can efficiently infect the early developing fetus, which does not have a competent immune system yet, might induce immunological tolerance and thereby avoid immune recognition. An example of the latter is infection of the pregnant female with rubella virus, which results in infection of the fetus in 80% of cases when it occurs within the first trimester of pregnancy.

Herpesviruses (e.g. herpes simplex virus (HSV), CMV or Epstein Barr virus (EBV)) have evolved a special form of co-existence with the host called latency. After clearance of primary infection, the viral genome is maintained in a few host cells as an episome (self-replicating circular virus genome) for the remainder of the host life span. With the exception of very few viral proteins, no viral antigens and infectious particles are produced and the virus thus hides from immune control. Specific stimuli can induce spontaneous production of virus particles (reactivation) limited in time, which then allows virus spread. In a similar manner, the integration of the adeno-associated viral genome into chromosome 19 circumvents expression of viral antigens if no helper virus is present.

3. CHANGE OF THE ANTIGENIC REPERTOIRE (ANTIGENIC VARIATION)

One of the first active immunoevasive mechanisms described for viruses was antigenic variation as a means of escape from the humoral response mediated by B-cells. This phenomenon is observed mainly with RNA viruses which use low fidelity RNA polymerases for replication and accumulate in average 10^{-4} to 10^{-5} mutations per replication cycle in their genomes, compared to 10^{-8} to 10^{-11} for DNA viruses. As a consequence, these mutations yield amino acid variations in immunogenic epitopes in the virus progeny, causing altered immunogenicity (antigenic drift). The altered proteins and epitopes may no longer be recognized by specific antibodies or B- and T-cell receptors. One prominent example for an antigenic drift is the accumulation of point mutations in the hemagglutinin and neuraminidase proteins of Influenza type 1 virus, a virus with a segmented RNA genome[2]. This virus produces a large number of virus progeny with structural variations in the hemagglutinin and the neuraminidase, resulting in a collection of differing viruses (quasispecies). Some of the mutants can escape from antibody recognition, since the epitopes that carry an amino acid exchange may no longer be recognized. An even more drastic change in immunogenicity of the Influenza type 1 virus can be caused by a second phenomenon, called antigenic shift. During infection of a secondary host, whole RNA segments of the viral genome can be exchanged between different viral strains, resulting in the creation of a new virus strain with major changes in the viral surface proteins. In these cases, antibodies directed against previous encountered virus strains are no longer protective and severe influenza pandemics are the result[2].

An example of antigenic shift that affects the recognition by specific cytotoxic T cells (CTLs) is HIV. HIV epitopes are processed and presented to CTLs by class I MHC molecules on the surface of infected cells. An epitope with a mutation can still be processed and presented to CTLs. But the epitope might no longer be recognized by the specific CTL. Another scenario is that mutated epitopes can still be recognized by CTLs that are directed against the cognate epitope sequence, but due to the mutation, these CTLs bail to be properly activated and are inactivated or "anergized"[3]. Hereby the mutant epitope inactivates a specific CTL that otherwise would have been able to eliminate virus variants that still carry the cognate epitope.

4. ACTIVE MANIPULATION OF HOST CELLS AND IMMUNE FUNCTIONS BY VIRAL PROTEINS

Every manipulation of the cell by a virus evokes an altered response of the cell to virus infection. This might also affect the ensuing immune response and therefore all these manipulations can be considered as immunomodulatory or immunoevasive mechanisms. There are many examples of viruses that interfere with the host cell cycle and gene expression. For instance, CMV leads to a G1 cell cycle arrest in infected cells. Another herpesvirus, HSV, expresses the virus host shut-off (vhs) protein which leads to a general arrest in cellular gene expression; coincidentally this also lowers the expression of, for example, class I MHC molecules on latently infected cells and leads to reduced CTL activation, allowing the virus to reactivate. Details on the interference of viruses with cellular transcription and translation are discussed in other chapters of this book.

We shall focus on viral proteins that actively influence or compromise host cell functions at the protein level.

4.1 Interference with cytokine and chemokine functions and related intracellular signaling

Viruses intensively interfere with cytokine and chemokine signaling, which plays a key role in initiation and regulation of the innate and adaptive immune responses. Viruses have evolved mechanisms to block synthesis, activity and signal transduction of cytokines and chemokines, and so affect immunosuppressive, anti- and pro-inflammatory and chemotactic processes. This is achieved by blocking cytokine expression, alteration of cytokine activity, interference with cytokine induced signal transduction and by viral mimicry of cytokines or chemokines and their receptors[4].

Other chapters of this book also discuss immunoevasive mechanisms concerning cytokines and chemokines.

4.2 Impairment of extracellular recognition of viruses by the humoral immune response

Variation of the antigenicity of viral proteins by mutation and genome rearrangement is an indirect strategy used mainly by small RNA viruses to escape recognition from virus-specific antibodies. But viruses also directly influence the humoral response and complement activation by expression of Fc receptor homologs and by inhibition of complement activation. The latter can be achieved through i) induction of the expression of cellular complement regulators or inhibitors, which may even be incorporated into virions, ii) viral expression of complement blocking proteins and iii) by viral mimicry of cellular complement regulators. For a detailed review on viral evasion of the humoral immune response see [5-7].

4.3 Impairment of host mediated killing of infected cells

Once the virus is inside a cell, the main strategy of the host is to limit virus replication and spread and to eliminate the infected cell from the body. This can be achieved by intracellular mediated killing of the infected cells by apoptosis or by extracellular killing through CTLs and NK cells.

4.3.1 Inhibition of apoptosis in infected cells

Apoptosis, also called programmed cell death, is induced by ligands of the TNF family, irradiation, DNA damage, cell cycle inhibitors and pathogens like viruses. To limit virus replication in infected cells, apoptosis is triggered as an innate cellular response upon infection. In addition, NK cells and CTLs kill virus-infected cells by induction of apoptosis via the secretion of perforins, granzymes, cytokines (e.g. TNF) or by activation of Fas. Therefore, many viruses express anti-apoptotic gene products to keep the infected cell alive and to sustain virus replication. To this end, diverse viral mechanisms evolved, such as inhibition of apoptosis inducing cellular proteins (e.g. p53), mimicry of the anti-apoptotic cellular Bcl-2 protein, modulation of TNF-R and Fas signaling or inhibition of caspases. Some viruses also actively induce apoptosis to their benefits. We refer to other reviews covering viral interference with apoptosis[7,8].

4.3.2 Impairment of class I and II MHC mediated antigen presentation in infected cells and protection from CTL killing

A major countermeasure by viruses is the impairment of class I and II MHC mediated antigen presentation of intracellular viral antigens to T cells at many different steps, at both the transcriptional and posttranslational level.

Viruses efficiently interfere with class I MHC antigen presentation at each step of this process and this includes interference with antigenic peptide generation by the proteasome in the cytosol, transport of the peptides to the endoplasmatic reticulum (ER) and loading on class I MHC molecules, trafficking of the antigen loaded class I MHC complexes to the cell surface and presentation to CTLs (Figure 1).

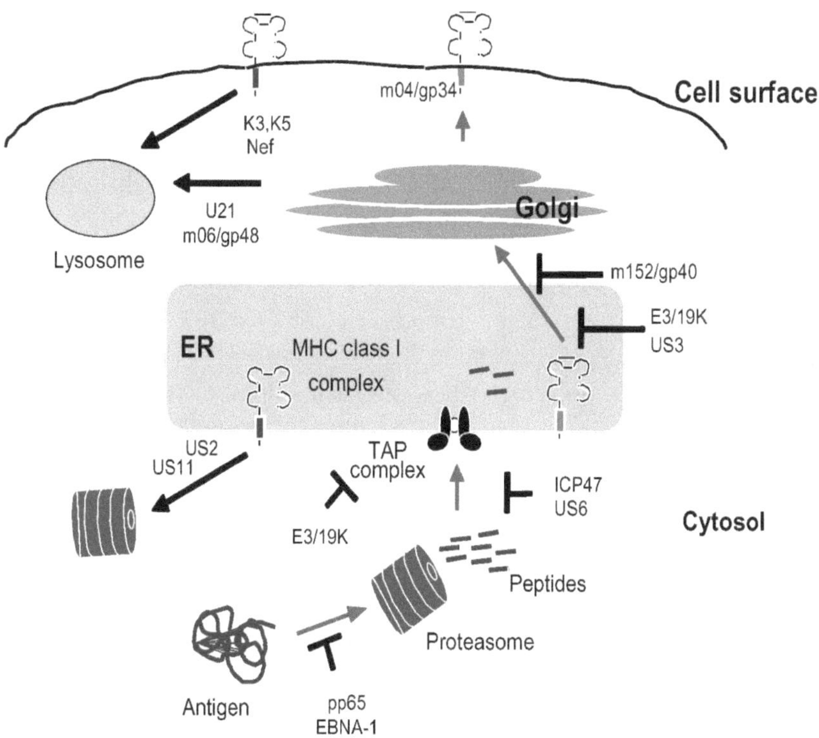

Figure 1. Interference of viral proteins with major histocompatibility complex (MHC) class I mediated antigen presentation. Proteins pp65 from human cytomegalovirus (HCMV) and EBNA-1 from Epstein-Barr virus (EBV) inhibit proteosomal antigen degradation. ICP47 from HSV and US6 from HCMV interfere with the transporter associated with antigen presentation (TAP)-mediated peptide translocation into the endoplasmic reticulum (ER). The adenoviral protein E3/19k can inhibit tapasin by binding to the TAP-complex and retains MHC class I molecules in the ER. The latter is also achieved by the HCMV protein US3. Similarly, the murine CMV (MCMV) protein m152/gp40 retains MHC class I molecules in the ER-Golgi intermediate compartment/cis-Golgi network. The HCMV proteins US2 and US11 target MHC class I molecules into the cytosol for proteasomal degradation. The MCMV protein m04/gp34 binds to MHC class I molecules and protects against cytotoxic T cell mediated

killing by a still unknown mechanism. The human herpes virus 8 proteins K3 and K5 induce enhanced internalization and lysosomal degradation of MHC class I molecules, as does the Nef protein from HIV. The U21 protein of human herpesvirus 7 and m06/gp48 from MCMV directly target class I molecules to lysosomes. Interfering viral proteins are indicated.

Interference with proteasomal cleavage of viral proteins

Viruses can change by mutation and this may result in a specific viral protein's ineffective proteolytic cleavage by the proteasome into antigenic peptides. An example of this is the impaired processing of an immunodominant CTL epitope of murine leukemia virus as the consequence of a single amino acid (aa) change in a proteolytic cleavage site[9]. Viruses can also actively prevent proteolytic processing of viral proteins by the proteasome by incorporation of unfavorable secondary structures into the viral proteins or by posttranslational modifications. The EBV-encoded nuclear antigen 1 (EBNA-1) contains a Gly-Ala co-repeat which inhibits ubiquitin-proteasome dependent proteolysis of this protein[10] and prevents generation of CTL epitopes[11]. The Gly-Ala co-repeat most likely assumes a beta-sheet conformation that does not easily unfold and so blocks entry into the proteasome[12]. An example of a posttranslational modification that prevents proteasomal degradation is phosphorylation of immediate early (IE) proteins at threonine residues by the HCMV protein pp65 (UL83)[13].

Class I MHC molecules

The class I MHC product is a complex composed of three subunits: a type one membrane glycoprotein heavy chain (HC) molecule, a β_2-microglobulin (β_2m) molecule which associates non-covalently with a HC, and a short peptide (8-12aa) that occupies the peptide binding groove present in the complex. Class I MHC molecules traffic to the cell surface and are involved in presenting a diverse selection of intracellular peptides to CD8+ T-cells, which can detect peptides derived from foreign proteins, e.g. from viruses, and specifically destroy cells harboring foreign antigens. Interfering with class I MHC antigen presentation allows viruses to avoid T-cell mediated elimination.

Antigen presentation by class I MHC molecules

Peptide presentation by class I MHC molecules is a complex but fascinating process and calls for many different protein complexes to act in concert to accomplish this task. The first step involves co-translational translocation of HC and β_2m into the ER lumen through the Sec61 complex (a protein channel involved in transporting proteins into the ER and secretary pathway), followed by chaperone-assisted folding and association of these two molecules. The HC-β_2m complex then associates with tapasin and is escorted to a transmembrane channel called the transporter associated with antigen presentation (TAP). TAP, a MHC-encoded heterodimer, transfers small peptides derived from proteasomal mediated degradation of intracellular proteins from the cytoplasm to the ER in an ATP-dependent manner. These peptides are suitable for binding to the class I MHC peptide binding groove, a process in which an MHC encoded chaperone, tapasin, is involved. Only peptide loaded class I MHC molecules dissociate from the TAP/tapasin complex and proceed through the secretory pathway from the ER to the Golgi and finally to the cell surface, where they present their cargo to CD8+ T-cells.

Viral interference with class I MHC mediated antigen presentation

The ability to evade the immune system by reducing surface expression of antigen presenting molecules has been described for many different viruses. The first observation was made with Adenovirus type 2[14]. The E3/19K gene in the E3 region of the Adenovirus type 2 expresses a 19 kDa protein which down-regulates surface expression of class I MHC molecules by mediating ER retention and also – for the fraction that may escape to the cell surface - by binding to the site necessary for interaction with T-cell receptors[15,16]. Herpes simplex viruses (HSV)-1 and 2 prevent loading of antigenic peptides on class I MHC molecules[17-20], while Human Herpesvirus 7 (HHV-7) expresses a protein called U21 which binds to class I MHC molecules and diverts them to lysosomal compartments, resulting in down regulation of surface expression of class I MHC molecules[21,22]. Human Herpesvirus 8 (HHV-8) and Human Immunodeficiency Virus (HIV) encode K3, K5 and the nef protein, respectively, which bind to class I MHC molecules and divert them to lysosomal compartments. The K3 and K5 genes of HHV-8 mediate enhanced internalization of class I MHC molecules after they reach the cell surface[23,24]. On the other hand, the nef protein of HIV misroutes the class I MHC molecules to lysosomes before they reach the cell surface by AP-1 and PACS complexes in clathrin-coated vesicles[25,26].

CMV, which is one of the best studied herpesviruses as far as interference with class I MHC biogenesis is concerned, employs several different proteins that all interfere with class I MHC surface expression at various stages of class I maturation[27-36]. These immunoevasions intercept class I MHC molecules at different stages of assembly and maturation and effectively down regulate their surface expression by different strategies. MCMV encodes 3 immunoevasions called m04/gp34, m06/gp48, and m152/gp60. m06/gp48 leads to lysosomal degradation of class I MHC molecules[32] while m152/gp60 retains class I MHC molecules in the ER[30,31]. m04/gp34 is a unique immunoevasion since it binds to class I MHC but does not alter its surface expression[33-35]. The exact mechanism that leads to evasion from CD8+ T-cells is still unknown, but it has been suggested that m04/gp34 may interfere either with peptide loading of class I MHC molecules in the ER, or it may interrupt the contact between TCR and the loaded class I MHC on the cell surface[34,35].

HCMV has four immunoevasive genes that are located at the unique short (US) region of the HCMV genome. For this reason, they are referred to as US2, US3, US6, and US11. US2 and US11 dislocate class I MHC molecules from the ER luman to the cytoplasm, resulting in the proteasomal mediated degradation of HC molecules. US6 binds to the TAP at the ER luminal side and prevents transport of the antigenic peptides into the ER, while US3 binds to the class I MHC molecules and retains them in the ER, preventing their progression to the cell surface. In addition, two other US genes, US8 and US10, have been shown to bind to class I MHC molecules[37,38]. Here, we will discuss each of the HCMV immunoevasions in more detail and analyze their mode of action separately. It should be kept in mind that down regulating surface expression of class I MHC molecules is the ultimate goal of all of these HCMV immunoevasions.

US6 interrupts transport of antigenic peptides into the ER

HCMV-encoded US6, an ER glycoprotein expressed at early times of infection, binds to the TAP complex at the ER-luminal side and prevents translocation of peptides from the cytosol into the ER. This results in retention of class I MHC molecules in the ER, since they are no longer loaded with antigenic peptides. US6 and TAP have similar localization in the ER, and US6-mediated TAP inhibition is independent of the presence of class I MHC and tapasin molecules[39-41]. TAP hydrolyzes ATP in order to transport antigenic peptides into the ER, where these are loaded on the class I MHC molecules, exported to the cell surface and displayed to CD8+ T-cells. US6 inactivates the whole TAP complex by preventing TAP1 from binding to ATP through a conformational effect, since ATP binding is inhibited at the cytoplasmic side of TAP1. US6 does not interfere with binding of the antigenic peptides to TAP, but may prevent the peptide-binding induced conformational rearrangement of the TAP complex, required for ATP binding[42,43]. Interestingly, cell surface expression of HLA-E, which inhibits NK cell mediated lysis of host cells, is not affected by US6, even though the delivery of HLA-E ligands is generally considered TAP dependent[44]. Therefore, the functional consequence of US6 mediated TAP inhibition is to hamper detection of HCMV infected cells by CD8+ T-cells through reduction of surface expression of class I molecules loaded with antigenic peptides.

US2 degrades assembled HC molecules

US2 is an ER resident glycoprotein that is expressed at early times of HCMV infection and can dislocate class I MHC molecules from the ER to the cytoplasm. Dislocation of class I MHC molecules results in dissociation of HC and β_2m, followed by their rapid proteasomal mediated degradation. The Sec61 complex may be involved in class I MHC dislocation, perhaps by a process that is reversal of translocation[45,46]. US2 mediated dislocation requires the cytoplasmic tail of class I MHC molecules, despite the fact that US2 maintains its capability to bind to class I molecules lacking the cytosolic domain. Hence, substrate recognition and dislocation by US2 are separable processes. This observation suggests that US2 binds to class I

MHC molecules and brings them to a protein channel or to the translocon (possibly Sec61), where the cytoplasmic tail of class I MHC molecules is recognized by a yet unknown protein(s) that mediate(s) their dislocation[47]. US2 binds through its ER-luminal domain to properly folded class I MHC molecules without significantly altering their conformation[48]. In addition, recombinant US2 only binds to HLA-A alleles but not HLA-B7, HLA-B27, HLA-Cw4, or HLA-E alleles. This strategy may result in selective down regulation of alleles that are involved in activation of CD8+ T-cells while preserving surface expression of NK inhibitory alleles[48,49]. The atomic structure of the luminal domain of US2 bound to the soluble and peptide-loaded luminal domain of the class I MHC molecule (HLA-A2 allele) reveals that US2 has a immunoglobulin (Ig)-like fold consisting of 7 beta sheets, forming a β-sandwich structure[50]. Several residues at the junction of the peptide binding region and the α_3 domain of the class I MHC molecule are involved in the interaction with US2[50]. This constitutes a unique US2 binding site, which does not overlap with regions of the class I molecule that are recognized by T-cell receptors or NK receptors. Sequence alignment of US gene products revealed that several additional HCMV genes may also have Ig-like folds[50,51]. Although US2 alone seems to efficiently down regulate surface expression of class I MHC molecules, HCMV utilizes a barrage of other proteins to prevent progression of class I molecules to the cell surface.

US11 Targets assembled and unassembled HC molecules for destruction

US11 is another early gene product of HCMV that is involved in degradation of class I MHC molecules. US11 is a type I transmembrane glycoporein capable of dislocating class I MHC molecules from the ER to the cytosol, a process that is promptly followed by proteasomal mediated degradation of dislocated HC molecules[45,46]. Although US11 and US2 seem to accomplish the same task, their substrate recognition and dislocation strategies utilize fundamentally different protein partners and complexes. Contrary to US2, which requires the intact and original cytoplasmic tail of class I molecules for their dislocation, class I molecules whose cytoplasmic tail are replaced by a random amino acid sequence are still substrates for US11 mediated dislocation (unpublished data). In addition, US2 is capable of dislocating HC molecules only when they are in complex with β_2m, while US11 can recognize and dislocate HC molecules to the cytosol regardless of their association with β_2m[52]. US11 binds to class I MHC molecules via its ER luminal domain, but its transmembrane domain is essential for dislocation of HC molecules. The transmembrane domain of US11 (specifically aa Q192) is important for interacting with the dislocation machinery in order to deliver HC molecules to this complex[53]. Recent studies have identified a novel protein called Derlin-1 which is recruited by the transmembrane domain of US11 and is directly involved in dislocation of HC molecules from the ER to the cytosol[54,55]. Derlin-1, however, is not involved in US2 mediated dislocation of HC molecules, suggesting that US2 and US11 employ different dislocation machineries to target HC molecules to the cytoplasm[54].

US3 leads to ER retention of class I MHC molecules

The US3 gene is an immediate early gene that encodes a type I membrane glycoprotein immediately following HCMV infection and is capable of efficient down regulation of surface expression of class I MHC molecules[56,57]. US3 binds to assembled class I MHC molecules through its transmembrane domain[58,59] while retaining itself in the ER through interaction of its luminal domain residues (aa S58, E63 and K64) with ER resident proteins[60]. US3-mediated binding and retention of class I MHC molecules in the ER, although efficient, is accomplished through transient association of US3 and class I molecules[61], hinting that dynamic interaction of few US3 molecules with one class I molecule may be required for efficient retention. The US3 luminal domain is capable of dynamic oligomerization, which can enhance retention of class I MHC molecules in the ER by a) formation of a larger complex which is dynamically less mobile, b) having more US3 transmembrane domains in the vicinity of class I molecules to prevent their escape, and c) increasing the capability of US3 molecules to retain themselves in ER by having more moieties capable of binding to the ER resident protein(s)[62]. Despite little sequence homology to US2, the luminal domain of US3 is structurally similar to US2. However, unlike US2, the luminal domain of US3 does not interact measurably with class I MHC molecules[62]. Besides binding to class I molecules, US3 binds directly to tapasin and TAP and inhibits tapasin-mediated peptide loading of class I MHC molecules. This allows US3 to selectively retain tapasin dependent class I MHC alleles in the ER and to discriminate between different alleles of the class I MHC[63]. TAP and tapasin may therefore be utilized by US3 to retain itself in the ER. Considering its small size, US3 perhaps achieves the means for simultaneous interactions with several different partner proteins, including other US3 molecules, through oligomerization and conformational changes. Taken together, these observations suggest that US3 only prevents surface expression of class I molecules that require tapasin for their peptide loading.

The above studies and observations testify to the amazing ability of HCMV to exploit different cellular pathways in order to evade the immune system by down regulating surface expression of class I MHC molecules. In addition to the functions described here, it was also reported that viral proteins can directly inhibit tapasin. The E3/19k protein of Adenovirus, which binds to class I MHC and leads to retention, can also bind to TAP and acts as a competitive inhibitor of tapasin[64].

Interference with Class II MHC molecule mediated antigen presentation

Class I MHC down regulation is by no means the only strategy by which viruses evade antigen presentation. Several viruses cause down-regulation of class II MHC expression, which affects antigen presentation to CD4+ T-cells. CD4+ T-cells play a key role in activation of CD8+ T-cells and in the development of B-cells. Examples for interference with class II MHC are the HIV nef protein[65,66] and the E5 protein of human papillomavirus type 16[67]. Also HCMV and MCMV down-regulate surface expression of class II MHC molecules on endothelial and epithelial cells[68-72]. In contrast to class I MHC molecules, this down-regulation was only described to be indirect as a consequence of inhibition of the transcriptional activation of class II MHC surface expression in infected cells. HCMV inhibits IFN-gamma induced class II MHC expression in endothelial cells by induction of IFN-beta[70]. For HCMV and MCMV, IFN-alpha/beta independent blocking of the IFN-gamma induced class II MHC up-regulation, for example by disruption of the Jak/STAT signaling pathway, has also been described[72-74]. Unfortunately, the exact mechanism of down regulation of class II MHC by CMV is still a matter of debate. Until recently, there were no reports on interaction of viral proteins with class II MHC molecules. But recent reports described that US2 and US3, when over-expressed by adenoviral vectors, are also capable of down regulating class II MHC molecules whose expression is induced artificially by the CIITA trans-activator[75-77]. However, these conditions may not be physiological since class II molecules are artificially induced in cells that do not normally express class II MHC. Further, the expression levels of US2 and US3 are very high in these cells and may reach levels not usually attained in CMV-infected cells. Further experiments are needed to clarify the open question of the effect of US2 and US3 on class II MHC proteins. MCMV mediated down-regulation of class II MHC products after infection of macrophages and dendritic cells (DCs) will be discussed below.

4.3.3 Evasion from natural killer (NK) cells

NK cells are an important component of the innate immune system and are crucial for the defense against certain viruses. They are lymphocytes which can cause direct lysis or killing of their target cells by releasing cytotoxic granzymes and perforins or by inducing receptor-mediated apoptosis. After activation they also secrete cytokines (such as IFN-γ and tumor necrosis factor ((TNF) α) during infection and inflammation. Activation of NK cells is regulated by interaction of NK-cell-activating and NK-cell-inhibiting receptors with specific cellular or viral ligands on target cells, and also by cytokines (e.g. Interleukin 2 (IL-2), IL-12, IL-15, IL-18,

IFN-α and IFN-β). Examples of activating receptors on NK cells are NKp30, -44, -46, CD2 and leukocyte function-associated antigen 1 (LFA-1) in human and Ly49D, –H and NKG2D in the mouse. The physiological ligands for most of these activating receptors are still not well defined. Inhibitory receptors are killer-cell immunoglobin-like receptors (KIR), immunoglobin-like inhibitory receptors (ILT), the lectin-like receptor CD94-NKG2A and in the mouse certain members of the Ly49 family. These inhibitory receptors mainly bind to self-class I MHC antigens as ligands on target cells which cause NK cell inhibition. Using this mechanism NK cells discriminate between normal and abnormal cells[78]. Since many viruses actively down-regulate MHC class I on the surface for protection against CTL lysis (see above), NK cells can use the MHC class I expression levels as a screening tool for recognition of infected and non-infected cells. Decreased MHC class I levels on target cells therefore lead to an increased susceptibility to NK cell killing ("missing self" hypothesis)[79]. Viruses of different families have evolved various means to evade the NK cell response (Figure 2). These mechanisms can be summarized as follows: a) expression of viral MHC class I homologs which can engage inhibitory NK cell receptors, b) preserving or inducing surface expression of distinct cellular MHC class I molecules that serve as ligands for NK-cell-inhibiting receptors, c) down-regulation of ligands binding to NK-cell-activating receptors, d) inhibition of intracellular signaling from NK-cell-activating receptors, e) inhibition of NK cells by direct infection, f) binding of viral proteins to non-class I NK-cell-inhibiting receptors, and g) inhibition or modulation of cytokines and chemokines involved in NK cell activation or migration. The latter mechanism is mediated indirectly by influencing cytokines and chemokines, and we therefore will not discuss it here.

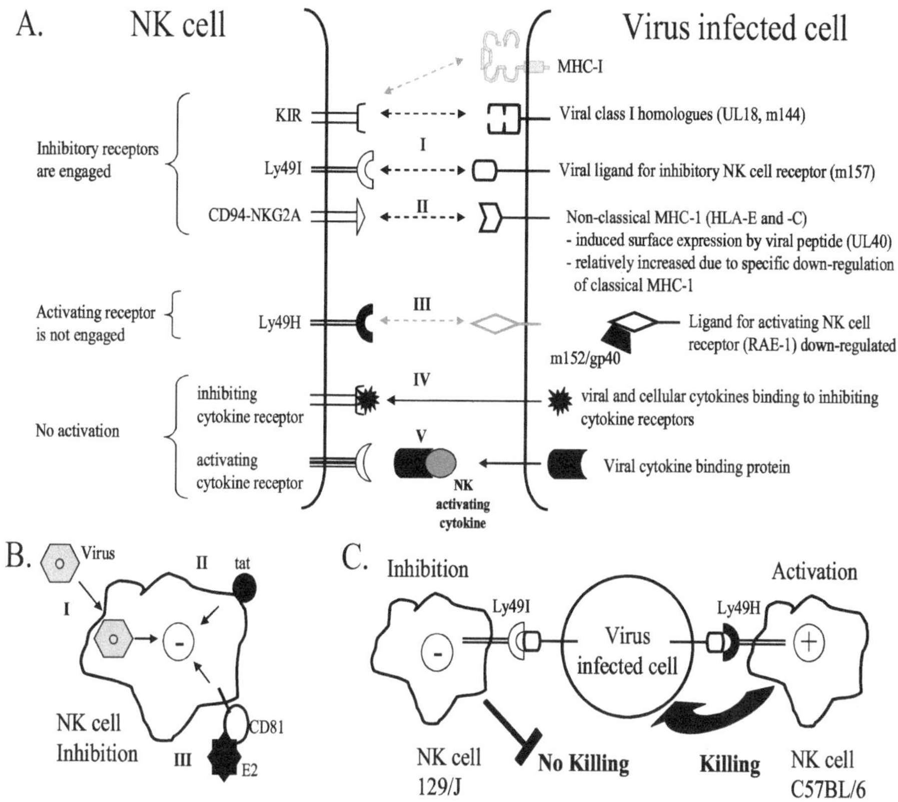

Figure 2. Viral evasion from natural killer (NK) cell-mediated lysis. A. Viral infection of cells in most cases leads to reduced classical major histocompatibility complex (MHC) class I surface expression. To compensate for interaction of MHC class I molecules with inhibitory receptors like the killer-cell immunoglobin-like receptors (KIR) on NK cells, viruses have evolved the following strategies: I) Viral class I homologs (UL18, m144 of CMV) and other viral proteins (m157 of MCMV) may substitute the lacking of down-regulated classical class I MHC and engage with inhibitory NK cell receptors (KIR, Ly49I, CD94-NKG2A) leading to NK cell inhibition. II) Indirect or direct up-regulation of non-classical class I MHC (HLA-E) by the virus is an additional strategy. III) Another strategy is to down-regulate activating ligands in infected cells. The MCMV protein m152/gp40 leads to down-regulation of ligands for activating NK cell receptors (e.g. RAE-1), consequently, the activating NK cell receptor (e.g. Ly49H) is not engaged and NK cells are not activated. IV) The viral expression or the up-regulation of cellular cytokines acting as NK cell inhibitors may also lead to NK cell inhibition. V) Viral cytokine binding proteins bind to and impair activating cytokines leading to evasion from NK cell activation. B. In addition, NK cells can be inhibited by direct contact with viruses. I) Direct virus infection (e.g. with HIV or HSV), II) binding of the HIV tat protein to NK cells leading to inhibition of L-type calcium channels and III) binding of viral proteins (e.g. E2 of hepatitis C virus) to non-class I-like NK-cell-inhibitory receptors (e.g. CD81) can lead to NK cell inhibition. C. A viral protein may have opposing outcomes for NK

cell activation. Expression of the MCMV protein m157 in infected cells can lead to NK cell inhibition or activation depending on the genetic background of the mouse. m157 is expressed in infected cells and engages with I) NK cell inhibiting receptors that are expressed in 129/J or BALB/C mice, leading to inhibition, or with II) NK cell activating receptors only expressed in C57BL/6 mice, leading to killing of the infected cell.

The viruses best studied concerning direct NK cell interference are the cytomegaloviruses[80]. Therefore we will discuss viral NK cell evasion mechanism in more detail, using CMV as an example (Figure 2). NK cells play an important role in anti-MCMV immunity[80-82]. Newborn or NK cell-deficient mice are highly susceptible to MCMV infection. In adult mice the susceptibility depends on the mouse strain. C57BL/6 mice are relatively resistant to MCMV infection in contrast to BALB/c, 129/J and DBA/2 mice, which are susceptible. The reason for this lies in the genetic differences on chromosome 6 of these mice: only C57BL/6 mice contain the *cmv-1* locus[83], which encodes the activating NK receptor Ly49H[84,85]. Ly49H belongs to the Ly49 gene family which encodes either activating or inhibitory receptors[86]. Activating Ly49 receptors associate with the DAP12 adaptor molecule which recruits the Syk and ZAP70 tyrosine kinases and which is essential for activating signal transduction[87]. DAP12-deficient mice are therefore susceptible to MCMV infection[88]. Surprisingly, the activating ligand for Ly49H in MCMV infected cells was identified as the product of the viral gene m157 which shares homologies to non-classical class I MHC molecules like CD1d[89,90]. This means that the virus itself is expressing a glycoprotein that acts as an activating ligand for NK cell mediated recognition, which is disadvantageous for the virus. In line with this unexpected observation are reports that a m157-deleted MCMV recombinant replicates to higher titers in C57BL/6 mice[91]. Very recent reports, surprisingly, showed that, similar to RNA viruses which escape the immune response by mutation, MCMV accumulates mutations in m157 during several passages in Ly49H+ wt mice or in mice without adaptive immune response. These m157 mutants can elude the Ly49H-dependent NK cell response[92]. Even more striking, French et al.[93] used MCMV infection of SCID-mice and showed for the first time the in vivo emergence of escape mutants of a DNA virus during the course of a single infection, which is driven by selective pressure from innate immunity.

Evasion of NK cell killing by expression of viral MHC class I homologs which can engage with inhibitory NK cell receptors

As mentioned above, infection with many viruses, including CMV, leads to a decreased class I MHC surface expression and a lack of engagement with inhibitory NK cell receptors. As a consequence, NK cells show increased cytotoxicity against these cells. To avoid this, CMVs express MHC-like surrogates that might replace cellular class I MHC molecules as ligands for inhibitory receptors.

The class I MHC-like protein m157 described above might be such a surrogate, since it was shown that it can also bind to the inhibitory Ly49I receptor in 129/J mice. This suggests that MCMV originally has acquired m157 to escape attack by NK cells in mice expressing Ly49I[90]. But since an m157 deletion mutant showed only a slight attenuation compared to wt MCMV in 129/J mice[91], the significance of an immunoevasive function of m157 in Ly49I+ mice strains awaits further confirmation.

MCMV also encodes the class I MHC homolog m144, which is a type I transmembrane glycoprotein that can associate with β_2m[94] and might bind to not yet identified inhibitory NK cell receptors. This suggestion is supported by the fact that an MCMV m144 deletion mutant is less virulent than wt MCMV in vivo and NK depletion rescued the phenotype[95].

Similarly, HCMV expresses the class I MHC homolog UL18[96], which is a glycosylated type I transmembrane glycoprotein that efficiently binds to the inhibitory receptor LIR-1 on monocytes and macrophages[97]. Interestingly, UL18 is not down-regulated by US2, US3, US6 and US11[98]. The hypothesis that UL18 therefore can protect against NK cell killing was confirmed by Reyburn et al.[99], who also reported that antibodies against CD94 abolish target cell protection mediated by UL18. But contrary to this report, Leong et al. showed that UL18 expression increased the susceptibility to NK cell mediated lysis[100]. A recent report even showed that UL18-LIR1 interaction allows recognition and non-MHC-restricted lysis of HCMV infected cells by CTLs[101]. Nevertheless, UL18 might also have an inhibitory role only for some specific NK cells subpopulations, or it engages with another yet unknown inhibitory receptor.

Preserving or inducing surface expression of distinct cellular MHC class I molecules which serve as ligands for NK-cell-inhibiting receptors

At first sight one might wonder why both MCMV and HCMV use several different class I MHC down-regulating proteins with seemingly redundant functions. Different expression kinetics could be one explanation, but recent reports also revealed that some of the m04/gp34, m06/gp48, m152/gp40 and the US2, US3, US6 and US11 proteins show a selective targeting of specific subsets of class I molecules[36,48,49,102]. Preserving the expression of class I MHC molecules that have a minor role in antigen presentation but that can serve as key ligands for inhibitory receptors (e.g. HLA-E and HLA-C) might allow CMV to escape from both T and NK cells. In this regard it is interesting that the HLA-E molecule, which mainly presents peptides derived from signal peptides, is not affected by US2 and US6[44,48] and that a peptide derived from the HCMV protein UL40 stabilizes HLA-E surface expression[102,103]. Consequently, infected cells could preserve HLA-E surface expression and prevent activation of CD94/NKG2A+ effector cells. Indeed, Wang et al.[104] showed that fibroblasts infected with a HCMV UL40 deletion mutant (strain AD169) were more efficiently killed by a CD94/NKGZA+ primary NK cell line than cells infected with wt

HCMV. In contrast to this report, Falk et al.[105] used HCMV mutants with deletions of UL40 and/or US2, US3, US6 and US11 and showed that UL40 was not sufficient to preserve HLA-E expression. To add even more complexity to this issue, Llano et al.[106] used a human B-cell lymphoma cell line to investigate the effect of US2, US6 and US11 on HLA-E and HLA class Ia expression and their influence on susceptibility to NK cell clones. This study showed that US6 down-regulated all class I molecules in contrast to US2 and US11, which preserved HLA-E. The discrepancies between these reports might be due to different experimental settings, virus strains and cell lines used and definitely needs further clarification. In summary, one can conclude that the interplay between the CMV immunoevasions, class I MHC molecules and NK cell receptors is extremely complex and reflects the sophisticated fine-tuning of immunoevasion mechanisms achieved during co-evolution of viruses and their hosts.

Down-regulation of cellular ligands binding to NK-cell-activating receptors

An additional strategy for NK cell evasion used by viruses is down-regulation of ligands on targets cells which are binding to activating NK cell receptors. The MCMV protein m152/gp40, which was previously shown to target class I MHC molecules[30,31], reduces the expression of ligands for the activating NKG2D receptor in MCMV infected cells[107,108]. NKG2D recognizes RAE-1, H60 and MULT-1 ligands in mice[109-111] and NK cells kill target cells expressing these ligands. It was shown that m152/gp40 specifically down-regulates RAE-1 proteins, which are up-regulated during MCMV infection[108]. An m152 deletion mutant is accordingly attenuated in vivo and virus titers can be rescued by treatment of mice with anti-NKG2D antibodies[107,108]. Since virus titers in mice infected with an MCMV deletion mutant lacking 14 additional genes besides m152 and m157 can be rescued by NK cell depletion using anti-NK1.1 antibodies[91], additional MCMV genes that affect expression of ligands for activating NK cell receptors should be discovered soon.

HCMV has evolved at least one viral protein that affects ligands for the NK cell activating receptors. The human activating NK cell receptor NKG2D recognizes MICA/B and ULBP ligands[112,113]. The early glycoprotein UL16 of HCMV binds to MICB, ULBP1 and ULBP2[113] and so prevents surface expression of these ligands in HCMV-infected cells[114-117]. Rolle and colleagues could even show that lack of UL-16 during HCMV infection increases the surface expression of ULPB ligands and also leads to increased NKG2D-mediated lysis. A recent report suggests that the protective function of UL16 is more likely due to a protective effect against different cytolytic mediators released by NK cells[118]. In addition to effects on NKG2D ligands, some HCMV strains also down-regulate the ligand LFA-3 for the activating receptor CD2, but the viral gene product responsible has not yet been identified[119].

Other viruses also use similar strategies. For instance, the K3 and K5 proteins of HHV-8 reduce ICAM-1 and CD86 by ubiquitination[120,121]. Both are ligands for the activating receptors LFA-1 and CD28.

Other viral mechanisms for NK cell escape

Several reports also described evasion from NK cell killing by inhibition of intracellular signaling from NK-cell-activating receptors. The tat protein of HIV-1 can block NK cell-mediated lysis of DCs after LFA-1 ligation by inhibition of L-type calcium channels, which are required for NK cell cytotoxicity[122,123]. Further viral proteins blocking intracellular signaling for NK cell cytotoxicity await discovery.

Interaction of NK-cell-activating receptors with their ligands can also be inhibited by virus-induced modifications of surface molecules by sialylation[124]. Some viruses can directly infect NK cells and inhibit their function. But even without an effective infection, viruses can actively inactivate NK cells. By binding of a viral protein to a non-class I-like NK-cell-inhibiting receptor on the cell surface, hepatitis C virus inhibits NK cell cytotoxicity[125,126]. The major envelope protein E2 binds to CD81 on NK cells and inhibits not only cytotoxicity but also IFN-gamma production and CD16-mediated activation[125,126].

Finally, viruses can evade the NK cell response by interacting with cytokines and chemokines involved in NK cell activation. We refer to other reviews describing viral proteins interfering with cytokines important for NK cell activity[127].

5. IMPAIRMENT OF THE FUNCTION OF IMMUNE CELLS BY DIRECT INFECTION

The majority of cells infected by viruses are non-immune cells. But some viruses can also directly infect cells of the immune system such as B cells, T-cells, NK cells, macrophages and DCs. Recently, more and more data is accumulating that shows that some viruses can impair immune cell functions by direct infection.

5.1 Functional paralysis of DCs after CMV infection

DCs are professional antigen presenting cells and have a crucial role in priming of naïve T-cells. Both HCMV and MCMV can infect DCs and lead to functional paralysis and reduced T-cell proliferation and cytotoxicity[128-131]. Infection leads to a down-regulation of class I and II MHC, CD83 and co-stimulatory molecules like CD40, CD80 and CD86. Infected DCs remain unresponsive to maturation stimuli and lose their capacity to secrete IL-2 and IL-12 and finally they are unable to prime an effective T-cell response[130].

5.2 Impairment of macrophage function after infection with MCMV

Recently, several groups have shown that CMV can cause decreased expression of MHC class II proteins by inhibition of IFN-gamma induced transcriptional up-regulation in epithelial and endothelial cells (see above). In addition, MCMV can also down-regulate class II MHC in infected macrophages[72]. This can be achieved by induction of cellular IL-10 expression by the virus[74] and by blocking IFN gamma-induced promoter assembly[132]. Also, class I MHC is specifically down-regulated in MCMV infected macrophages[133].

A recent report showed that HCMV infection of monocytes can inhibit cytokine-induced macrophage differentiation[134], which adds a new potential viral trick for manipulation of the effectiveness of the immune control.

5.3 Infection and inhibition of NK cells

Few viruses can directly infect NK cells and lead to inhibition of their function or to reduced viability. It was reported that HIV can lead to NK cell death by in vitro infection[135] and that HSV infection of NK cells inhibited NK cell cytotoxicity[136]. The mechamisms for both observations are still unknown.

5.4 Down-regulation of CD4 and the TCR/CD3 by infection of T-cells

Stimulation of T-cells by antigens presented by the MHC complex requires co-expression of CD4 or CD8 on the cell surface. Some viruses can directly infect T-cells and modulate CD4 and TCR/CD3 expression levels. The HIV proteins env, vpu and nef have been implicated in such a down-regulation[137,138]. Env binds to CD4 and retains it in the ER[139], vpu translocates CD4 to the cytosol and leads to its proteasomal degradation, whereas nef down-regulates, in addition to class I MHC[140], CD4[141,142] and TCR/CD3[143]. Human herpesvirus 6A can also down-regulate the TCR/CD3 expression on the transcriptional level by infection of T-cells[144].

6. CONCLUSIONS

Viruses have co-evolved an amazing array of viral functions that actively manipulate the host immune response. In the past 10 years, an exciting insight into the functional mechanisms of immunoevasion, especially concerning viral proteins interfering with class I MHC mediated antigen presentation and T-cell activation, has been gained. In addition to the MHC-TCR interaction, co-stimulatory molecules expressed in antigen presenting cells (e.g. CD80 and CD86) are likewise essential for T-cell activation. Therefore it is likely that viruses also target these molecules with manifold functions to evade T-cell activation. Several reports showed already that CMV infection of DCs leads to down-regulation of CD40, CD80 and CD86 (see above). By identification of the viral proteins responsible, new mechanistic insights in the interference with trafficking of these surface molecules will be gained in future.

The recent discovery of viral proteins that inhibit activating NK cell receptors or their ligands should reveal new exiting viral interactions also with the innate immune system. It is increasingly clear that the pattern recognition receptors, in particular the Toll-like receptors (TLRs), not only recognize bacterial but also viral components, and therefore interference of viruses with TLR-mediated immune responses is very likely. Indeed, the vaccinia virus proteins A46R and A52R antagonize TLR signaling[145,146].

Until now, biologists focused on protein-encoding genes when studying virus-host interactions and immune evasion strategies. But this might change due to the finding of virally expressed *miRNAs* by Pfeffer et al.[147]. miRNAs are endogenously transcribed small RNA molecules that can inhibit *mRNA* translation or induce *mRNA* degradation of target genes, leading to a knock-down of gene expression. Pfeffer and colleagues cloned five viral *miRNAs* from B-cells infected with *EBV*, a DNA virus belonging to the herpesvirus family. These findings extend the group of organisms dealing with *miRNAs* from plants and animals to viruses. Initially described to be important in early plant development, *miRNA*-mediated effects are now recognized as a new epigenetic gene silencing mechanism with many different functions: e.g. developmental-timing, regulation of cell proliferation and cancer, apoptosis, fat metabolism and lymphocyte development[148]. It seems that *EBV* is taking advantage of the cellular *RNAi* machinery and that it has evolved its own *miRNAs*. Amongst the predicted target genes is the virally encoded DNA polymerase, but also 16 cellular genes, such as the *bcl-2* gene involved in apoptosis, the *E2F1* transcription factor, the *p53* tumor suppressor gene and the tumor necrosis factor (TNF) receptor gene *TNFRSF1A*[147]. Despite the lack of experimental confirmation of down-regulation of the predicted targets, the finding by Pfeffer and colleagues suggests that at least large DNA viruses might use *miRNAs* as non-protein-based tools for manipulation of cells and potentially the host immune response. This adds a completely new mechanism to the list of clever viral strategies for manipulating host cells and it might initiate a new research direction for investigation of virus-host cell interactions.

ACKNOWLEDGEMENTS

M.W. is supported by a Human Frontier Science Program Organization long-term fellowship.

REFERENCES

1. Dunn W, Chou C, Li H, Hai R, Patterson D, Stolc V, Zhu H, Liu F. 2003. Functional profiling of a human cytomegalovirus genome. *Proc Natl Acad Sci USA* **100:**14223-8
2. Gorman OT, Bean WJ, Webster RG. 1992. Evolutionary processes in influenza viruses: divergence, rapid evolution, and stasis. *Curr Top Microbiol Immunol* **176:**75-97
3. Klenerman P, Meier UC, Phillips RE, McMichael AJ. 1995. The effects of natural altered peptide ligands on the whole blood cytotoxic T lymphocyte response to human immunodeficiency virus. *Eur J Immunol* **25:**1927-31
4. Liston A, McColl S. 2003. Subversion of the chemokine world by microbial pathogens. *Bioessays* **25:**478-88
5. Vossen MT, Westerhout EM, Soderberg-Naucler C, Wiertz EJ. 2002. Viral immune evasion: a masterpiece of evolution. *Immunogenetics* **54:**527-42
6. Favoreel HW, Van de Walle GR, Nauwynck HJ, Pensaert MB. 2003. Virus complement evasion strategies. *J Gen Virol* **84:**1-15
7. Alcami A, Koszinowski UH. 2000. Viral mechanisms of immune evasion. *Trends Microbiol* **8:**410-8
8. Benedict CA, Norris PS, Ware CF. 2002. To kill or be killed: viral evasion of apoptosis. *Nat Immunol* **3:**1013-8
9. Ossendorp F, Eggers M, Neisig A, Ruppert T, Groettrup M, Sijts A, Mengede E, Kloetzel PM, Neefjes J, Koszinowski U, Melief C. 1996. A single residue exchange within a viral CTL epitope alters proteasome-mediated degradation resulting in lack of antigen presentation. *Immunity* **5:**115-24
10. Levitskaya J, Sharipo A, Leonchiks A, Ciechanover A, Masucci MG. 1997. Inhibition of ubiquitin/proteasome-dependent protein degradation by the Gly-Ala repeat domain of the Epstein-Barr virus nuclear antigen 1. *Proc Natl Acad Sci USA* **94:**12616-21
11. Levitskaya J, Coram M, Levitsky V, Imreh S, Steigerwald-Mullen PM, Klein G, Kurilla MG, Masucci MG. 1995. Inhibition of antigen processing by the internal repeat region of the Epstein-Barr virus nuclear antigen-1. *Nature* **375:**685-8
12. Sharipo A, Imreh M, Leonchiks A, Branden C, Masucci MG. 2001. cis-Inhibition of proteasomal degradation by viral repeats: impact of length and amino acid composition. *FEBS Lett* **499:**137-42
13. Gilbert MJ, Riddell SR, Plachter B, Greenberg PD. 1996. Cytomegalovirus selectively blocks antigen processing and presentation of its immediate-early gene product. *Nature* **383:**720-2
14. Burgert HG, Kvist S. 1985. An adenovirus type 2 glycoprotein blocks cell surface expression of human histocompatibility class I antigens. *Cell* **41:**987-97

15. Burgert HG, Maryanski JL, Kvist S. 1987. "E3/19K" protein of adenovirus type 2 inhibits lysis of cytolytic T lymphocytes by blocking cell-surface expression of histocompatibility class I antigens. *Proc Natl Acad Sci USA* **84:**1356-60
16. Burgert HG, Kvist S. 1987. The E3/19K protein of adenovirus type 2 binds to the domains of histocompatibility antigens required for CTL recognition. *Embo J* **6:**2019-26
17. Hill A, Jugovic P, York I, Russ G, Bennink J, Yewdell J, Ploegh H, Johnson D. 1995. Herpes simplex virus turns off the TAP to evade host immunity. *Nature* **375:**411-5
18. Ahn K, Meyer TH, Uebel S, Sempe P, Djaballah H, Yang Y, Peterson PA, Fruh K, Tampe R. 1996. Molecular mechanism and species specificity of TAP inhibition by herpes simplex virus ICP47. *Embo J* **15:**3247-55
19. Tomazin R, Hill AB, Jugovic P, York I, van Endert P, Ploegh HL, Andrews DW, Johnson DC. 1996. Stable binding of the herpes simplex virus ICP47 protein to the peptide binding site of TAP. *Embo J* **15:**3256-66
20. York IA, Roop C, Andrews DW, Riddell SR, Graham FL, Johnson DC. 1994. A cytosolic herpes simplex virus protein inhibits antigen presentation to CD8+ T lymphocytes. *Cell* **77:**525-35
21. Hudson AW, Blom D, Howley PM, Ploegh HL. 2003. The ER-lumenal domain of the HHV-7 immunoevasin U21 directs class I MHC molecules to lysosomes. *Traffic* **4:**824-37
22. Hudson AW, Howley PM, Ploegh HL. 2001. A human herpesvirus 7 glycoprotein, U21, diverts major histocompatibility complex class I molecules to lysosomes. *J Virol* **75:**12347-58
23. Ishido S, Wang C, Lee BS, Cohen GB, Jung JU. 2000. Downregulation of major histocompatibility complex class I molecules by Kaposi's sarcoma-associated herpesvirus K3 and K5 proteins. *J Virol* **74:**5300-9
24. Coscoy L, Ganem D. 2000. Kaposi's sarcoma-associated herpesvirus encodes two proteins that block cell surface display of MHC class I chains by enhancing their endocytosis. *Proc Natl Acad Sci USA* **97:**8051-6
25. Le Gall S, Erdtmann L, Benichou S, Berlioz-Torrent C, Liu L, Benarous R, Heard JM, Schwartz O. 1998. Nef interacts with the mu subunit of clathrin adaptor complexes and reveals a cryptic sorting signal in MHC I molecules. *Immunity* **8:**483-95
26. Piguet V, Wan L, Borel C, Mangasarian A, Demaurex N, Thomas G, Trono D. 2000. HIV-1 Nef protein binds to the cellular protein PACS-1 to downregulate class I major histocompatibility complexes. *Nat Cell Biol* **2:**163-7
27. Beersma MF, Bijlmakers MJ, Ploegh HL. 1993. Human cytomegalovirus down-regulates HLA class I expression by reducing the stability of class I H chains. *J Immunol* **151:**4455-64
28. Yamashita Y, Shimokata K, Saga S, Mizuno S, Tsurumi T, Nishiyama Y. 1994. Rapid degradation of the heavy chain of class I major histocompatibility complex antigens in the endoplasmic reticulum of human cytomegalovirus-infected cells. *J Virol* **68:**7933-43
29. Jones TR, Hanson LK, Sun L, Slater JS, Stenberg RM, Campbell AE. 1995. Multiple independent loci within the human cytomegalovirus unique short region down-regulate expression of major histocompatibility complex class I heavy chains. *J Virol* **69:**4830-41
30. Ziegler H, Muranyi W, Burgert HG, Kremmer E, Koszinowski UH. 2000. The luminal part of the murine cytomegalovirus glycoprotein gp40 catalyzes the retention of MHC class I molecules. *Embo J* **19:**870-81
31. Ziegler H, Thale R, Lucin P, Muranyi W, Flohr T, Hengel H, Farrell H, Rawlinson W, Koszinowski UH. 1997. A mouse cytomegalovirus glycoprotein retains MHC class I complexes in the ERGIC/cis-Golgi compartments. *Immunity* **6:**57-66
32. Reusch U, Muranyi W, Lucin P, Burgert HG, Hengel H, Koszinowski UH. 1999. A cytomegalovirus glycoprotein re-routes MHC class I complexes to lysosomes for degradation. *Embo J* **18:**1081-91

33. Kleijnen MF, Huppa JB, Lucin P, Mukherjee S, Farrell H, Campbell AE, Koszinowski UH, Hill AB, Ploegh HL. 1997. A mouse cytomegalovirus glycoprotein, gp34, forms a complex with folded class I MHC molecules in the ER which is not retained but is transported to the cell surface. *Embo J* **16:**685-94
34. Kavanagh DG, Koszinowski UH, Hill AB. 2001. The murine cytomegalovirus immune evasion protein m4/gp34 forms biochemically distinct complexes with class I MHC at the cell surface and in a pre-Golgi compartment. *J Immunol* **167:**3894-902
35. Kavanagh DG, Gold MC, Wagner M, Koszinowski UH, Hill AB. 2001. The multiple immune-evasion genes of murine cytomegalovirus are not redundant: m4 and m152 inhibit antigen presentation in a complementary and cooperative fashion. *J Exp Med* **194:**967-78
36. Wagner M, Gutermann A, Podlech J, Reddehase MJ, Koszinowski UH. 2002. Major histocompatibility complex class I allele-specific cooperative and competitive interactions between immune evasion proteins of cytomegalovirus. *J Exp Med* **196:**805-16
37. Tirabassi RS, Ploegh HL. 2002. The human cytomegalovirus US8 glycoprotein binds to major histocompatibility complex class I products. *J Virol* **76:**6832-5
38. Furman MH, Dey N, Tortorella D, Ploegh HL. 2002. The human cytomegalovirus US10 gene product delays trafficking of major histocompatibility complex class I molecules. *J Virol* **76:**11753-6
39. Hengel H, Koopmann JO, Flohr T, Muranyi W, Goulmy E, Hammerling GJ, Koszinowski UH, Momburg F. 1997. A viral ER-resident glycoprotein inactivates the MHC-encoded peptide transporter. *Immunity* **6:**623-32
40. Ahn K, Gruhler A, Galocha B, Jones TR, Wiertz EJ, Ploegh HL, Peterson PA, Yang Y, Fruh K. 1997. The ER-luminal domain of the HCMV glycoprotein US6 inhibits peptide translocation by TAP. *Immunity* **6:**613-21
41. Lehner PJ, Karttunen JT, Wilkinson GW, Cresswell P. 1997. The human cytomegalovirus US6 glycoprotein inhibits transporter associated with antigen processing-dependent peptide translocation. *Proc Natl Acad Sci USA* **94:**6904-9
42. Hewitt EW, Gupta SS, Lehner PJ. 2001. The human cytomegalovirus gene product US6 inhibits ATP binding by TAP. *Embo J* **20:**387-96
43. Kyritsis C, Gorbulev S, Hutschenreiter S, Pawlitschko K, Abele R, Tampe R. 2001. Molecular mechanism and structural aspects of transporter associated with antigen processing inhibition by the cytomegalovirus protein US6. *J Biol Chem* **276:**48031-9
44. Ulbrecht M, Hofmeister V, Yuksekdag G, Ellwart JW, Hengel H, Momburg F, Martinozzi S, Reboul M, Pla M, Weiss EH. 2003. HCMV glycoprotein US6 mediated inhibition of TAP does not affect HLA-E dependent protection of K-562 cells from NK cell lysis. *Hum Immunol* **64:**231-7
45. Wiertz EJ, Jones TR, Sun L, Bogyo M, Geuze HJ, Ploegh HL. 1996. The human cytomegalovirus US11 gene product dislocates MHC class I heavy chains from the endoplasmic reticulum to the cytosol. *Cell* **84:**769-79.
46. Wiertz EJ, Tortorella D, Bogyo M, Yu J, Mothes W, Jones TR, Rapoport TA, Ploegh HL. 1996. Sec61-mediated transfer of a membrane protein from the endoplasmic reticulum to the proteasome for destruction. *Nature* **384:**432-8.
47. Story CM, Furman MH, Ploegh HL. 1999. The cytosolic tail of class I MHC heavy chain is required for its dislocation by the human cytomegalovirus US2 and US11 gene products. *Proc Natl Acad Sci USA* **96:**8516-21
48. Gewurz BE, Wang EW, Tortorella D, Schust DJ, Ploegh HL. 2001. Human cytomegalovirus US2 endoplasmic reticulum-lumenal domain dictates association with major histocompatibility complex class I in a locus-specific manner. *J Virol* **75:**5197-204
49. Barel MT, Ressing M, Pizzato N, van Leeuwen D, Le Bouteiller P, Lenfant F, Wiertz EJ. 2003. Human cytomegalovirus-encoded US2 differentially affects surface expression of MHC class I locus products and targets membrane-bound, but not soluble HLA-G1 for degradation. *J Immunol* **171:**6757-65

50. Gewurz BE, Gaudet R, Tortorella D, Wang EW, Ploegh HL, Wiley DC. 2001. Antigen presentation subverted: Structure of the human cytomegalovirus protein US2 bound to the class I molecule HLA-A2. *Proc Natl Acad Sci USA* **98:**6794-9
51. Gewurz BE, Gaudet R, Tortorella D, Wang EW, Ploegh HL. 2001. Virus subversion of immunity: a structural perspective. *Curr Opin Immunol* **13:**442-50
52. Blom D, Hirsch C, Stern P, Tortorella D, Ploegh HL. 2004. A glycosylated type I membrane protein becomes cytosolic when peptide: N-glycanase is compromised. *Embo J* **23:**650-8
53. Lilley BN, Tortorella D, Ploegh HL. 2003. Dislocation of a type I membrane protein requires interactions between membrane-spanning segments within the lipid bilayer. *Mol Biol Cell* **14:**3690-8
54. Lilley BN, Ploegh HL. 2004. A membrane protein required for dislocation of misfolded proteins from the ER. *Nature* **429:**834-40
55. Ye Y, Shibata Y, Yun C, Ron D, Rapoport TA. 2004. A membrane protein complex mediates retro-translocation from the ER lumen into the cytosol. *Nature* **429:**841-7
56. Ahn K, Angulo A, Ghazal P, Peterson PA, Yang Y, Fruh K. 1996. Human cytomegalovirus inhibits antigen presentation by a sequential multistep process. *Proc Natl Acad Sci USA* **93:**10990-5
57. Jones TR, Wiertz EJ, Sun L, Fish KN, Nelson JA, Ploegh HL. 1996. Human cytomegalovirus US3 impairs transport and maturation of major histocompatibility complex class I heavy chains. *Proc Natl Acad Sci USA* **93:**11327-33
58. Lee S, Yoon J, Park B, Jun Y, Jin M, Sung HC, Kim IH, Kang S, Choi EJ, Ahn BY, Ahn K. 2000. Structural and functional dissection of human cytomegalovirus US3 in binding major histocompatibility complex class I molecules. *J Virol* **74:**11262-9
59. Zhao Y, Biegalke BJ. 2003. Functional analysis of the human cytomegalovirus immune evasion protein, pUS3(22kDa). *Virology* **315:**353-61
60. Lee S, Park B, Ahn K. 2003. Determinant for endoplasmic reticulum retention in the luminal domain of the human cytomegalovirus US3 glycoprotein. *J Virol* **77:**2147-56
61. Gruhler A, Peterson PA, Fruh K. 2000. Human cytomegalovirus immediate early glycoprotein US3 retains MHC class I molecules by transient association. *Traffic* **1:**318-25
62. Misaghi S, Sun ZY, Stern P, Gaudet R, Wagner G, Ploegh H. 2004. Structural and functional analysis of human cytomegalovirus US3 protein. *J Virol* **78:**413-23
63. Park B, Kim Y, Shin J, Lee S, Cho K, Fruh K, Ahn K. 2004. Human cytomegalovirus inhibits tapasin-dependent peptide loading and optimization of the MHC class I peptide cargo for immune evasion. *Immunity* **20:**71-85
64. Bennett EM, Bennink JR, Yewdell JW, Brodsky FM. 1999. Cutting edge: adenovirus E19 has two mechanisms for affecting class I MHC expression. *J Immunol* **162:**5049-52
65. Stumptner-Cuvelette P, Morchoisne S, Dugast M, Le Gall S, Raposo G, Schwartz O, Benaroch P. 2001. HIV-1 Nef impairs MHC class II antigen presentation and surface expression. *Proc Natl Acad Sci USA* **98:**12144-9
66. Schindler M, Wurfl S, Benaroch P, Greenough TC, Daniels R, Easterbrook P, Brenner M, Munch J, Kirchhoff F. 2003. Down-modulation of mature major histocompatibility complex class II and up-regulation of invariant chain cell surface expression are well-conserved functions of human and simian immunodeficiency virus nef alleles. *J Virol* **77:**10548-56
67. Zhang B, Li P, Wang E, Brahmi Z, Dunn KW, Blum JS, Roman A. 2003. The E5 protein of human papillomavirus type 16 perturbs MHC class II antigen maturation in human foreskin keratinocytes treated with interferon-gamma. *Virology* **310:**100-8
68. Buchmeier NA, Cooper NR. 1989. Suppression of monocyte functions by human cytomegalovirus. *Immunology* **66:**278-83

69. Ng-Bautista CL, Sedmak DD. 1995. Cytomegalovirus infection is associated with absence of alveolar epithelial cell HLA class II antigen expression. *J Infect Dis* **171:**39-44
70. Sedmak DD, Guglielmo AM, Knight DA, Birmingham DJ, Huang EH, Waldman WJ. 1994. Cytomegalovirus inhibits major histocompatibility class II expression on infected endothelial cells. *Am J Pathol* **144:**683-92
71. Heise MT, Virgin HWt. 1995. The T-cell-independent role of gamma interferon and tumor necrosis factor alpha in macrophage activation during murine cytomegalovirus and herpes simplex virus infections. *J Virol* **69:**904-9
72. Heise MT, Connick M, Virgin HWt. 1998. Murine cytomegalovirus inhibits interferon gamma-induced antigen presentation to CD4 T cells by macrophages via regulation of expression of major histocompatibility complex class II-associated genes. *J Exp Med* **187:**1037-46
73. Miller DM, Rahill BM, Boss JM, Lairmore MD, Durbin JE, Waldman JW, Sedmak DD. 1998. Human cytomegalovirus inhibits major histocompatibility complex class II expression by disruption of the Jak/Stat pathway. *J Exp Med* **187:**675-83
74. Redpath S, Angulo A, Gascoigne NR, Ghazal P. 1999. Murine cytomegalovirus infection down-regulates MHC class II expression on macrophages by induction of IL-10. *J Immunol* **162:**6701-7
75. Tomazin R, Boname J, Hegde NR, Lewinsohn DM, Altschuler Y, Jones TR, Cresswell P, Nelson JA, Riddell SR, Johnson DC. 1999. Cytomegalovirus US2 destroys two components of the MHC class II pathway, preventing recognition by CD4+ T cells. *Nat Med* **5:**1039-43
76. Hegde NR, Tomazin RA, Wisner TW, Dunn C, Boname JM, Lewinsohn DM, Johnson DC. 2002. Inhibition of HLA-DR assembly, transport, and loading by human cytomegalovirus glycoprotein US3: a novel mechanism for evading major histocompatibility complex class II antigen presentation. *J Virol* **76:**10929-41
77. Chevalier MS, Johnson DC. 2003. Human cytomegalovirus US3 chimeras containing US2 cytosolic residues acquire major histocompatibility class I and II protein degradation properties. *J Virol* **77:**4731-8
78. Ravetch JV, Lanier LL. 2000. Immune inhibitory receptors. *Science* **290:**84-9
79. Ljunggren HG, Karre K. 1990. In search of the 'missing self': MHC molecules and NK cell recognition. *Immunol Today* **11:**237-44
80. Biron CA, Nguyen KB, Pien GC, Cousens LP, Salazar-Mather TP. 1999. Natural killer cells in antiviral defense: function and regulation by innate cytokines. *Annu Rev Immunol* **17:**189-220
81. Bukowski JF, Warner JF, Dennert G, Welsh RM. 1985. Adoptive transfer studies demonstrating the antiviral effect of natural killer cells in vivo. *J Exp Med* **161:**40-52
82. Welsh RM, Brubaker JO, Vargas-Cortes M, O'Donnell CL. 1991. Natural killer (NK) cell response to virus infections in mice with severe combined immunodeficiency. The stimulation of NK cells and the NK cell-dependent control of virus infections occur independently of T and B cell function. *J Exp Med* **173:**1053-63
83. Scalzo AA, Fitzgerald NA, Wallace CR, Gibbons AE, Smart YC, Burton RC, Shellam GR. 1992. The effect of the Cmv-1 resistance gene, which is linked to the natural killer cell gene complex, is mediated by natural killer cells. *J Immunol* **149:**581-9
84. Lee SH, Girard S, Macina D, Busa M, Zafer A, Belouchi A, Gros P, Vidal SM. 2001. Susceptibility to mouse cytomegalovirus is associated with deletion of an activating natural killer cell receptor of the C-type lectin superfamily. *Nat Genet* **28:**42-5
85. Brown MG, Dokun AO, Heusel JW, Smith HR, Beckman DL, Blattenberger EA, Dubbelde CE, Stone LR, Scalzo AA, Yokoyama WM. 2001. Vital involvement of a natural killer cell activation receptor in resistance to viral infection. *Science* **292:**934-7
86. Yokoyama WM, Plougastel BF. 2003. Immune functions encoded by the natural killer gene complex. *Nat Rev Immunol* **3:**304-16

87. Lanier LL, Bakker AB. 2000. The ITAM-bearing transmembrane adaptor DAP12 in lymphoid and myeloid cell function. *Immunol Today* **21:**611-4
88. Sjolin H, Tomasello E, Mousavi-Jazi M, Bartolazzi A, Karre K, Vivier E, Cerboni C. 2002. Pivotal role of KARAP/DAP12 adaptor molecule in the natural killer cell-mediated resistance to murine cytomegalovirus infection. *J Exp Med* **195:**825-34
89. Smith HR, Heusel JW, Mehta IK, Kim S, Dorner BG, Naidenko OV, Iizuka K, Furukawa H, Beckman DL, Pingel JT, Scalzo AA, Fremont DH, Yokoyama WM. 2002. Recognition of a virus-encoded ligand by a natural killer cell activation receptor. *Proc Natl Acad Sci USA* **99:**8826-31
90. Arase H, Mocarski ES, Campbell AE, Hill AB, Lanier LL. 2002. Direct recognition of cytomegalovirus by activating and inhibitory NK cell receptors. *Science* **296:**1323-6
91. Bubi I, Wagner M, Krmpoti A, Saulig T, Kim S, Yokoyama WM, Jonji S, Koszinowski UH. 2004. Gain of virulence caused by loss of a gene in murine cytomegalovirus. *J Virol* **78:**7536-44
92. Voigt V, Forbes CA, Tonkin JN, Degli-Esposti MA, Smith HR, Yokoyama WM, Scalzo AA. 2003. Murine cytomegalovirus m157 mutation and variation leads to immune evasion of natural killer cells. *Proc Natl Acad Sci USA* **100:**13483-8
93. French AR, Pingel JT, Wagner M, Bubic I, Yang L, Kim S, Koszinowski U, Jonjic S, Yokoyama WM. 2004. Escape of mutant double-stranded DNA virus from innate immune control. *Immunity* **20:**747-56
94. Chapman TL, Bjorkman PJ. 1998. Characterization of a murine cytomegalovirus class I major histocompatibility complex (MHC) homolog: comparison to MHC molecules and to the human cytomegalovirus MHC homolog. *J Virol* **72:**460-6
95. Farrell HE, Vally H, Lynch DM, Fleming P, Shellam GR, Scalzo AA, Davis-Poynter NJ. 1997. Inhibition of natural killer cells by a cytomegalovirus MHC class I homologue in vivo. *Nature* **386:**510-4
96. Beck S, Barrell BG. 1988. Human cytomegalovirus encodes a glycoprotein homologous to MHC class-I antigens. *Nature* **331:**269-72
97. Cosman D, Fanger N, Borges L, Kubin M, Chin W, Peterson L, Hsu ML. 1997. A novel immunoglobulin superfamily receptor for cellular and viral MHC class I molecules. *Immunity* **7:**273-82
98. Park B, Oh H, Lee S, Song Y, Shin J, Sung YC, Hwang SY, Ahn K. 2002. The MHC class I homolog of human cytomegalovirus is resistant to down-regulation mediated by the unique short region protein (US)2, US3, US6, and US11 gene products. *J Immunol* **168:**3464-9
99. Reyburn HT, Mandelboim O, Vales-Gomez M, Davis DM, Pazmany L, Strominger JL. 1997. The class I MHC homologue of human cytomegalovirus inhibits attack by natural killer cells. *Nature* **386:**514-7
100. Leong CC, Chapman TL, Bjorkman PJ, Formankova D, Mocarski ES, Phillips JH, Lanier LL. 1998. Modulation of natural killer cell cytotoxicity in human cytomegalovirus infection: the role of endogenous class I major histocompatibility complex and a viral class I homolog. *J Exp Med* **187:**1681-7
101. Saverino D, Ghiotto F, Merlo A, Bruno S, Battini L, Occhino M, Maffei M, Tenca C, Pileri S, Baldi L, Fabbi M, Bachi A, De Santanna A, Grossi CE, Ciccone E. 2004. Specific recognition of the viral protein UL18 by CD85j/LIR-1/ILT2 on CD8+ T cells mediates the non-MHC-restricted lysis of human cytomegalovirus-infected cells. *J Immunol* **172:**5629-37
102. Tomasec P, Braud VM, Rickards C, Powell MB, McSharry BP, Gadola S, Cerundolo V, Borysiewicz LK, McMichael AJ, Wilkinson GW. 2000. Surface expression of HLA-E, an inhibitor of natural killer cells, enhanced by human cytomegalovirus gpUL40. *Science* **287:**1031

103. Ulbrecht M, Martinozzi S, Grzeschik M, Hengel H, Ellwart JW, Pla M, Weiss EH. 2000. Cutting edge: the human cytomegalovirus UL40 gene product contains a ligand for HLA-E and prevents NK cell-mediated lysis. *J Immunol* **164:**5019-22
104. Wang EC, McSharry B, Retiere C, Tomasec P, Williams S, Borysiewicz LK, Braud VM, Wilkinson GW. 2002. UL40-mediated NK evasion during productive infection with human cytomegalovirus. *Proc Natl Acad Sci USA* **99:**7570-5
105. Falk CS, Mach M, Schendel DJ, Weiss EH, Hilgert I, Hahn G. 2002. NK cell activity during human cytomegalovirus infection is dominated by US2-11-mediated HLA class I down-regulation. *J Immunol* **169:**3257-66
106. Llano M, Guma M, Ortega M, Angulo A, Lopez-Botet M. 2003. Differential effects of US2, US6 and US11 human cytomegalovirus proteins on HLA class Ia and HLA-E expression: impact on target susceptibility to NK cell subsets. *Eur J Immunol* **33:**2744-54
107. Krmpotic A, Busch DH, Bubic I, Gebhardt F, Hengel H, Hasan M, Scalzo AA, Koszinowski UH, Jonjic S. 2002. MCMV glycoprotein gp40 confers virus resistance to CD8+ T cells and NK cells in vivo. *Nat Immunol* **3:**529-35
108. Lodoen M, Ogasawara K, Hamerman JA, Arase H, Houchins JP, Mocarski ES, Lanier LL. 2003. NKG2D-mediated natural killer cell protection against cytomegalovirus is impaired by viral gp40 modulation of retinoic acid early inducible 1 gene molecules. *J Exp Med* **197:**1245-53
109. Cerwenka A, Bakker AB, McClanahan T, Wagner J, Wu J, Phillips JH, Lanier LL. 2000. Retinoic acid early inducible genes define a ligand family for the activating NKG2D receptor in mice. *Immunity* **12:**721-7
110. Diefenbach A, Jamieson AM, Liu SD, Shastri N, Raulet DH. 2000. Ligands for the murine NKG2D receptor: expression by tumor cells and activation of NK cells and macrophages. *Nat Immunol* **1:**119-26
111. Carayannopoulos LN, Naidenko OV, Fremont DH, Yokoyama WM. 2002. Cutting edge: murine UL16-binding protein-like transcript 1: a newly described transcript encoding a high-affinity ligand for murine NKG2D. *J Immunol* **169:**4079-83
112. Bauer S, Groh V, Wu J, Steinle A, Phillips JH, Lanier LL, Spies T. 1999. Activation of NK cells and T cells by NKG2D, a receptor for stress-inducible MICA. *Science* **285:**727-9
113. Cosman D, Mullberg J, Sutherland CL, Chin W, Armitage R, Fanslow W, Kubin M, Chalupny NJ. 2001. ULBPs, novel MHC class I-related molecules, bind to CMV glycoprotein UL16 and stimulate NK cytotoxicity through the NKG2D receptor. *Immunity* **14:**123-33
114. Wu J, Chalupny NJ, Manley TJ, Riddell SR, Cosman D, Spies T. 2003. Intracellular retention of the MHC class I-related chain B ligand of NKG2D by the human cytomegalovirus UL16 glycoprotein. *J Immunol* **170:**4196-200
115. Rolle A, Mousavi-Jazi M, Eriksson M, Odeberg J, Soderberg-Naucler C, Cosman D, Karre K, Cerboni C. 2003. Effects of human cytomegalovirus infection on ligands for the activating NKG2D receptor of NK cells: up-regulation of UL16-binding protein (ULBP)1 and ULBP2 is counteracted by the viral UL16 protein. *J Immunol* **171:**902-8
116. Dunn C, Chalupny NJ, Sutherland CL, Dosch S, Sivakumar PV, Johnson DC, Cosman D. 2003. Human cytomegalovirus glycoprotein UL16 causes intracellular sequestration of NKG2D ligands, protecting against natural killer cell cytotoxicity. *J Exp Med* **197:**1427-39
117. Welte SA, Sinzger C, Lutz SZ, Singh-Jasuja H, Sampaio KL, Eknigk U, Rammensee HG, Steinle A. 2003. Selective intracellular retention of virally induced NKG2D ligands by the human cytomegalovirus UL16 glycoprotein. *Eur J Immunol* **33:**194-203
118. Odeberg J, Browne H, Metkar S, Froelich CJ, Branden L, Cosman D, Soderberg-Naucler C. 2003. The human cytomegalovirus protein UL16 mediates increased resistance to natural killer cell cytotoxicity through resistance to cytolytic proteins. *J Virol* **77:**4539-45

119. Fletcher JM, Prentice HG, Grundy JE. 1998. Natural killer cell lysis of cytomegalovirus (CMV)-infected cells correlates with virally induced changes in cell surface lymphocyte function-associated antigen-3 (LFA-3) expression and not with the CMV-induced down-regulation of cell surface class I HLA. *J Immunol* **161:**2365-74
120. Coscoy L, Ganem D. 2001. A viral protein that selectively downregulates ICAM-1 and B7-2 and modulates T cell costimulation. *J Clin Invest* **107:**1599-606
121. Ishido S, Choi JK, Lee BS, Wang C, DeMaria M, Johnson RP, Cohen GB, Jung JU. 2000. Inhibition of natural killer cell-mediated cytotoxicity by Kaposi's sarcoma-associated herpesvirus K5 protein. *Immunity* **13:**365-74
122. Zocchi MR, Rubartelli A, Morgavi P, Poggi A. 1998. HIV-1 Tat inhibits human natural killer cell function by blocking L-type calcium channels. *J Immunol* **161:**2938-43
123. Poggi A, Carosio R, Spaggiari GM, Fortis C, Tambussi G, Dell'Antonio G, Dal Cin E, Rubartelli A, Zocchi MR. 2002. NK cell activation by dendritic cells is dependent on LFA-1-mediated induction of calcium-calmodulin kinase II: inhibition by HIV-1 Tat C-terminal domain. *J Immunol* **168:**95-101
124. Zheng ZY, Zucker-Franklin D. 1992. Apparent ineffectiveness of natural killer cells vis-a-vis retrovirus-infected targets. *J Immunol* **148:**3679-85
125. Crotta S, Stilla A, Wack A, D'Andrea A, Nuti S, D'Oro U, Mosca M, Filliponi F, Brunetto RM, Bonino F, Abrignani S, Valiante NM. 2002. Inhibition of natural killer cells through engagement of CD81 by the major hepatitis C virus envelope protein. *J Exp Med* **195:**35-41
126. Tseng CT, Klimpel GR. 2002. Binding of the hepatitis C virus envelope protein E2 to CD81 inhibits natural killer cell functions. *J Exp Med* **195:**43-9
127. Orange JS, Fassett MS, Koopman LA, Boyson JE, Strominger JL. 2002. Viral evasion of natural killer cells. *Nat Immunol* **3:**1006-12
128. Raftery MJ, Schwab M, Eibert SM, Samstag Y, Walczak H, Schonrich G. 2001. Targeting the function of mature dendritic cells by human cytomegalovirus: a multilayered viral defense strategy. *Immunity* **15:**997-1009
129. Moutaftsi M, Mehl AM, Borysiewicz LK, Tabi Z. 2002. Human cytomegalovirus inhibits maturation and impairs function of monocyte-derived dendritic cells. *Blood* **99:**2913-21
130. Andrews DM, Andoniou CE, Granucci F, Ricciardi-Castagnoli P, Degli-Esposti MA. 2001. Infection of dendritic cells by murine cytomegalovirus induces functional paralysis. *Nat Immunol* **2:**1077-84
131. Mathys S, Schroeder T, Ellwart J, Koszinowski UH, Messerle M, Just U. 2003. Dendritic cells under influence of mouse cytomegalovirus have a physiologic dual role: to initiate and to restrict T cell activation. *J Infect Dis* **187:**988-99
132. Popkin DL, Watson MA, Karaskov E, Dunn GP, Bremner R, Virgin HWt. 2003. Murine cytomegalovirus paralyzes macrophages by blocking IFN gamma-induced promoter assembly. *Proc Natl Acad Sci USA* **100:**14309-14
133. LoPiccolo DM, Gold MC, Kavanagh DG, Wagner M, Koszinowski UH, Hill AB. 2003. Effective inhibition of K(b)- and D(b)-restricted antigen presentation in primary macrophages by murine cytomegalovirus. *J Virol* **77:**301-8
134. Gredmark S, Tilburgs T, Soderberg-Naucler C. 2004. Human cytomegalovirus inhibits cytokine-induced macrophage differentiation. *J Virol* **78:**10378-89
135. Chehimi J, Bandyopadhyay S, Prakash K, Perussia B, Hassan NF, Kawashima H, Campbell D, Kornbluth J, Starr SE. 1991. In vitro infection of natural killer cells with different human immunodeficiency virus type 1 isolates. *J Virol* **65:**1812-22
136. York IA, Johnson DC. 1993. Direct contact with herpes simplex virus-infected cells results in inhibition of lymphokine-activated killer cells because of cell-to-cell spread of virus. *J Infect Dis* **168:**1127-32
137. Fujita K, Maldarelli F, Silver J. 1996. Bimodal down-regulation of CD4 in cells expressing human immunodeficiency virus type 1 Vpu and Env. *J Gen Virol* **77 (Pt 10):**2393-401

138. Chen BK, Gandhi RT, Baltimore D. 1996. CD4 down-modulation during infection of human T cells with human immunodeficiency virus type 1 involves independent activities of vpu, env, and nef. *J Virol* **70:**6044-53
139. Piguet V, Schwartz O, Le Gall S, Trono D. 1999. The downregulation of CD4 and MHC-I by primate lentiviruses: a paradigm for the modulation of cell surface receptors. *Immunol Rev* **168:**51-63
140. Swann SA, Williams M, Story CM, Bobbitt KR, Fleis R, Collins KL. 2001. HIV-1 Nef blocks transport of MHC class I molecules to the cell surface via a PI 3-kinase-dependent pathway. *Virology* **282:**267-77
141. Na YS, Yoon K, Nam JG, Choi B, Lee JS, Kato I, Kim S. 2004. Nef from a primary isolate of human immunodeficiency virus type 1 lacking the EE(155) region shows decreased ability to down-regulate CD4. *J Gen Virol* **85:**1451-61
142. Levesque K, Finzi A, Binette J, Cohen EA. 2004. Role of CD4 receptor down-regulation during HIV-1 infection. *Curr HIV Res* **2:**51-9
143. Munch J, Janardhan A, Stolte N, Stahl-Hennig C, Ten Haaft P, Heeney JL, Swigut T, Kirchhoff F, Skowronski J. 2002. T-cell receptor:CD3 down-regulation is a selected in vivo function of simian immunodeficiency virus Nef but is not sufficient for effective viral replication in rhesus macaques. *J Virol* **76:**12360-4
144. Lusso P, Malnati M, De Maria A, Balotta C, DeRocco SE, Markham PD, Gallo RC. 1991. Productive infection of CD4+ and CD8+ mature human T cell populations and clones by human herpesvirus 6. Transcriptional down-regulation of CD3. *J Immunol* **147:**685-91
145. Bowie A, Kiss-Toth E, Symons JA, Smith GL, Dower SK, O'Neill LA. 2000. A46R and A52R from vaccinia virus are antagonists of host IL-1 and toll-like receptor signaling. *Proc Natl Acad Sci USA* **97:**10162-7
146. Harte MT, Haga IR, Maloney G, Gray P, Reading PC, Bartlett NW, Smith GL, Bowie A, O'Neill LA. 2003. The poxvirus protein A52R targets Toll-like receptor signaling complexes to suppress host defense. *J Exp Med* **197:**343-51
147. Pfeffer S, Zavolan M, Grasser FA, Chien M, Russo JJ, Ju J, John B, Enright AJ, Marks D, Sander C, Tuschl T. 2004. Identification of virus-encoded microRNAs. *Science* **304:**734-6
148. Bartel DP. 2004. MicroRNAs: genomics, biogenesis, mechanism, and function. *Cell* **116:**281-97

Chapter 5

INTERFERON ANTAGONISTS OF INFLUENZA VIRUSES

ADOLFO GARCÍA-SASTRE
Department of Microbiology, Mount Sinai School of Medicine, New York, NY, USA

1. INTRODUCTION

It has become apparent that as a result of co-evolution of viruses with their hosts, most if not all mammalian viruses have acquired gene products that dampen the induction of the antiviral mechanisms mediated by the type I interferon (IFN) system, allowing them to replicate in an otherwise very hostile environment. Interestingly, most viruses encode IFN antagonists that are unique for their specific groups and that share no similarity among other viruses. In this chapter, I will review our current understanding on how specific viral proteins encoded by the group of the orthomyxoviruses, which include the influenza viruses, attenuate the induction of the type I IFN system during viral infection, contributing to efficient viral replication in their mammalian hosts.

2. INFLUENZA VIRUSES

The orthomyxoviruses include the influenza viruses and the less characterized tick-born thogotoviruses and the fish isaviruses. The influenza viruses are further divided into three types, A, B and C, according to the antigenic differences of their internal structural components. Yearly epidemics of influenza A or B viruses cause significant morbidity and mortality in humans, with 30,000 estimated yearly deaths caused only in the

P. Palese (ed.), Modulation of Host Gene Expression and Innate Immunity by Viruses, 95-114.

US[1]. In addition, influenza A viruses are also significant animal pathogens of poultry, horses and pigs, and multiple antigenically diverse strains exist in a aquatic wild bird reservoir[2]. Pandemic influenza occurs when a human influenza A virus acquires novel antigenic determinants from an avian influenza virus and now can infect and propagate in humans in the absence of pre-existing immunity. Although influenza B viruses do not cause pandemics, during some epidemic years they are known to cause more significant mortality and morbidity than influenza A viruses. By contrast, influenza C viruses are believed to cause mild or asymptomatic respiratory infections in humans.

All orthomyxoviruses are characterized by having a segmented negative strand RNA genome that replicates in the nucleus of the infected cell. Both influenza A and B virus genomes consist of eight RNA segments, whereas influenza C viruses contain seven segments[3,4]. Each RNA segment encodes one or two viral proteins (Figure 1). The RNA segments are encapsidated by the viral nucleoprotein in the form of ribonucleoproteins, which are also known to associate with the viral RNA-dependent RNA polymerase, consisting of three subunits (PB2, PB1 and PA in the case of influenza A and B viruses). The structure of the virion is completed by a layer of matrix (M1 protein) that is surrounded by the viral envelope, in which the viral glycoproteins hemagglutinin (HA), responsible for viral attachment and entry, and neuraminidase (NA), with receptor destroying activity, are anchored. Influenza C viruses lack an NA protein, and all attachment, entry and receptor destroying activities are performed by a single glycoprotein, the hemagglutinin-esterase-fusion (HEF) protein[4]. Also anchored in the viral envelope, influenza A, B and possible C viruses contain an ion channel protein, M2[5], BM2[6], and CM2[7], respectively. The influenza B viruses encode one more transmembrane protein, or NB, of unknown function, and most strains of influenza A virus have recently been shown to express a pro-apoptotic factor, known as PB1-F2, in infected cells[8]. In addition, all three types of viruses encode a non structural protein, NS1, which is expressed at high levels in infected cells, and a nuclear export protein (NEP, previously referred as NS2 protein) that is a minor structural component of the viral core and that mediates nucleo-cytoplasmic trafficking of the viral genome[9]. This chapter will focus on the known IFN antagonist properties of the NS1 protein.

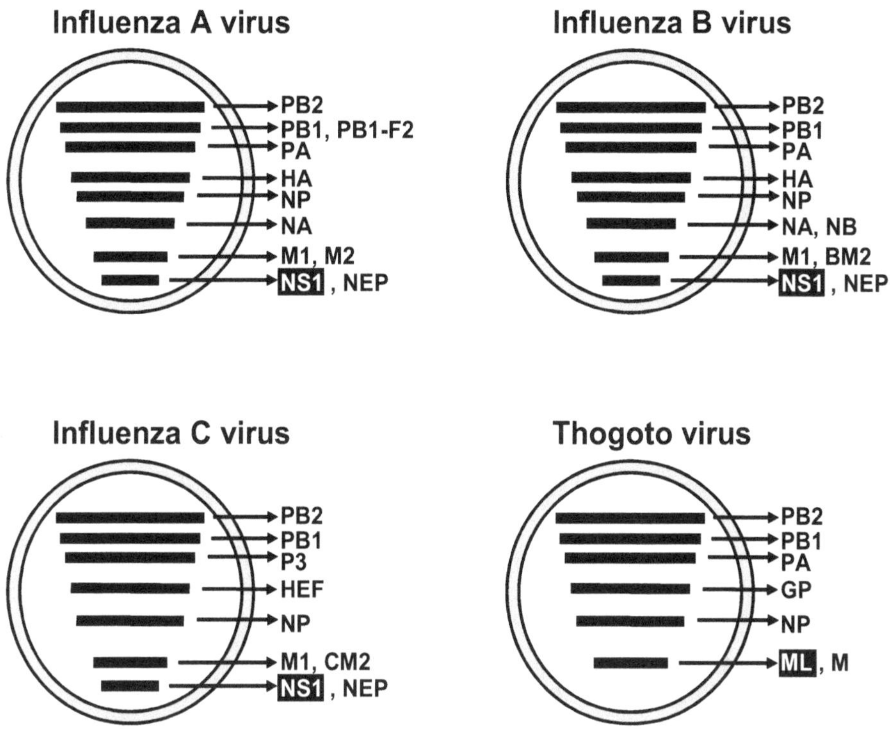

Figure 1. Schematic representation of genes and gene products encoded by influenza and Thogoto viruses. The NS1 and ML IFN antagonist gene products are highlighted.

3. TRIGGERING OF THE TYPE I IFN RESPONSE BY INFLUENZA VIRUSES

Complex organisms, such as mammals, have developed multiple primary barriers to prevent infections by pathogens. However, once a virus overcomes these barriers and invades its host, the host relies on the ability to recognize at the molecular level that it is being infected in order to put in place appropriate defense mechanisms to block the infection. Among the host defense mechanisms, the type I IFN system is a major component of the innate immune response against viruses. Type I IFN is secreted in response to virus infection by the activation of IRF-3 in combination with NF-κB and AP-1 transcription factors[10-14]. Activation of IRF-3 is mediated by phosphorylation through the action of the virus-activated kinases TBK1 and IKK-ε[15,16]. Once secreted, type I IFN interacts with its receptor, IFNAR, mediating the activation of a JAK/STAT pathway that results in transcriptional stimulation of more than one hundred genes[17] and in the induction of the antiviral state (for more details in these processes see chapters 2 and 3 in this book). The type I IFN system plays a critical role in influenza virus pathogenicity, as it was found that in the absence of the type I IFN receptor (IFNAR) or of STAT1, multiple tissues, in addition to the respiratory tissue, became highly susceptible to influenza virus infection[18]. Type I IFN is also important to promote an efficient adaptive immune response against influenza virus, as there was a biased Th2 response after influenza virus infection in STAT1 and IFNAR -/- mice[19].

Ideally for the host, virus recognition leading to type I IFN secretion should occur very early on during infection. Toll-like receptors (TLR) are molecules that are dedicated to detect invasion by pathogenic organisms. The TLRs are expressed at the cell surface and/or in the endosomes of specialized cells and sample the extracellular and endosomal environments for the presence of specific molecular patterns present in pathogens[20,21]. Upon binding of a TLR to a pathogen specific molecule, there is an activation of specific intracellular signaling events leading to the transcriptional induction of a subset of cellular genes that will directly attempt to stop the replication of the pathogen and that will alert the rest of the organism of an ongoing infection and attract immune cells to the site, resulting in further amplification of innate immune responses and in promoting the initiation of the adaptive immune response. It is now known that many TLRs recognize specific bacterial molecules, such as LPS, which are not present in eukaryotic cells. However, viruses rely on the host cell for its production, and therefore, strictly speaking, viruses are composed of cellular components, making it more challenging for a host to recognize specific viral products different from its own cellular products. Nevertheless, several TLRs appear to recognize viral glycoproteins. For instance, the hemagglutinin protein of measles virus[22] and the fusion protein of respiratory syncytial virus[23], two negative-strand RNA viruses, have been found to stimulate TLR2- and TLR4-mediated pathways, respectively. However, no clear interactions between TLRs and the glycoproteins of influenza virus have been identified so far. On the other hand, TLR3, that recognizes dsRNA[24], and TLR7 and TLR8, that recognize ssRNA[25,26], might be involved in the recognition of influenza virus and of RNA viruses in general, most likely at an endosomal compartment, leading to the production of type I IFN[25].

While TLRs might participate in the detection of influenza virus infection by specialized cells, it is also likely that intracellular cytoplasmic or even nuclear sensors are involved in recognizing specific viral products and in subsequent activation of signaling events leading to the production of type I IFN. In this respect, RIG-I, a cytoplasmic RNA helicase that has recently been found to activate type I IFN synthesis in response to dsRNA[27], or some other unidentified intracellular dsRNA sensors, might be critical factors for the recognition of structured viral RNA products generated during influenza virus infection, resulting in the transcriptional induction of type I IFN. In addition, other viral components, such as viral nucleocapsids, might also trigger type I IFN synthesis[28,29]. However, the specific molecular events leading to the activation of the type I IFN system by influenza viruses are still unknown. In the case of influenza A virus, viral mRNA synthesis is most likely required for the stimulation of IRF-3, while influenza B virus can stimulate IRF-3 and therefore IFN production prior to viral mRNA synthesis[30]. In addition, it is clear that influenza virus infection induces several signaling pathways within the infected cell, some of which are likely to participate in the regulation of the type I IFN response[31].

4. THE NS1 PROTEIN MEDIATES EVASION OF THE TYPE I IFN RESPONSE BY INFLUENZA A VIRUSES

IFNs were first described by Isaacs and Lindenmann as secreted cellular factors induced by heat-inactivated influenza A viruses that interfere with viral replication[32]. Intriguingly, Isaacs and Burke subsequently found that live influenza A virus infection is a poor stimuli of IFN production[33]. Lindemann also described in 1960 that infection with live viruses inhibits the induction of IFN by inactivated viruses[34]. These very early observations clearly suggest that influenza A virus encodes an inhibitor of IFN synthesis that is lost upon heat inactivation of the virus. It was not until 1998 that it was found that this inhibitor is the viral NS1 protein[35]. This was possible through the use of reverse genetic techniques that allowed the introduction of mutations in the NS gene of influenza virus[36], encoding both NS1 and NEP proteins (Figure 2). Using these techniques, it was possible to generate an influenza A virus lacking the NS1 gene[35]. These experiments not only demonstrated that the NS1 gene is not essential for influenza A virus infectivity, but also allowed for the first time to determine the functional role of the NS1 protein during viral infection by investigating the phenotype of an NS1 knock-out (delNS1) influenza A virus.

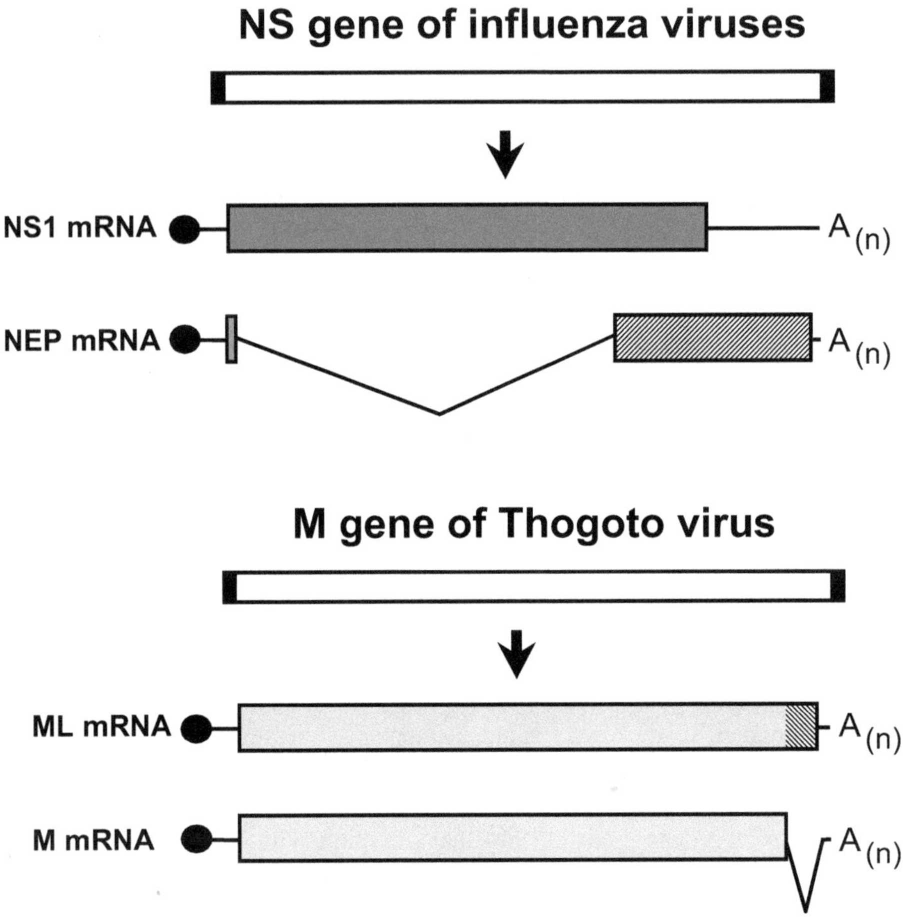

Figure 2. Coding strategy of the NS and M genes of influenza and Thogoto viruses. The NS and M genes are transcribed into mRNAs by the viral polymerase. Alternative splicing of the mRNAs results in expression of two proteins, NS1 and NEP (influenza virus NS gene) and ML and M (Thogoto virus M gene).

The delNS1 influenza virus exhibits a host range phenotype. This virus grows to titers close to wild-type influenza virus in IFN deficient substrates, such as Vero cells, that lack the type I IFN genes[37] and 6 to 7-day-old chicken embryos, that have an immature type I IFN system[38]. However, delNS1 virus replicates poorly in substrates that have a functional type I IFN response, such as MDCK cells and 10-day-old chicken embryos[35,39]. More striking are the properties of this virus in STAT1-/- mice. The parental wild-type influenza A virus replicates to high titers in lungs of wild-type CD1, BALB/c or Black/6 mice, and induces significant morbidity and mortality in these mice after intranasal infection. By contrast, the delNS1 virus replicates very poorly and it is completely attenuated in these mice. However, both wild-type and delNS1 viruses are lethal in STAT1-/- mice and they both grow to high titers in lungs. These observations clearly indicate that the NS1 protein is dispensable for viral replication in IFN deficient systems, while it is required in IFN competent substrates and hosts[35], and this strongly suggests that the main role of the NS1 protein during influenza virus infection is the inhibition of the IFN response.

Additional data was obtained demonstrating that the NS1 protein of influenza A virus is an IFN antagonist that inhibits the production of IFN during viral infection. The delNS1 virus is a high inducer of type I IFN, both IFNα and IFNβ, while the wild-type parental virus is a poor inducer[40-43]. The inhibition of IFN production by infection with wild-type influenza virus correlates with low levels of IFNα/β mRNA induced in infected cells, and with a poor activation of transcription factors involved in the transcriptional stimulation of the IFN genes, including IRF-3, NF-κB and AP-1. In contrast, delNS1 virus infection results in unimpeded activation of IRF-3, NF-κB and AP-1 transcription factors and in stimulation of the IFNβ promoter[40,41,44]. Expression of the NS1 protein in the absence of any other influenza viral protein resulted in inhibition of IFNβ promoter activation in response to dsRNA or to infection with an IFN-inducing virus, such as Sendai virus[40,41]. These observations demonstrate that the NS1 protein of influenza A virus directly inhibits the activation of transcription factors involved in the stimulation of the IFNβ promoter.

Although the mechanism of action on how the NS1 protein of influenza A virus inhibits the activation of transcription factors associated with IFN production is not fully elucidated, this activity is mainly mediated by its first 73 N-terminal amino acids[45] containing an unconventional dsRNA-binding domain[46,47]. Nevertheless, deletion of C-terminal amino acids from the NS1 results in mutant influenza viruses with intermediate phenotypes between wild-type and delNS1 virus, indicating that the C-terminal region of the NS1 also contributes to efficient inhibition of the IFN system[39,45,48,49]. Interestingly, the C-terminal amino acids of the NS1 protein can be functionally substituted by short unrelated dimerization domains[45]. These results indicate that the C-terminal region of the NS1 plays a structural role and contributes to efficient dimerization of a functional IFN antagonist protein. Since dsRNA binding by the NS1 protein depends on dimerization[50], and since mutations in amino acid residues R38 and K41 that play a key role in dsRNA binding result in the loss of the IFN antagonist properties of the NS1[39,51], it is highly likely that the inhibition of IFN synthesis by the NS1 is at least in part due to its ability to bind to dsRNA. This ability might result in sequestration from cellular sensors of viral dsRNA and/or of structured viral RNA generated as byproducts of viral RNA replication and transcription. However, it is also likely that additional mechanisms contribute to optimal IFN antagonistic functions of the NS1 protein. For instance, the ability of the NS1 to inhibit the IFN response appears to be host-specific[52], and this would suggest that a cellular factor whose sequence varies among different host species is a target for the NS1 protein. Moreover, a mutant influenza A virus expressing a dsRNA binding defective NS1 protein was able to partially gain the ability to inhibit IFN induction by a compensatory mutation within the N-terminal of the NS1 that did not confer dsRNA binding properties[51]. It is then likely that a cellular factor that participates in the IFN induction pathway is inhibited by the NS1. Although many cellular factors have been identified that interact with the NS1 protein of influenza A virus[53-59], none of them have been implicated in the induction of IFN, and therefore, the identity of such a factor remains unknown.

Not only does the NS1 protein of influenza A virus inhibit the type I IFN response by preventing the activation of transcription factors involved in the induction of IFN, but this viral protein has been described to inhibit the IFN-mediated antiviral response at multiple steps. The carboxy-terminal region of the NS1 protein appears to be implicated in inhibition of cellular mRNA processing by specifically interacting with CPSF, a critical component of the cellular polyadenylation machinery, resulting in reduced levels of cellular polydenylated messages in infected cells[55]. Moreover, either due to direct interaction of the N-terminal domain of the NS1 with cellular snRNPs[60], or indirectly due to inhibition of cellular mRNA polyadenylation[61], the NS1 protein also appears to affect mRNA splicing[62,63]. Finally, the last amino acids of the NS1 protein of influenza A/Udorn/72 virus have been described to interact with and block the function of PABII, a cellular protein that participates in the nucleo-cytoplasmic transport of polyadenylated cellular mRNA[57]. This interaction might be specific of some strains of influenza A viruses, since the majority of the influenza A viruses encode NS1 proteins between 219 and 230 amino acids long, instead of the 237 amino acids long NS1 of influenza A/Udorn/72 virus. Binding to PABII might not be required for inhibition of nucleo-cytoplasmic mRNA transport by the NS1 protein, since this inhibition can also be seen in the absence of the C-terminal domain of the NS1[64]. Nevertheless, inhibition of cellular RNA processing and transport by the NS1 could lead to a significant reduction of cytoplasmic levels of newly synthesized cellular mRNA[65]. Since the induction of the IFN response requires de novo synthesis of IFN as well as of IFN-induced genes, the inhibition of cellular mRNA processing by the NS1 might also contribute to the inhibition of the IFN response. In fact mutations that affect the interaction of the NS1 with CPSF have been shown to attenuate viral growth of influenza A/Udorn/72 virus[66]. Although this domain appears to be dispensable at least in some other viral strains[45,48], the NS1 protein might achieve a more potent inhibition of the IFN system by inhibiting the IFN response at both pre-transcriptional and post-transcriptional levels.

Another component of the type I IFN response that appears to also be inhibited by the NS1 protein is the PKR-mediated antiviral pathways. PKR is a cellular kinase that is expressed at low levels but whose expression dramatically increases in response to type I IFN. This kinase is usually present in an inactive state, but becomes activated by binding to dsRNA[67]. Once activated, PKR mediates phosphorylation of several proteins, eIF-2α being one of the most critical substrates. Phosphorylation of eIF-2α results in inhibition of cellular translation. PKR is one of the best characterized antiviral effectors induced by IFN. Once activated by dsRNA present in virus-infected cells, it stops viral replication by inducing a general translational shut-off. The fact that many viruses encode inhibitors of the PKR pathway illustrates the critical role of PKR as a cellular inhibitor of virus replication[68]. The NS1 protein of influenza A virus appears not only to prevent the transcriptional induction of PKR due to its ability to inhibit IRF-3 activation and IFN synthesis, but also to directly prevent PKR activation by sequestering the activator of PKR, dsRNA[69-72]. The critical role that the NS1 plays in inhibiting the PKR pathway has been demonstrated. Thus, replication of delNS1 virus is enhanced in the presence of PKR inhibitors and in PKR-/- mice[73]. Inhibition of PKR might also be responsible for the NS1-mediated enhancement of viral growth in cells pretreated with IFN[74-76].

The ability of the NS1 protein to bind to dsRNA, to prevent the activation of transcription factors involved in IFN synthesis and to inhibit the activation of PKR are reminiscent of the E3L protein of vaccinia virus[77-79]. It is highly remarkable that two completely unrelated viruses: vaccinia virus, a large dsDNA virus that replicates in the cytoplasm, and influenza A virus, a small negative strand RNA virus that replicates in the nucleus, have come up with two different proteins, the E3L and NS1 proteins, with no significant sequence identity, that nevertheless inhibit the type I IFN response by very similar mechanisms. Also remarkable is the fact that IFN antagonist proteins from different viruses are to a large extent functionally interchangeable. For example, replication of delNS1 influenza A virus could be complemented by expression of the IFN antagonist protein of Ebola virus VP35[80]. VP35 of Ebola virus was subsequently found to inhibit IRF3 activation[81]. Replication of delNS1 virus could also be complemented by expression of the herpesvirus protein ICP34.5, an inhibitor of PKR-mediated translational repression[82]. Functional replacement of the ICP34.5 protein by the NS1 protein was also possible in the context of infectious herpes simplex virus[83]. In addition, the NS1 protein of influenza A virus could functionally substitute for the V protein of a paramyxovirus[84], a viral protein that has been implicated in inhibition of IFN signaling and of IFN production[85,86]. In all cases, these viral proteins share no sequence identity and most likely antagonize the IFN system through different mechanisms.

While the NS1 protein of influenza A virus is clearly implicated in antagonizing the type I IFN response of the host, they might be additional mechanisms employed by the virus that could directly or indirectly contribute to inhibition of the IFN system. For instance, viral infection is known to induce the translational repression of cellular protein synthesis[87,88]. This protein synthesis shut-off is independent of the presence of the NS1 protein[72,89] and is likely to reduce the levels of expression of IFN and IFN-inducible genes during viral infection. Influenza A viruses also appear to inhibit PKR activation by a second, NS1-independent mechanism that involves the activation of the cellular PKR inhibitor p58[90,91]. It is actually not uncommon for a virus to encode multiple mechanisms of inhibition of the IFN response[92], and this underscores the high selective pressure that the IFN response has exerted in the evolution of viruses.

5. INTERFERON ANTAGONISTS OF INFLUENZA B VIRUS AND OF THOGOTO VIRUS

All influenza viruses have an NS gene that encodes two proteins due to mRNA alternative splicing: the NS1 protein from the unspliced mRNA and the NEP protein from the spliced mRNA[9]. While little is known with respect to the NS1 protein of influenza C virus, it was recently demonstrated that elimination of the NS1 gene from a recombinant influenza B virus lead to the generation of a high IFN-inducing virus, similarly to the situation with influenza A virus[43]. It was subsequently found that the NS1 protein of influenza B virus also prevents the activation of the IRF3 transcription factor during viral infection, resulting in the prevention of transcriptional activation of the IFNβ promoter[93]. Although the NS1 proteins of influenza A and B viruses have only approximately 20% amino acid sequence identity, both proteins contain N-terminal dsRNA binding domains that are predicted to be structurally related[94,95]. Like the A/NS1 protein, the dsRNA binding domain of the B/NS1 protein was able to prevent IRF3 and IFNβ promoter activation when expressed in the absence of any other viral component. However, the carboxy-terminal domain of the B/NS1 protein was also able to prevent IRF3 activation, suggesting that both dsRNA binding-dependent and -

independent mechanisms are responsible for the IFN antagonism of the B/NS1[93]. As predicted by these observations, influenza B viruses lacking the carboxy-terminal domains of their NS1 proteins were attenuated in IFN-competent systems[39].

As opposed to the A/NS1 protein, the B/NS1 protein lacks the ability to inhibit cellular mRNA processing[94]. However, the B/NS1 protein, like the A/NS1 protein, has been reported to inhibit PKR activation[94]. In addition, the B/NS1 protein, and not the A/NS1 protein has the ability to inhibit the activity of ISG15 that, like PKR, is an IFN-inducible protein[96]. This inhibitory activity has been mapped within the N-terminal region of the B/NS1 protein but can be separated from the dsRNA binding activity[95]. ISG15 is a protein related to ubiquitinin that is conjugated to cellular proteins after IFN stimulation by the actions of the E1 and E2 enzymes Ube1L/E1^{ISG15} and UbcH8, respectively[96,97]. Although the identity of the target proteins and the functional consequences of ISG15 conjugation are not well understood, ISGylation has been found to promote IFN signaling, and it might represent a positive feed-back mechanism of the IFN system[98]. Inhibition of PKR and ISG15 activity by the B/NS1 protein is likely to contribute to the IFN antagonism functions of this viral protein.

Among orthomyxoviruses, Thogoto virus represents an interesting virus with respect to the inhibition of the IFN system due to the absence of an NS gene, encoding NS1 and NEP proteins, in its viral genome. While the functional homologue of the NEP protein of Thogoto virus is not known, the M gene of this virus has recently been described to encode an IFN antagonist protein, the ML protein[99]. Alternative splicing of the mRNA encoded by the M gene of Thogoto virus results in expression of the matrix M protein from the spliced mRNA, and of the ML protein from the unspliced mRNA[100] (Figure 2). The ML protein is identical to the M protein except for a carboxy-terminal extension that confers to the protein the ability to inhibit IFN production[99]. An ML knock-out Thogoto virus induces higher levels of IFN and shows attenuation in IFN competent mice containing a functional Mx1 gene[101]. Like the NS1 protein of influenza virus, expression of the ML protein suppresses IFN synthesis by blocking the transcriptional activation of the IFNβ promoter by IRF-3. However, in contrast to the NS1 protein, the ML protein does not inhibit phosphorylation and nuclear translocation of IRF-3 but prevents its dimerization and interaction with the coactivator CBP in the nucleus[102].

6. MULTIPLE FUNCTIONS ASSOCIATED WITH THE NS1 PROTEIN OF INFLUENZA A VIRUS

In addition to specifically inhibiting the type I IFN system, the NS1 protein of the influenza A virus has been shown to be involved in several other functions, some of which may or may not be related to its IFN antagonistic properties. The NS1 protein functions as a translational enhancer[103,104], and although this property might be in part explained by its PKR inhibitory activity[72], a direct interaction between the NS1 protein and components of the cellular translational machinery might be responsible for specific enhancement of viral mRNA translation[58,59]. A role of the NS1 protein in viral RNA synthesis has also been proposed, since mutations in the NS1 protein have resulted in reduced viral RNA replication[105].

Specific cellular processes different from the IFN response have also been found to be affected by the NS1 protein. Both pro-apoptotic[106,107] as well as apoptotic inhibitory effects have been associated with expression of the NS1[84,107,108]. Inhibition of apoptosis by the NS1 might be explained by an inhibition of pro-apototic pathways induced by IFN. Intriguingly, the NS1 protein also inhibits RNA silencing when expressed in insect and plant cells[109-111]. Whether this property reflects a convergence of RNA silencing and IFN pathways between different kingdoms and/or the existence of RNA silencing innate antiviral pathways in mammalian hosts that are inhibited by influenza virus remains to be determined.

7. ROLE OF THE INFLUENZA A VIRUS NS1 PROTEIN IN VIRAL PATHOGENICITY

Due to the central role that the NS1 protein of influenza virus plays in inhibition of innate immune responses mediated by IFN, it is likely that this protein also influences viral pathogenicity in different hosts. For example, the NS1 protein of the unusually virulent strain of human influenza A virus that caused the 1918-1919 viral pandemia was found to attenuate virulence of a mouse-adapted influenza virus in mice[112], but to efficiently inhibit the type I IFN response in human cells[52]. The NS1 proteins of the avian H5N1 influenza A viruses isolated in Hong Kong in 1997 that that have caused limited infections in humans associated with high levels of mortality, appear to be very efficient at mediating viral resistance against the antiviral effects of type I IFN and TNF-α, contributing to enhanced virulence in a pig animal model[76,113]. This property appears to be dependent on the presence of a glutamic acid residue at position 92 of the NS1[76]. Interestingly, resistance to the action of IFN and TNF-α also correlates with enhanced expression of these cytokines as well as other proinflammatory cytokines in infected macrophages as well as humans, and this might have contributed to the high lethality of these viruses[114]. Thus, H5N1 viral replication appears to trigger a cytokine storm that does not prevent further virus replication leading to more proinflammatory cytokine secretion resulting in immunopathological damage of the host and acute respiratory distress syndrome.

The role of NS1 in virulence can be exploited to our own benefit. Inhibition of NS1 function is predicted to lead to an unimpeded innate immune response during viral infection, resulting in fast viral clearance. While specific inhibitors of NS1 function remain to be found, influenza virus strains expressing mutated NS1 proteins are likely to be attenuated in vivo and therefore might be good live vaccine candidates against influenza. In this respect, different truncations in the NS1 gene have been associated with different degrees of attenuation, so it might be possible to select NS1 mutations resulting in optimal levels of attenuation and immunogenicity[39]. Such strains would replicate in the host to levels enough to induce a potent immune response but not sufficient to cause disease. Attenuated viruses containing truncated NS1 proteins might also be used to express antigens as vaccine vectors against HIV or tumor cells[115-117]. Interestingly, loss of NS1 function appears to lead to selected specificity for viral growth and killing of tumor cells, many of which have deficiencies in the type I IFN response[118,119]. Therefore, it might be possible to use in the future NS1-mutant influenza viruses as oncolytic agents for the therapeutic treatment of cancer.

8. CONCLUSIONS

Although our understanding on how influenza and Thogoto viruses inhibit the IFN response has greatly increased in the last years, there are still many interesting unanswered questions in this field. It will be important to determine the precise mechanisms of action of the NS1 and ML proteins of these viruses, as well as their contribution to host and tissue tropism and to virulence. Research in this area also requires a better knowledge on the cell processes that result in activation of the IFN system and in the induction of the IFN-mediated antiviral state. In addition, development of vaccines and antivirals against influenza virus might be possible by targeting the NS1 protein. If successful, these approaches would also represent a proof-of-concept that can be applied to many other viruses known to contain IFN antagonist genes.

ACKNOWLEDGEMENTS

Work in the laboratory of AG-S has been partly supported by grants from the NIH.

REFERENCES

1. Thompson, W.W., Shay, D.K., Weintraub, E., Brammer, L., Cox, N., Anderson, L.J. & Fukuda, K., 2003, Mortality associated with influenza and respiratory syncytial virus in the United States. *JAMA* **289:**179-186
2. Webster, R.G., Bean, W.J., Gorman, O.T., Chambers, T.M. & Kawaoka, Y., 1992, Evolution and ecology of influenza A viruses. *Microbiol. Rev.* **56:**152-179
3. Palese, P., 1977, The genes of influenza virus. *Cell* **10:**1-10
4. Lamb, R.A. & Krug, R.M., 2001. In Fields Virology (Knipe, D.M. et al. eds.) Lippincott-Raven, Philadelphia, pp1487-1531.
5. Pinto, L.H., Holsinger, L.J. & Lamb, R.A., 1992, Influenza virus M2 protein has ion channel activity. *Cell* **69:**517-528
6. Mould, J.A., Paterson, R.G., Takeda, M., Ohigashi, Y., Venkataraman, P., Lamb, R.A. & Pinto, L.H., 2003, Influenza B virus BM2 protein has ion channel activity that conducts protons across membranes. *Dev. Cell.* **5:**175-184
7. Hongo, S., Ishii, K., Mori, K., Takashita, E., Muraki, Y., Matsuzaki, Y. & Sugawara, K., 2004, Detection of ion channel activity in Xenopus laevis oocytes expressing Influenza C virus CM2 protein. *Arch. Virol.* **149:**35-50
8. Chen, W., Calvo, P.A., Malide, D., Gibbs, J., Schubert, U., Bacik, I., Basta, S., O'Neill, R., Schickli, J., Palese, P., Henklein, P., Bennink, J.R. & Yewdell, J.W., 2001, A novel influenza A virus mitochondrial protein that induces cell death. *Nat. Med.* **7:**1306-1312
9. Paragas, J., Talon, J., O'Neill, R.E., Anderson, D.K., García-Sastre, A. & Palese, P., 2001, Influenza B and C virus NEP (NS2) proteins possess nuclear export activities. *J. Virol.* **75:**7375-7383
10. Wathelet, M.G., Lin, C.H., Parekh, B.S., Ronco, L.V., Howley, P.M. & Maniatis, T., 1998, Virus infection induces the assembly of coordinately activated transcription factors on the IFN-beta enhancer in vivo. *Mol. Cell* **1:**507-518
11. Sato, M., Tanaka, N., Hata, N., Oda, E. & Taniguchi, T., 1998, Involvement of the IRF family transcription factor IRF-3 in virus-induced activation of the IFN-β gene. *FEBS Lett.* **425:**112-116
12. Juang, Y., Lowther, W., Kellum, M., Au, W.C., Lin, R., Hiscott, J. & Pitha, P.M., 1998, Primary activation of interferon A and interferon B gene transcription by interferon regulatory factor 3. *Proc. Natl. Acad. Sci. (USA)* **95:**9837-9842
13. Yoneyama, M., Suhara, W., Fukuhara, Y., Fukuda, M., Nishida, E. & Fujita, T., 1998, Direct triggering of the type I interferon system by virus infection: activation of a transcription factor complex containing IRF-3 and CBP/p300. *EMBO J.* **17:**1087-1095
14. Weaver, B.K., Kumar, K.P. & Reich, N.C., 1998, Interferon regulatory factor 3 and CREB-binding protein/p300 are subunits of double-stranded RNA-activated transcription factor DRAF1. *Mol. Cell. Biol.* **18:**1359-1368
15. Sharma, S., TenOever, B.R., Grandvaux, N., Zhou, G.P., Lin, R. & Hiscott, J., 2003, Triggering the interferon antiviral response through an IKK-related pathway. *Science* **300:**1148-1151

16. Fitzgerald, K.A., McWhirter, S.M., Faia, K.L., Rowe, D.C., Latz, E., Golenbock, D.T., Coyle, A.J., Liao, S.M. & Maniatis, T., 2003, IKKepsilon and TBK1 are essential components of the IRF3 signaling pathway. *Nat. Immunol.* **4:**491-496
17. Der, S.D., Zhou, A., Williams, B.R. & Silverman, R.H., 1998, Identification of genes differentially regulated by interferon alpha, beta, or gamma using oligonucleotide arrays. *Proc. Natl. Acad. Sci. (USA)* **95:**15623-15628
18. García-Sastre, A., Durbin, R.K., Zheng, H., Palese, P., Gertner, R., Levy, D.E. & Durbin, J.E., 1998, The role of interferon in the tropism of influenza virus. *J. Virol.* **72:**8550-8558
19. Durbin, J.E., Fernandez-Sesma, A., Lee, C.K., Rao, T.D., Frey, A.B., Moran, T.M., Vukmanovic, S., García-Sastre, A. & Levy, D.E., 2000, Type I IFN modulates innate and specific antiviral immunity. *J. Immunol.* **164:**4220-4228
20. Pasare, C. & Medzhitov, R., 2004, Toll-like receptors and acquired immunity. *Semin. Immunol.* **16:**23-26
21. Akira, S. & Takeda, K., 2004, Toll-like receptor signalling. *Nat. Rev. Immunol.* **4:**499-511
22. Bieback, K., Lien, E., Klagge, I.M., Avota, E., Schneider-Schaulies, J., Duprex, W.P., Wagner, H., Kirschning, C.J., Ter Meulen, V. & Schneider-Schaulies, S., 2002, Hemagglutinin protein of wild-type measles virus activates toll-like receptor 2 signaling. *J. Virol.* **76:**8729-8736
23. Kurt-Jones, E.A., Popova, L., Kwinn, L., Haynes, L.M., Jones, L.P., Tripp, R.A., Walsh, E.E., Freeman, M.W., Golenbock, D.T., Anderson, L.J. & Finberg, R.W., 2000, Pattern recognition receptors TLR4 and CD14 mediate response to respiratory syncytial virus. *Nat. Immunol.* **1:**398-401
24. Alexopoulou, L., Holt, A.C., Medzhitov, R. & Flavell, R.A., 2001, Recognition of double-stranded RNA and activation of NF-kappaB by Toll-like receptor 3. *Nature* **413:**732-738
25. Diebold, S.S., Kaisho, T., Hemmi, H., Akira, S. & Reis e Sousa, C., 2004, Innate antiviral responses by means of TLR7-mediated recognition of single-stranded RNA. *Science* **303:**1529-1531
26. Heil, F., Hemmi, H., Hochrein, H., Ampenberger, F., Kirschning, C., Akira, S., Lipford, G., Wagner, H. & Bauer, S., 2004, Species-specific recognition of single-stranded RNA via toll-like receptor 7 and 8. *Science* **303:**1526-1529
27. Yoneyama, M., Kikuchi, M., Natsukawa, T., Shinobu, N., Imaizumi, T., Miyagishi, M., Taira, K., Akira, S. & Fujita, T., 2004, The RNA helicase RIG-I has an essential function in double-stranded RNA-induced innate antiviral responses. *Nat. Immunol.* **5:**730-7
28. tenOever, B.R., Servant, M.J., Grandvaux, N., Lin, R. & Hiscott, J., 2002, Recognition of the measles virus nucleocapsid as a mechanism of IRF-3 activation. *J. Virol.* **76:**3659-3669
29. tenOever, B.R., Sharma, S., Zou, W., Sun, Q., Grandvaux, N., Julkunen, I., Hemmi, H., Yamamoto, M., Akira, S., Yeh, W.C., Lin, R. & Hiscott, J., 2004, Activation of TBK1 and IKKvarepsilon kinases by vesicular stomatitis virus infection and the role of viral ribonucleoprotein in the development of interferon antiviral immunity. *J. Virol.* **78:**10636-10649
30. Kim, M.J., Latham, A.G. & Krug, R.M., 2002, Human influenza viruses activate an interferon-independent transcription of cellular antiviral genes: Outcome with influenza A virus is unique. *Proc. Natl. Acad. Sci. (USA)* **99:**10096-10101
31. Ludwig, S., Planz, O., Pleschka, S. & Wolff, T., 2003, Influenza-virus-induced signaling cascades: targets for antiviral therapy? *Trends Mol. Med.* **9:**46-52
32. Isaacs, A. & Lindenmann, J., 1957, Virus interference. 1. The interferon. *Proc. R. Soc. Lond. B.* **147:**258-267
33. Isaacs, A. & Burke, D.C., 1958, Mode of action of interferon. *Nature* **4642:**1073-1076
34. Lindenmann, J., 1960, Interferon und inverse Interferenz. *Zeitschr. f. Hygiene* **146:**287-309

35. García-Sastre, A., Egorov, A., Matassov, D., Brandt, S., Levy, D.E., Durbin, J.E., Palese, P. & Muster, T., 1998, Influenza A virus lacking the NS1 gene replicates in interferon-deficient systems. *Virology* **252:**324-330
36. Enami, M. & Palese, P., 1991, High-efficiency formation of influenza virus transfectants. *J. Virol.* **65:**2711-2713
37. Diaz, M.O., Ziemin, S., Le Beau, M.M., Pitha, P., Smith, S.D., Chilcote, R.R. & Rowley, J.D., 1988, Homozygous deletion of the alpha- and beta 1-interferon genes in human leukemia and derived cell lines. *Proc. Natl. Acad. Sci. USA* **85:**5259-5263
38. Sekellick, M.J., Biggers, W.J. & Marcus, P.I., 1990, Development of the interferon system. I. In chicken cells development *in ovo* continues on time *in vitro. In Vitro Cell. Dev. Biol.* **26:**997-1003
39. Talon, J., Salvatore, M., O'Neill, R.E., Nakaya, Y., Zheng, H., Muster, T., García-Sastre, A. & Palese, P., 2000, Influenza A and B viruses expressing altered NS1 proteins: a vaccine approach. *Proc. Natl. Acad. Sci. (USA)* **97:**4309-4314
40. Talon, J., Horvath, C.M., Polley, R., Basler, C.F., Muster, T., Palese, P. & García-Sastre, A., 2000, Activation of interferon regulatory factor 3 is inhibited by the influenza A virus NS1 protein. *J. Virol.* **74:**7989-7996
41. Wang, X., Li, M., Zheng, H., Muster, T., Palese, P., Beg, A.A. & García-Sastre, A., 2000, Influenza A virus NS1 protein prevents the activation of NF-κB and induction of type I IFN. *J. Virol.* **74:**11566-11573
42. Diebold, S.S., Montoya, M., Unger, H., Alexopoulou, L., Roy, P., Haswell, L.E., Al-Shamkhani, A., Flavell, R., Borrow, P. & Reis e Sousa, C., 2003, Viral infection switches non-plasmacytoid dendritic cells into high interferon producers. *Nature* **424:**324-328
43. Dauber, B., Heins, G. & Wolff, T., 2004, The influenza B virus nonstructural NS1 protein is essential for efficient viral growth and antagonizes beta interferon induction. *J. Virol.* **78:**1865-1872
44. Ludwig, S., Wang, X., Ehrhardt, C., Zheng, H., Donelan, N., Planz, O., Pleschka, S., García-Sastre, A., Heins, G. & Wolff, T., 2002, The influenza A virus NS1 protein inhibits activation of Jun N-terminal Kinase and AP-1 transcription factors. *J. Virol.* **76:**11166-11171
45. Wang, X., Basler, C.F., Williams, B.R.G., Silverman, R.H., Palese, P. & García-Sastre, A., 2002, Functional replacement of the carboxy-terminal two thirds of the influenza A virus NS1 protein with short heterologous dimerization domains. *J. Virol.* **76:**12951-12962
46. Hatada, E. & Fukuda, R., 1992, Binding of influenza A virus NS1 protein to dsRNA in vitro. *J. Gen. Virol.* **73:**3325-3329
47. Chien, C.Y., Tejero, R., Huang, Y., Zimmerman, D.E., Rios, C.B., Krug, R.M. & Montelione, G.T., 1997, A novel RNA-binding motif in influenza A virus non-structural protein 1. *Nat. Struct. Biol.* **4:**891-895
48. Kittel, C., Sereinig, S., Ferko, B., Stasakova, J., Romanova, J., Wolkerstorfer, A., Katinger, H. & Egorov, A., 2004, Rescue of influenza virus expressing GFP from the NS1 reading frame. *Virology* **324:**67-73
49. Enami, M. & Enami, K., 2000, Characterization of influenza virus NS1 protein by using a novel helper-virus-free reverse genetic system. *J.Virol.* **74:**5556-5561
50. Wang, W., Riedel, K., Lynch, P., Chien, C.Y., Montelione, G.T. & Krug, R.M., 1999, RNA binding by the novel helical domain of the influenza virus NS1 protein requires its dimer structure and a small number of specific basic amino acids. *RNA* **5:**195-205
51. Donelan, N., Basler, C.F. & García-Sastre, A., 2003, A recombinant influenza A virus expressing an RNA-binding defective NS1 protein induces high levels of IFN-β and is attenuated in mice. *J. Virol.* **77:**13257-13266
52. Geiss, G.K., Salvatore, M., Tumpey, T.M., Carter, V.S., Wang, X., Basler, C.F., Taubenberger, J.K., Bumgarner, R.E., Palese, P., Katze, M.G. & García-Sastre, A., 2002, Cellular transcriptional profiling in influenza A virus infected lung epithelial cells: the

role of the nonstructural NS1 protein in the evasion of the host innate defense and its potential contribution to pandemic influenza. *Proc. Natl. Acad. Sci. (USA)* **99:**10736-10741

53. Wolff, T., O'Neill, R.E. & Palese, P., 1996, Interaction cloning of NS1-I, a human protein that binds to the nonstructural NS1 proteins of influenza A and B viruses. *J. Virol.* **70:**5363-5372
54. Wolff, T., O'Neill, R.E. & Palese, P., 1998, NS1-binding protein (NS1-BP): A novel human protein that interacts with the influenza A virus nonstructural NS1 protein is relocalized in the nucleus of infected cells. *J. Virol.* **72:**7170-7180
55. Nemeroff, M.E., Barabino, S.M., Li, Y., Keller, W. & Krug, R.M., 1998, Influenza virus NS1 protein interacts with the cellular 30 kDa subunit of CPSF and inhibits 3'end formation of cellular pre-mRNAs. *Mol. Cell* **1:**991-1000
56. Falcon, A.M., Fortes, P., Marion, R.M., Beloso, A. & Ortín, J., 1999, Interaction of influenza virus NS1 protein and the human homologue of Staufen *in vivo* and *in vitro*. *Nucleic Acids Res.* **27:**2241-2247
57. Chen, Z., Li, Y. & Krug, R.M., 1999, Influenza A virus NS1 protein targets poly(A)-binding protein II of the cellular 3'-end processing machinery. *EMBO J.* **18:**2273-2283
58. Aragón, T., de La Luna, S., Novoa, I., Carrasco, L., Ortín, J. & Nieto, A., 2000, Eukaryotic translation initiation factor 4GI is a cellular target for NS1 protein, a translational activator of influenza virus. *Mol. Cell. Biol.* **20:**6259-6268
59. Burgui, I., Aragon, T., Ortin, J. & Nieto, A., 2003, PABP1 and eIF4GI associate with influenza virus NS1 protein in viral mRNA translation initiation complexes. *J. Gen. Virol.* **84:**3263-3274
60. Qiu, Y., Nemeroff, M. & Krug, R.M., 1995, The influenza virus NS1 protein binds to a specific region in human U6 snRNA and inhibits U6-U2 and U6-U4 snRNA interactions during splicing. *RNA* **1:**304-316
61. Li, Y., Chen, Z.Y., Wang, W., Baker, C.C. & Krug, R.M., 2001, The 3'-end-processing factor CPSF is required for the splicing of single-intron pre-mRNAs in vivo. *RNA* **7:**920-931
62. Fortes, P., Beloso, A. & Ortín, J., 1994, Influenza virus NS1 protein inhibits pre-mRNA splicing and blocks mRNA nucleocytoplasmic transport. *EMBO J.* **13:**704-712
63. Lu, Y., Qian, X.Y. & Krug, R.M., 1994, The influenza virus NS1 protein: a novel inhibitor of pre-mRNA splicing. *Genes Dev.* **8:**1817-1828
64. Marión, R.M., Aragón, T., Beloso, A., Nieto, A. & Ortín, J., 1997, The N-terminal half of the influenza virus NS1 protein is sufficient for nuclear retention of mRNA and enhancement of viral mRNA translation. *Nucleic Acids Res.* **25:**4271-4277
65. Shimizu, K., Iguchi, A., Gomyou, R. & Ono, Y., 1999, Influenza virus inhibits cleavage of the HSP70 pre-mRNAs at the polyadenylation site. *Virology* **254:**213-219
66. Noah, D.L., Twu, K.Y. & Krug, R.M., 2003, Cellular antiviral responses against influenza A virus are countered at the posttranscriptional level by the viral NS1A protein via its binding to a cellular protein required for the 3' end processing of cellular pre-mRNAS. *Virology* **307:**386-395
67. Williams, B.R., 1999, PKR; a sentinel kinase for cellular stress. *Oncogene* **18:**6112-6120
68. Gale, M.J. & Katze, M.G., 1998, Molecular mechanisms of interferon resistance mediated by viral-directed inhibition of PKR, the interferon-induced protein kinase. *Pharmacol. Ther.* **78:**29-46
69. Lu, Y., Wambach, M., Katze, M.G. & Krug, R.M., 1995, Binding of the influenza virus NS1 protein to double-stranded RNA inhibits the activation of the protein kinase that phosphorylates the eIF-2 translation initiation factor. *Virology* **214:**222-228
70. Tan, S.L. & Katze, M.G., 1998, Biochemical and genetic evidence for complex formation between the influenza A virus NS1 protein and the interferon-induced PKR protein kinase. *J. Interferon Cytokine Res.* **18:**757-766

71. Hatada, E., Saito, S. & Fukuda, R., 1999, Mutant influenza viruses with a defective NS1 protein cannot block the activation of PKR in infected cells. *J. Virol.* **73:**2425-2433
72. Salvatore, M., Basler, C.F., Parisien, J.-P., Horvath, C.M., Bourmakina, S., Zheng, H., Muster, T., Palese, P. & García-Sastre, A., 2002, Effects of influenza A virus NS1 protein on protein expression: the NS1 protein enhances translation and is not required for shutoff of host protein synthesis. *J. Virol.* **76:**1206-1212
73. Bergmann, M., García-Sastre, A., Carnero, E., Pehamberger, H., Wolff, K., Palese, P. & Muster, T., 2000, Influenza virus NS1 protein counteracts PKR-mediated inhibition of replication. *J. Virol.* **74:**6203-6206
74. Dubrovina, T.I., Egorov, A.I., Ivanova, I.A., Pokhil'ko, A.V. & Poliak, R.I., 1995, The effect of mutation in the NS gene on the biological properties of the influenza virus. *Zh. Mikrobiol. Epidemiol. Immunobiol.* **3:**75-79
75. Sekellick, M.J., Carra, S.A., Bowman, A., Hopkins, D.A. & Marcus, P.I., 2000, Transient resistance of influenza virus to interferon action attributed to random multiple packaging and activity of NS genes. *J. Interferon Cytokine Res.* **20:**963-970
76. Seo, S.H., Hoffmann, E. & Webster, R.G., 2002, Lethal H5N1 influenza viruses escape host antiviral cytokine responses. *Nat. Med.* **8:**950-954
77. Chang, H.W., Watson, J.C. & Jacobs, B.L., 1992, The E3L gene of vaccinia virus encodes an inhibitor of the interferon-induced, double-stranded RNA-dependent protein kinase. *Proc. Natl. Acad. Sci. USA* **89:**4825-4829
78. Smith, E.J., Marié, I., Prakash, A., García-Sastre, A. & Levy, D.E., 2001, IRF3 and IRF7 phosphorylation in virus-infected cells does not require double-stranded RNA-dependent protein kinase R or IκB kinase but is blocked by vaccinia virus E3L protein. *J. Biol. Chem.* **276:**8951-8957
79. Xiang, Y., Condit, R.C., Vijaysri, S., Jacobs, B., Williams, B.R. & Silverman, R.H., 2002, Blockade of interferon induction and action by the E3L double-stranded RNA binding proteins of vaccinia virus. *J. Virol.* **76:**5251-5259
80. Basler, C.F., Wang, X., Mühlberger, E., Volchkov, V., Paragas, J., Klenk, H.-D., García-Sastre, A. & Palese, P., 2000, The Ebola virus VP35 protein functions as a type I interferon antagonist. *Proc. Natl. Acad. Sci. (USA)* **97:**12289-12294
81. Basler, C.F., Mikulasova, A., Martinez-Sobrido, L., Paragas, J., Muhlberger, E., Bray, M., Klenk, H.D., Palese, P. & García-Sastre, A., 2003, The Ebola virus VP35 protein inhibits activation of interferon regulatory factor 3. *J. Virol.* **77:**7945-7956
82. He, B., Gross, M. & Roizman, B., 1997, The γ(1) 34.5 protein of herpes simplex virus 1 complexes with protein phosphatase 1α to dephosphorylate the α subunit of the eukaryotic translation initiation factor 2 and preclude the shutoff of protein synthesis by double-stranded RNA-activated protein kinase. *Proc. Natl. Acad. Sci. (USA)* **94:**843-848
83. Jing, X., Cerveny, M., Yang, K. & He, B., 2004, Replication of herpes simplex virus 1 depends on the $\gamma_1$34.5 functions that facilitate virus response to interferon and egress in the different stages of productive infection. *J. Virol.* **78:**7653-7666
84. Park, M.S., García-Sastre, A., Cros, J.F., Basler, C.F. & Palese, P., 2003, Newcastle disease virus V protein is a determinant of host range restriction. *J. Virol.* **77:**9522-9532
85. Didcock, L., Young, D.F., Goodbourn, S. & Randall, R.E., 1999, The V protein of simian virus 5 inhibits interferon signalling by targeting STAT1 for proteasome-mediated degradation. *J.Virol.* **73:**9928-9933
86. Poole, E., He, B., Lamb, R.A., Randall, R.E. & Goodbourn, S., 2002, The V proteins of simian virus 5 and other paramyxoviruses inhibit induction of interferon-beta. *Virology* **303:**33-46
87. Katze, M.G., DeCorato, D. & Krug, R.M., 1986, Cellular mRNA translation is blocked at both initiation and elongation after infection by influenza virus or adenovirus. *J. Virol.* **60:**1027-1039
88. Feigenblum, D. & Schneider, R.J., 1993, Modification of eukaryotic initiation factor 4F during infection by influenza virus. *J. Virol.* **67:**3027-3035

89. Zürcher, T., Marión, R.M. & Ortín, J., 2000, Protein synthesis shut-off induced by influenza virus infection is independent of PKR activity. *J. Virol.* **74:**8781-8784
90. Lee, T.G., Tomita, J., Hovanessian, A.G. & Katze, M.G., 1990, Purification and partial characterization of a cellular inhibitor of the interferon-induced protein kinase of Mr 68,000 from influenza virus-infected cells. *Proc. Natl. Acad. Sci. (USA)* **87:**6208-6212
91. Lee, T.G., Tomita, J., Hovanessian, A.G. & Katze, M.G., 1992, Characterization and regulation of the 58,000-dalton cellular inhibitor of the interferon-induced, dsRNA-activated protein kinase. *J. Biol. Chem.* **267:**14238-14243
92. Levy, D.E. & García-Sastre, A., 2001, The virus battles: IFN induction of the antiviral state and mechanisms of viral evasion. *Cytokine Growth Factor Rev.* **12:**143-156.
93. Donelan, N.R., Dauber, B., Wang, X., Basler, C.F., Wolff, T. & García-Sastre, A., 2004, The N- and C-terminal domains of the NS1 protein of influenza B virus can independently inhibit IRF-3 and IFNβ promoter activation. *J. Virol.***:** in press
94. Wang, W. & Krug, R.M., 1996, The RNA-binding and effector domains of the viral NS1 protein are conserved to different extents among influenza A and B viruses. *Virology* **223:**41-50
95. Yuan, W., Aramini, J.M., Montelione, G.T. & Krug, R.M., 2002, Structural basis for ubiquitin-like ISG 15 protein binding to the NS1 protein of influenza B virus: A protein-protein interaction function that is not shared by the corresponding N-terminal domain of the NS1 protein of influenza A virus. *Virology* **304:**291-301
96. Yuan, W. & Krug, R.M., 2001, Influenza B virus NS1 protein inhibits conjugation of the interferon (IFN)-induced ubiquitin-like ISG15 protein. *EMBO J.* **20:**362-371
97. Zhao, C., Beaudenon, S.L., Kelley, M.L., Waddell, M.B., Yuan, W., Schulman, B.A., Huibregtse, J.M. & Krug, R.M., 2004, The UbcH8 ubiquitin E2 enzyme is also the E2 enzyme for ISG15, an IFN-alpha/beta-induced ubiquitin-like protein. *Proc. Natl. Acad. Sci. (USA)* **101:**7578-7582
98. Malakhova, O.A., Yan, M., Malakhov, M.P., Yuan, Y., Ritchie, K.J., Kim, K.I., Peterson, L.F., Shuai, K. & Zhang, D.E., 2003, Protein ISGylation modulates the JAK-STAT signaling pathway. *Genes Dev.* **17:**455-460
99. Hagmaier, K., Jennings, S., Buse, J., Weber, F. & Kochs, G., 2003, Novel gene product of thogoto virus segment 6 codes for an interferon antagonist. *J. Virol.* **77:**2747-2752
100. Kochs, G., Weber, F., Gruber, S., Delvendahl, A., Leitz, C. & Haller, O., 2000, Thogoto virus matrix protein is encoded by a spliced mRNA. *J. Virol.* **74:**10785-10789
101. Pichlmair, A., Buse, J., Jennings, S., Haller, O., Kochs, G. & Staeheli, P., 2004, Thogoto Virus Lacking Interferon antagonistic Protein ML Is Strongly Attenuated in Newborn Mx1-Positive but Not Mx1-Negative Mice. *J. Virol.* **78:**11422-11424
102. Jennings, S., Martínez-Sobrido, L., García-Sastre, A., Weber, F. & Kochs, G., Thogoto virus ML protein suppresses IRF3 function. submitted
103. Enami, K., Sato, T.A., Nakada, S. & Enami, M., 1994, Influenza virus NS1 protein stimulates translation of the M1 protein. *J. Virol.* **68:**1432-1437
104. de la Luna, S., Fortes, P., Beloso, A. & Ortín, J., 1995, Influenza virus NS1 protein enhances the rate of translation initiation of viral mRNAs. *J. Virol.* **69:**2427-2433
105. Falcon, A.M., Marion, R.M., Zurcher, T., Gomez, P., Portela, A., Nieto, A. & Ortin, J., 2004, Defective RNA replication and late gene expression in temperature-sensitive influenza viruses expressing deleted forms of the NS1 protein. *J. Virol.* **78:**3880-3888
106. Schultz-Cherry, S., Dybdahl-Sissoko, N., Neumann, G., Kawaoka, Y. & Hinshaw, V.S., 2001, Influenza virus NS1 protein induces apoptosis in cultured cells. *J. Virol.* **75:**7875-7881
107. Morris, S.J., Smith, H. & Sweet, C., 2002, Exploitation of the Herpes simplex virus translocating protein VP22 to carry influenza virus proteins into cells for studies of apoptosis: direct confirmation that neuraminidase induces apoptosis and indications that other proteins may have a role. *Arch. Virol.* **147:**961-979

108. Zhirnov, O.P., Konakova, T.E., Wolff, T. & Klenk, H.D., 2002, NS1 protein of influenza A virus down-regulates apoptosis. *J. Virol.* **76:**1617-1625
109. Li, W.X., Li, H., Lu, R., Li, F., Dus, M., Atkinson, P., Brydon, E.W., Johnson, K.L., García-Sastre, A., Ball, L.A., Palese, P. & Ding, S.W., 2004, Interferon antagonist proteins of influenza and vaccinia viruses are suppressors of RNA silencing. *Proc. Natl. Acad. Sci. (USA)* **101:**1350-13555
110. Bucher, E., Hemmes, H., de Haan, P., Goldbach, R. & Prins, M., 2004, The influenza A virus NS1 protein binds small interfering RNAs and suppresses RNA silencing in plants. *J. Gen. Virol.* **85:**983-991
111. Delgadillo, M.O., Saenz, P., Salvador, B., Garcia, J.A. & Simon-Mateo, C., 2004, Human influenza virus NS1 protein enhances viral pathogenicity and acts as an RNA silencing suppressor in plants. *J. Gen. Virol.* **85:**993-999
112. Basler, C.F., Reid, A.H., Dybing, J.K., Janczewski, T.A., Fanning, T.G., Zheng, H., Salvatore, M., Perdue, M.L., Swayne, D.E., García-Sastre, A., Palese, P. & Taubenberger, J.K., 2001, Sequence of the 1918 pandemic influenza virus nonstructural gene (NS) segment and characterization of recombinant viruses bearing the 1918 NS genes. *Proc. Natl. Acad. Sci. (USA)* **98:**2746-2751
113. Seo, S.H., Hoffmann, E. & Webster, R.G., 2004, The NS1 gene of H5N1 influenza viruses circumvents the host antiviral cytokine responses. *Virus Res.* **103:**107-113
114. Cheung, C.Y., Poon, L.L., Lau, A.S., Luk, W., Lau, Y.L., Shortridge, K.F., Gordon, S., Guan, Y. & Peiris, J.S., 2002, Induction of proinflammatory cytokines in human macrophages by influenza A (H5N1) viruses: a mechanism for the unusual severity of human disease? *Lancet* **360:**1831-1837
115. Ferko, B., Stasakova, J., Sereinig, S., Romanova, J., Katinger, D., Niebler, B., Katinger, H. & Egorov, A., 2001, Hyperattenuated recombinant influenza A virus nonstructural-protein-encoding vectors induce human immunodeficiency virus type 1 Nef-specific systemic and mucosal immune responses in mice. *J Virol* **75:**8899-908
116. Takasuka, N., Enami, M., Itamura, S. & Takemori, T., 2002, Intranasal inoculation of a recombinant influenza virus containing exogenous nucleotides in the NS segment induces mucosal immune response against the exogenous gene product in mice. *Vaccine* **20:**1579-1585
117. Efferson, C.L., Schickli, J., Ko, B.K., Kawano, K., Mouzi, S., Palese, P., García-Sastre, A. & Ioannides, C.G., 2003, Activation of tumor antigen-specific cytotoxic T lymphocytes (CTLs) by human dendritic cells infected with an attenuated influenza A virus expressing a CTL epitope derived from the HER-2/neu proto-oncogene. *J Virol* **77:**7411-7424
118. Bergmann, M., Romirer, I., Sachet, M., Fleischhacker, R., García-Sastre, A., Palese, P., Wolff, K., Pehamberger, H., Jakesz, R. & Muster, T., 2001, A genetically engineered influenza A virus with ras-dependent oncolytic properties. *Cancer Res.* **61:**8188-8193
119. Muster, T., Rajtarova, J., Sachet, M., Unger, H., Fleischhacker, R., Romirer, I., Grassauer, A., Url, A., García-Sastre, A., Wolff, K., Pehamberger, H. & Bergmann, M., 2004, Interferon resistance promotes oncolysis by influenza virus NS1-deletion mutants. *Int. J. Cancer* **110:**15-21

Chapter 6

THE ANTI-INTERFERON MECHANISMS OF PARAMYXOVIRUSES

NICOLA STOCK*, STEPHEN GOODBOURN#, and RICHARD E. RANDALL*

**School of Biology, Biomolecular Sciences Building, North Haugh, University of St. Andrews, Fife, Scotland, UK; #Department of Biochemistry and Immunology, St. George's Hospital Medical School, University of London, London, UK*

1. INTRODUCTION

This chapter is divided into three parts. The first part briefly introduces the reader to paramyxoviruses, the second deals with general concepts of how these viruses interact with the interferon (IFN) response and the consequences of these interactions, whilst the third part deals with the specifics of how individual members of this group of viruses counteract the interferon response.

2. THE PARAMYXOVIRUSES

The family *Paramyxoviridae* is part of the virus order *Mononegavirales*, which includes all viruses with non-segmented negative strand RNA genomes and contains two other families, the *Filoviridae* and the *Rhabdoviridae*. The *Paramyxoviridae* family is divided into two sub-families, *Paramyxovirinae* and *Pneumovirinae,* which are then further sub-divided into genera according to characteristics such as genome organization, virus morphology, protein characteristics and relatedness of protein sequence, as shown in Table 1.

P. Palese (ed.), Modulation of Host Gene Expression and Innate Immunity by Viruses, 115-139.

Table 1. Family Paramyxoviridae.

Family *Paramyxoviridae*	
Subfamily *Paramyxovirinae*	Subfamily *Pneumovirinae*
Genus *Respirovirus*	Genus *Pneumovirus*
Sendai virus	**Human respiratory syncytial virus**
Human Parainfluenzavirus 1 & 3	Bovine respiratory syncytial virus
Genus *Rubulavirus*	Genus *Metapneumovirus*
Mumps virus	**Turkey rhinotracheitis virus**
Human Parainfluenzavirus 2	
Simian virus 5	
Genus *Morbillivirus*	
Measles virus	
Canine distemper virus	
Rinderpest virus	
Genus *Henipavirus*	
Hendra virus	
Nipah virus	
Genus *Avulavirus*	
Newcastle disease virus	
Avian Parainfluenzavirus 2, 3, 4 & 5	
Genus *"TPMV-like Viruses"*	
Tupaia virus	

The *Paramyxoviridae* family includes a number of important disease-causing viruses, including measles virus (MeV), mumps virus (MuV), the human parainfluenza viruses (HPIV) and human respiratory syncytial virus (HRSV) of man, as well as Newcastle disease virus (NDV), bovine respiratory syncytial virus (BRSV), rinderpest virus, turkey rhinotracheitis virus and Sendai virus (SeV) of mammals and birds. Certain paramyxoviruses also have zoonotic potential[1], observed during outbreaks of the newly emergent Hendra virus (HeV) and Nipah virus (NiV), which appear to have a natural reservoir in fruit bats but have also infected farm animals, domestic animals and humans.

Paramyxoviruses are small enveloped viruses with a single-stranded negative sense genome of 15 to 19kb (reviewed in [2]). The complete genome sequences of nearly all known paramyxoviruses are currently available and the number of identified genes ranges from six to ten, depending on the virus. However, the number of proteins encoded by viruses within the subfamily *Paramyxovirinae* is larger than the number of genes contained in their genome, as their P/V/C genes have overlapping open reading frames (ORFs) that give rise to multiple, distinct gene products (Figure 1), some of which allow these viruses to circumvent the IFN response. The P genes of viruses within the *Pneumovirinae* subfamily do not encode more than one protein, but rather some of these viruses, including HRSV and BRSV, have two extra genes, NS1 and NS2, the products of which act as IFN antagonists.

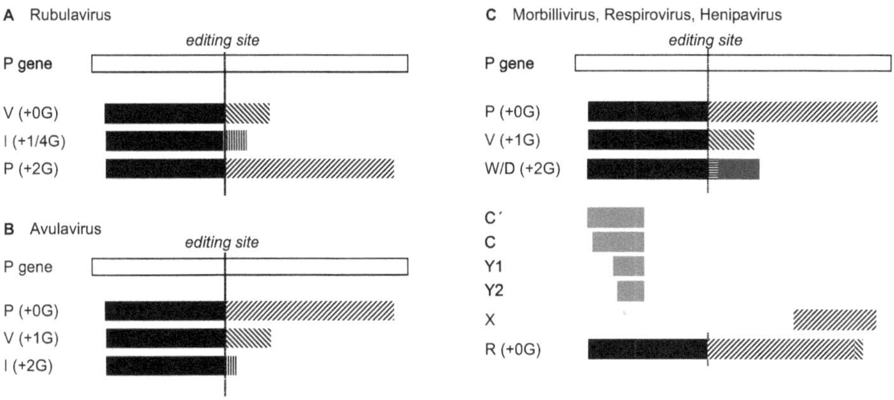

Figure 1. Illustration of the coding strategy for the multiple proteins encoded by Paramyxovirus P genes.

3. GENERAL CONCEPTS ARISING FROM THE INTERACTION OF PARAMYXOVIRUSES WITH THE INTERFERON SYSTEM

In general, viruses interfere with the IFN response by i) reducing/preventing the production of IFN, ii) inhibiting the expression of IFN responsive genes, iii) blocking the activity of cellular enzymes capable of inhibiting virus replication or iv) by having a replication strategy that is not sensitive to the intracellular "antiviral state" induced by IFN (for reviews see [3-11]). Viruses within the subfamilies *Pneumovirinae* and *Paramyxovirinae* use a variety of these strategies to antagonize the IFN response (for reviews see also [12,13]).

3.1 IFN evasion strategies of paramyxoviruses

All members of the *Paramyxovirinae* so far examined inhibit IFN signaling, thereby inhibiting the expression of IFN-stimulated genes (see Figure 2 for details of IFN signaling pathways). Interestingly, the molecular mechanisms by which the different viruses within this subfamily achieve this are very distinct and are discussed in detail in section three. In human cells, for example, simian virus 5 (SV5) blocks IFN signaling by targeting STAT1 for degradation, parainfluenza virus type 2 (HPIV2) primarily targets STAT2, whilst mumps virus (MuV) targets both STAT1 and STAT3. For these viruses it is the V protein that blocks IFN signaling. By contrast, the Respirovirus Sendai virus blocks IFN signaling by preventing STAT1 phosphorylation, and it is another set of proteins encoded by the P/V/C gene, namely the C proteins, which are responsible for this aspect of IFN antagonism (for proteins synthesised by the P/V/C gene of different viruses, see Figure 1).

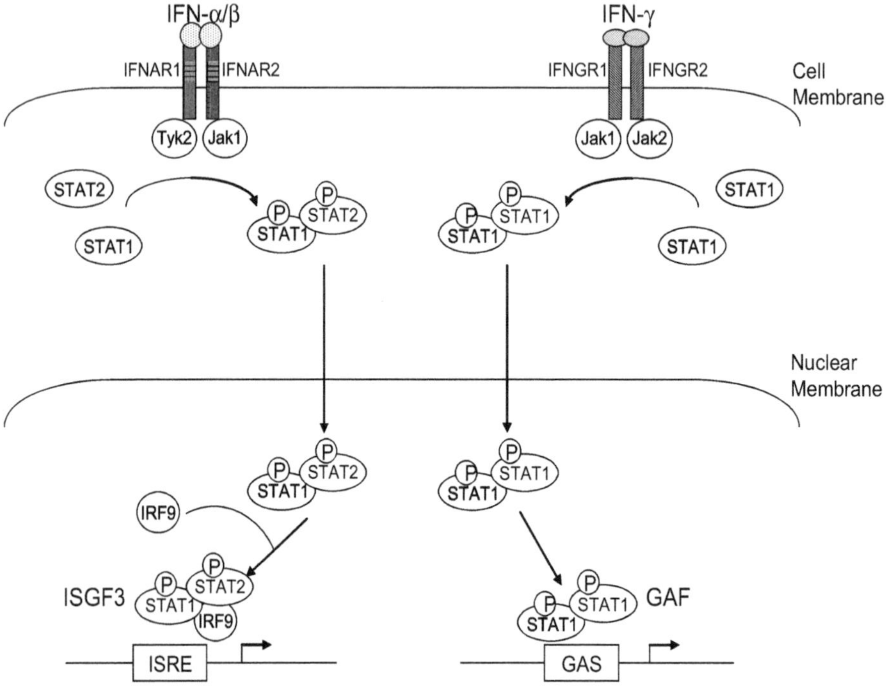

Figure 2. Schematic representation of the IFN-α/β and IFN-γ signaling pathways.

However, blocking IFN signaling alone is not sufficient to allow these viruses to fully circumvent the IFN response because even if a virus blocks IFN signaling, infected cells may still respond to infection by releasing IFN. This would induce an antiviral state in surrounding uninfected cells, making it difficult for the virus to spread from the initial foci of infection[14]. This is clearly the case for SV5, HPIV2 and MuV, which do not replicate efficiently in cells that are already in an IFN-induced antiviral state. These viruses overcome this problem by also specifically limiting IFN production in a mechanism that requires the highly conserved cysteine-rich, carboxy-terminal domain of the V proteins[15]. In contrast to the diverse mechanisms used to block IFN signaling, all members of the *Paramyxovirinae* subfamily, including those which use the viral C proteins to block IFN signaling such as SeV, may use the same method to block IFN production[15]. The importance of inhibiting IFN production is reinforced by the observation that the Pneumoviruses HRSV and BRSV block IFN production, although their ability to inhibit IFN signaling remains a matter of controversy (see Section 4).

To date there is no specific evidence that any paramyxovirus can block the function of an antiviral enzyme, such as PKR, MxA or 2'-5'oligoadenylate synthetase, although this possibility is one that might be expected and should be explored. Furthermore, there is a need to examine in greater detail the replication of paramyxoviruses in cells that are in an "antiviral state," especially as it has been suggested that HRSV replication may be naturally resistant to the "antiviral state" induced by IFN[16].

3.2 Consequences of IFN antagonism for viral pathogenesis

An intriguing question that drives much of the work in this area is why a given virus has evolved a particular strategy for circumventing the IFN response and what consequences this has for the biology of the virus and the type of disease it causes. One consideration is that the method used by viruses in general to block the IFN response may lead to either cell death or cell survival. Clearly viruses that inhibit cellular processes in a gross way, for example by blocking host cell transcription and/or translation, are likely to kill the infected cell, whereas viruses like SV5, which block both IFN signaling and IFN production, do so without necessarily inducing cell death (these viruses may also have specific mechanisms to block apoptosis, but this remains to be firmly established). This might be important if the virus naturally establishes persistent or prolonged infections. Another consideration is that if a virus establishes a persistent infection whilst interfering with important cellular functions, such as the IFN response, the resulting infection may lead to chronic disease. Although the role of paramyxoviruses in chronic human disease such as Paget's bone disease[17-19] remains highly controversial, a greater understanding of the effect of these viruses on the IFN response and the role of IFN in the control of normal cellular function (such as osteoclast function[20]) may lead to a more rational basis for their involvement in chronic disease and this is an area worthy of further study. Recent work on SV5, NDV and BRSV has also suggested that the ability to antagonize the IFN system may be an important factor in the determination of paramyxovirus host range[21-24].

3.3 IFN antagonism and the development of novel vaccines and antiviral drugs

Studies of how paramyxoviruses, and viruses in general, circumvent innate immune responses are important not only because they help to explain the molecular pathogenesis of virus infections and define important intracellular signaling pathways, but also because such studies may lead to advances in vaccine manufacturing and the development of novel medicines. Thus, viruses engineered to be IFN-sensitive may be developed as attenuated virus vaccines; such viruses are likely to be non-pathogenic, but may still induce strong, protective CTL and antibody responses[25-27]. However, since the virus proteins which are involved in blocking the IFN response are often multifunctional proteins, it seems likely that if such an approach is to be successful, it will be necessary to isolate or engineer mutant viruses in which the ability to circumvent the IFN response has been modified without interfering with other important virus functions. This is because the simple deletion of genes involved in blocking the IFN response may over-attenuate virus replication and thus impair the ability to induce vigorous immune responses. A further problem that arises from such an approach is that it may be difficult to grow IFN-sensitive viruses in tissue culture cells such as MRC5 cells, which are licensed for use in vaccine manufacturing, as many of these cells can produce and respond to IFN as a consequence of virus infection. However, a simple solution to this problem is to engineer tissue culture cells such that they continuously express a virus protein, such as the V protein of SV5, so that they can no longer respond to IFN[26].

In addition to potentially influencing vaccine design and manufacturing, elucidation of the molecular mechanisms that viruses have evolved to circumvent the IFN response, may also lead to the development of novel medicines, such as antiviral drugs that prevent the virus from circumventing the IFN response. Furthermore, since abnormal STAT signaling has been reported in a variety of diseases including cancer and certain types of chronic inflammation (reviewed in [28]), it may be that a greater understanding of how paramyxoviruses interfere with STAT function may lead to treatments for such diseases. For example, it has been suggested that the ability of MuV or its V protein to target STAT3 for degradation may have practical therapeutic applications in diseases characterised by overactive STAT3, including certain cancers[13,29]. However, for such a process to be successful it may be necessary to isolate a mutant of the mumps V protein that specifically targets STAT3 for degradation but not STAT1, and to date, it is not clear whether this can be achieved.

4. MECHANISMS OF IFN EVASION BY PARAMYXOVIRUSES

In the following section, the IFN evasion strategies of five major genera of the *Paramyxoviridae* are discussed with reference to the viral proteins involved and their modes of action as far as is currently known.

4.1 Rubulavirus: Targeted degradation of STATs via the V protein

The findings that paramyxoviruses can specifically block IFN signaling[25,30] and IFN production[15,31,32] via the activities of their V proteins were firstly clearly demonstrated for Simian virus 5 (SV5), the prototype Rubulavirus. SV5 blocks both IFN-α/β and IFN-γ signaling by targeting STAT1 (a component of both signaling pathways) for proteasome-mediated degradation[30]. Proteins targeted for proteasome-mediated degradation are usually first poly-ubiquitinated (reviewed in [33,34]) and indeed, unstable, ubiquitinated STAT1 degradation intermediates have been identified by expression of low levels of exogenous SV5 V protein[35].

Subsequent studies showed that other Rubulaviruses have similar IFN evasion strategies, including MuV and Simian virus 41 (SV41), which also target STAT1 for degradation[36-38], and human parainfluenza virus type 2 (HPIV2), which in human cells preferentially targets STAT2 for degradation and therefore only antagonizes IFN-α/β signaling[39,40]. NDV, formerly a member of the *Rubulavirus* genus but recently reclassified as the type member of the *Avulavirus* genus, also targets STAT1 for degradation via the activity of its V protein[41]. In addition to the degradation of STAT1, MuV V also targets STAT3 for degradation, interfering with cytokine signaling, including IL-6, although the biological consequences of this have yet to be fully appreciated[29]. Similarly, the significance and mechanisms by which different strains of SV5 affect the secretion of chemokines, including IL8, has yet to be evaluated[42].

4.1.1 Requirement for STAT proteins in the degradation process

Studies in cells lacking STAT1 and STAT2, as a result of either genetic mutation or the stable expression of Rubulavirus V proteins, revealed that the degradation of STAT1 or STAT2 by SV5 V and HPIV2 V respectively, requires the presence of the other STAT protein such that STAT1 degradation cannot occur in cells lacking STAT2 and *vice versa*[43,44]. Both studies also demonstrated that STAT1 degradation is independent of IFN signaling and STAT phosphorylation. Furthermore, the studies of Parisien *et al*[43] showed that degradation does not require any components of the IFN signaling pathway other than STAT1 and STAT2 and that cell lines lacking STAT1 or STAT2 can be complemented with the expression of exogenous STATs, with no requirement for the presence of tyrosine phosphorylation residues or SH2 domains. Chimeric STAT1/STAT2 proteins and truncations of STAT2 showed that the first 578 residues of the amino-terminus of STAT2 are required for SV5 V to degrade STAT1[43]. Other studies indicate that STAT degradation only occurs when the relative amounts of V and STAT proteins are properly balanced. This was demonstrated by the use of UV-inactivated SV5 in both naïve and IFN pre-treated cells, where the small amount of V protein present in the inactivated virions was sufficient to target the relatively low levels of STAT1 in naïve cells for degradation but could not completely degrade the higher levels of STAT1 in IFN-treated cells[30]. However, SV5 V can degrade high levels of exogenously expressed STAT1 in human cells, but only in the presence of sufficient STAT2 provided by exogenous expression[44].

A role for STAT2 in STAT1 degradation is also supported by the finding that STAT2 can act as a host range determinant for SV5. Thus expression of human STAT2 in murine cells, known to be non-permissive for SV5 replication due to the sensitivity of SV5 to the murine IFN response[25,45], allowed the degradation of murine STAT1 by SV5 V[22]. A similar phenomenon was also observed for HPIV2 V, whereby complementation of murine cells with human STAT2 led to the degradation of the human STAT2, but not the murine STAT2 and furthermore to a loss of degradation fidelity, with some degradation of murine STAT1. STAT2 is also required for the antagonism of IFN signaling by MuV V, but in contrast to SV5 and HPIV2 V proteins, MuV V can utilize murine STAT2 in order to degrade STAT1 and STAT3 in murine cells[29]. The apparent inability of SV5 V to utilize murine STAT2 for the degradation of STAT1 can be reversed by the introduction of a single amino acid change in the amino terminus of V, N100D, as found in a murine-adapted isolate of SV5, suggesting that a single amino acid change can overcome the species barrier for IFN antagonism and potentially extend the host range of SV5[24]. The finding that STAT2 can act as a host range determinant is also supported by the observation that the addition of human STAT2 to rabbit reticulocyte extracts is required to enable the *in vitro* ubiquitination and degradation of human STAT1 that is dependent upon the V proteins of SV5 and HPIV2 (our unpublished observations).

Protein:protein interactions between Rubulavirus V proteins and STAT1, STAT2 and, in the case of MuV V, STAT3 have been observed, although whether these interactions are direct or indirect still remains to be established[29,43,46]. Studies using MuV V suggested that the interactions with STATs may be independent of the conserved carboxy-terminal cysteine residues of MuV V, but require the presence of a conserved tryptophan motif in this region, also found in HPIV2 V[46]. However, this motif is not found in SV5 V and a second study of MuV V and STATs suggests that the cysteine residues are required for the interaction with STAT1[47], so currently the residues of MuV V and those of other Rubulavirus Vs required for interaction with STATs are unknown.

4.1.2 DDB1 and the "V degradation complex"

Before it had been established that the V proteins of Rubulaviruses acted as IFN antagonists, they had been shown to interact with viral NP[48], bind zinc via cysteine residues in the highly conserved V-unique carboxy terminus[49] and bind ssRNA through the shared N-terminal domain of V and P[50]. The V protein had also been shown to interact with DDB1, the 127kDa subunit of the cellular damage-specific DNA binding protein[51]. The interaction of SV5 V with DDB1 was reported to slow the cell cycle of HeLa T4 cells infected with SV5, as expression of additional DDB1 in cells expressing V could partially restore normal cell cycle progression[52]. Subsequent studies have clearly established that the interaction with DDB1 is also essential for SV5 V to target STAT1 for proteasome-mediated degradation[35,44].

The evidence that DDB1 is required for STAT degradation is threefold: firstly, an isolate of SV5, termed CPI-, which fails to bind DDB1 also fails to degrade STAT1; secondly, mutations that abolish the binding of V to DDB1 also prevent V from blocking IFN signaling; thirdly, the use of siRNA to knock down DDB1 expression adversely affects the ability of V to degrade STAT1[44]. It was also demonstrated that V can target STAT1 for degradation in cells derived from patients with the disease Xeroderma pigmentosum which lack the normal interaction between the two DDB subunits, DDB1 and DDB2, indicating that DDB2 is not required in the V-mediated degradation process, and recently the binding of SV5 V to DDB1 has been shown to displace DDB2[53]. Regions in both the amino and carboxy termini of V are important for its interaction with DDB1. Mutations of any of the conserved cysteine residues within the carboxy terminus of V result in a loss of DDB1 binding and an inability to block IFN signaling. Studies of a recombinant SV5 with a V protein lacking the cysteine-rich C-terminus (rSV5VΔC) underlined the requirement for this region in IFN antagonism as rSV5VΔC did not block IFN-α/β signaling or the formation of ISGF3 in infected cells, or target STAT1 for degradation[31]. On the other hand, there are only three amino acid differences between the V proteins of two closely related dog isolates of SV5, termed CPI+ and CPI-, all of which are located in the amino terminus of V. However in contrast to CPI+, CPI- fails to bind DDB1 or target STAT1 for degradation.

The relevance of DDB1 in the degradation of STAT1 was suspected as DDB1 has been known for some time to interact with cullin 4a (Cul4a) in normal cells. Cullins are a family of proteins that form part of E3 ligases that target substrates for ubiquitin-dependent degradation by the 26S proteasome (for a review of cullin-based ubiquitin ligases see [54]). Furthermore, a cellular ubiquitin ligase complex that regulates the half-life of c-jun and contains

DDB1 and Cul4a has recently been identified[55]. The evidence that Cul4a is required for the targeted degradation of STAT1 by the V protein of SV5 relies primarily on the use of siRNA to knock down the levels of Cul4a within cells, but this only reduced the efficiency of STAT1 degradation by SV5 V by 10-20%, despite causing a substantial decrease in Cul4a levels[35]. One possible explanation of this is that other cullins may substitute for Cul4a in the degradation complex. It has been suggested that the acronym VDC should be the term used for the degradation complex, as it can refer to V/DDB1/CUL4A, Virus Degradation Complex and V-dependent Degradation Complex[35]. Similar degradation complexes have been suggested for HPIV2 V and MuV V, although the exact composition of the MuV V complex appears to differ somewhat from that formed by SV5 V, presumably reflecting the ability of MuV V to target both STAT1 and STAT3 for degradation[29].

To further define the role of V in the targeted degradation of STAT1, Ulane & Horvath[35] devised an *in vitro* E3 ligase assay, in which proteins acting as E3 ligases are auto-ubiquitinated. In this assay GST-SV5 V was mono-ubiquitinated in the presence of ATP and E1 and E2 ligases, suggesting that Rubulavirus V proteins represent a new class of viral ubiquitin ligase enzymes that satisfy a minimal definition of E3 enzymatic activity. Additionally, the requirement for the E2 ligase was not absolute, indicating that V may also have some E2 ligase activity, but both these activities have to be firmly established. Indeed, it is unlikely that the specific molecular mechanism of STAT degradation will be defined until *in vitro* assays have been developed using purified proteins in which the specificity of STAT degradation can be reproduced.

4.1.3 Mumps V and RACK1

In addition to its interactions with STATs and DDB1, MuV V has been reported to interact with the cellular protein RACK1, which may mediate the interaction between the IFN-α receptor and STAT1[56,57]. This interaction involves the carboxy terminus of MuV V, but in contrast to the DDB1 interaction, it does not require the presence of the conserved carboxy-terminal cysteine residues[58]. GST fusions of part of the cytoplasmic domain of the IFN-α receptor subunit IFN-αRβL show that infection with MuV disrupts the normal association between IFN-αRβL, RACK1 and STAT1, but has no effect on their interactions with STAT2. It has been suggested that the interaction of MuV V and RACK1 may result in the dissociation of STAT1 from the IFN-α receptor and contribute to its subsequent poly-ubiquitination and degradation in MuV-infected cells. However, there is currently no evidence suggesting that other Rubulavirus V proteins interact with RACK1.

4.1.4 Suppression of IFN production by Rubulaviruses

Although SV5 can target STAT1 for degradation in cells already in an IFN-induced antiviral state, the pre-existing antiviral state delays viral replication[30]. This can be seen in infections of IFN pre-treated monolayers of cultured cells, in which the cell-to-cell spread of SV5 is severely slowed. However, since SV5 plaques efficiently in cells that are capable of producing and responding to IFN, it was suspected that the virus not only antagonizes IFN signaling but must also limit the amount of IFN produced by infected cells[14]. More recent studies have demonstrated that infection of cells with SV5 blocks the production of IFN and that this is a property of the V protein[15,31].

Cells infected with wild-type SV5 release very little IFN-β, in contrast to rSV5VΔC-infected cells, which produce large amounts (rSV5VΔC is a recombinant virus that lacks the unique carboxy-terminal domain of the V protein). This was also reflected in the relative amounts of IFN-β mRNA produced in infected cells, indicating that the carboxy-terminus of V is involved in blocking the production of IFN-β during wild-type SV5 infections[15]. The ability of SV5 to inhibit IFN production was confirmed by assays showing that full-length SV5 V blocks the activation of the IFN-β promoter by the synthetic dsRNA, poly(I):poly(C). Similar assays showed that deletions of up to 126aa from the amino-terminus of V are tolerated, but not deletions from the carboxy-terminus. In contrast to blocking IFN signaling, expression of the V-unique region alone blocks IFN-β promoter activation as efficiently as full-length SV5 V. Furthermore, point mutations of the carboxy-terminal conserved cysteine residues also abrogate the ability of V to block the activation of the IFN-β promoter, confirming the importance of these residues. The V proteins of HPIV2 and SeV also suppress the activation of the IFN-β promoter, presumably in a similar manner to SV5 V, via their highly conserved cysteine-rich carboxy termini[15].

Investigations of the mechanism of the block of IFN-β production have shown that the SV5 V protein blocks the activation of both IRF-3 and NF-κB by wild-type viral infection and by poly(I):poly(C), and that conversely, rSV5VΔC infection leads to the activation of IRF-3 and NF-κB[15]. The block is probably acting at the level of signal transduction since IRF-3 remains in the cytoplasm when the V protein is expressed[31], although the target of the viral block remains unknown. Interestingly, higher concentrations of V are required to effectively block IFN production than to inhibit IFN signaling[15].

4.2 Respirovirus: STAT sequestration, hypo/hyper-phosphorylation and degradation

Although there is no evidence that Sendai virus (SeV) infects wild mice populations, it is able to efficiently infect laboratory animals and has been widely studied as a model for virus pathogenicity. Studies on SeV strains generated by reverse genetics have allowed the V and C proteins to be identified as accessory proteins. SeV strains that either cannot edit their transcript to make a V protein or have stop codons that prevent its translation, replicate efficiently in cell culture but are attenuated *in vivo*[59-62], implying that the V protein plays a role in maintaining viability in the face of host immunity. SeV C proteins are made as a nested set of co-carboxy terminal proteins (called C', C, Y1 and Y2) and inactivation of these rendered SeV non-pathogenic[63], although in contrast to the V gene deleted viruses, inactivation of the C proteins also impaired virus growth in eggs and in cell culture[63,64]. Modifications to a subset of the C proteins generated viruses with intermediate properties, suggesting a complex set of interactions with the host's immune system.

The demonstration that SeV could antagonize both IFN-α/β and IFN-γ signaling in human and murine cells[25] offered a suggestion as to how the accessory proteins might function in evading innate immunity. Initial observations of the molecular events associated with the signaling inhibition showed that unlike infection by Rubulaviruses, SeV infection did not appear to lead to STAT degradation, but rather, at late times post infection caused an increase in the accumulation of tyrosine pY(701)-phosphorylation of STAT1. Despite this apparent activation, STAT1 could not be detected in a form that bound to DNA[39]. Furthermore, the expected phosphorylation of STAT1 on serine 727 (Ser727), by a kinase with MAP-like activity and which is required for the transactivating activity of STAT1, was not observed in SeV-infected cells[39].

Further research has shown that the situation appears to be more complicated, with at least three distinct mechanisms operating. These events will be discussed in more detail below and are i) an early event in which IFN signaling is blocked by sequestering STAT1 into a complex, ii) a late event in which STAT1 becomes hyper-phosphorylated and iii) the targeted degradation of STAT1. As predicted by the pathogenesis studies discussed above, the SeV C proteins are responsible for the antagonism of IFN signaling and play a role in all three of these mechanisms.

In contrast to the detection of activated STAT1 reported by Young *et al*[39], SeV infection actually blocks the tyrosine phosphorylation of STAT1 at early times post infection[39,65-69]. This block requires a direct interaction between STAT1 and the C-terminal 106 amino acids of the C protein (and hence all four SeV C proteins can bind STAT1 in an interaction that also requires the F170 residue), which sequesters STAT1 into a high molecular weight complex of more than 2MDa such that STAT1 can neither be phosphorylated nor bind to DNA[67,69]. The phosphorylation of STAT2 is also blocked at early times post infection, and this appears to be a function of interaction with the SeV C proteins, probably via the intermediacy of STAT1[65,70].

At later times in infection, the block of tyrosine phosphorylation is not as pronounced and levels of pY(701)-STAT1 increase. It has been reported that this is associated with a C-mediated block to pY(701)-STAT1 de-phosphorylation, but this has not been observed in all cell types studied[69,71]. The increase in pY(701)-STAT1 has been specifically associated with the longer forms of C[71] and it may be significant that only the longer C' and C proteins interact with pY(701)-STAT1[67,70-72]. However, it is currently difficult to see how STAT1 becomes phosphorylated if IFN signaling is blocked; in this context it is interesting to note that Garcin *et al*[71] reported that pY(701)-STAT1 could be detected in SeV infected cells that are defective in IFN signaling (including cells that lack Jak1), implying that a novel phosphorylation component is involved. The functional significance of a virally-directed increase in pY(701)-STAT1 levels is currently unclear.

Although bulk STAT1 levels do not appear to decrease during SeV infections in many cell types[39,65,67,73-75], this is not universal, since in some cases (e.g. mouse embryonal fibroblasts) SeV infection results in the mono-ubiquitination and instability of STAT1[66,72,76]. This is a property of the C and C' proteins only and is independent of interferon signaling[71,72]. The relative contribution to viral replication of STAT1 turnover versus direct IFN signal blocking remains to be determined.

In addition to blocking IFN signaling, SeV can limit the production of type I IFN. Comparison of wild-type SeV and strains that do not produce either intact V or C proteins shows that the defective viruses induce more IFN-β[77]. These results can be explained either by the proteins acting to limit the production of an IFN inducer (such as replicative dsRNA intermediates) or by directly blocking the production of IFN. In this context, it is of note that the SeV V protein has been shown to block the activation of the IFN-β promoter by synthetic dsRNA[15], but this has yet to be tested for the C protein.

A second member of the *Respirovirus* genus, human Parainfluenza virus type 3 (HPIV3), can also antagonize IFN-α/β and IFN-γ signaling[39]. The C protein of HPIV3, similarly to that of SeV, has been linked to viral pathogenesis and mutation of residue F164, the equivalent of SeV C^{F170}, attenuated HPIV3 *in vivo*[78,79]. Recent studies have shown that the C and L proteins of HPIV3 and SeV can interact and that HPIV3 C is an inhibitor of viral transcription, as is SeV C, indicating that despite their low sequence homology (38%), the C proteins of HPIV3 and SeV may be functionally equivalent, and thus HPIV3 C may be involved in IFN antagonism[80,81]. Despite these similarities and in contrast to SeV infection, homodimers of pY(701)-STAT1 capable of binding DNA have been observed in cells infected with HPIV3, although these are unable to activate transcription[39]. Since both SeV and HPIV3 infections block ISGF3 formation, this distinct behavior with respect to GAS site binding represents a subtle difference in action between these Respiroviruses.

4.3 Morbillivirus: IFN antagonism via the V and C proteins

Although it is well-established that MeV antagonizes the IFN response, the experimental data is conflicting as to the extent of this antagonism and the mechanism by which it occurs. These conflicts may be a result of strain differences, but it is possible that differences in experimental approach have also contributed.

Similar to other paramyxoviruses, the P gene of MeV encodes both V and C proteins in addition to the P protein (Figure 1). Recombinant MeV of the Edmonston vaccine strain (MeVEd) engineered to be defective in the expression of either the V or C proteins are less pathogenic in mouse model systems than the parental virus[82-84], but grow to similar titres as parental virus in IFN non-producing Vero cells[85,86]. Whilst the reason for these results was not clear at the time, it now seems likely that the attenuated phenotype of the MeV V(-) and MeV C(-) viruses may at least be partially explained by their inability to circumvent the IFN response. Similarly, experiments using cells persistently infected with MeVEd showed that in some cases, the expression of the IFN-stimulated gene 2'-5' oligoadenylate synthetase was suppressed[87], again consistent with the ability of MeV to block IFN signaling.

Expression of the V protein of MeVEd has since been demonstrated to block both IFN-α/β and IFN-γ signaling, and also to disrupt IL-6 and v-src signaling, both of which depend on STAT3. Furthermore, a recombinant MeVEd C(-) virus was shown to be IFN-sensitive and expression of the viral C protein blocked IFN-α/β signaling and reduced IFN-γ signaling by 50% in Vero cells. It is currently unclear how the V and C proteins of MeVEd antagonize IFN signaling, but MeVEd does not appear to target STAT1 or STAT2 for degradation or block their tyrosine phosphorylation in response to IFN-α[88]. Complexes that include MeVEd V, STAT1, STAT2, STAT3 and IRF-9 have been isolated from cells expressing MeVEd V and in such cells, STAT1 and STAT2 do not translocate to the nucleus in response to IFN-α, so it has been suggested that the V protein of MeVEd interferes with signal transduction at a point downstream of STAT tyrosine phosphorylation but upstream of their nuclear import. It is not currently clear whether the C protein of MeVEd is involved in these complexes.

In contrast, experiments using wild strains of MeV have shown that infection enables the replication of an IFN-sensitive virus following IFN-α but not IFN-γ stimulation and that in such cells the IFN-α/β signaling pathway, but not the IFN-γ signaling pathway was blocked[88]. Expression of wild MeV V alone has been demonstrated to block the induction of an antiviral state by IFN-α but not IFN-γ[89], but the expression of the C protein had no equivalent effect. These experiments found no evidence for the degradation of STAT1, STAT2 or IRF-9, but did show a block of the phosphorylation of STAT1 and Jak1 in response to IFN-α[88]. Immune precipitations from cells infected with wild MeV, using an antibody specific to part of the IFN-α/β receptor, demonstrated the presence of a complex including the IFN-α/β receptor, MeV V and C proteins, STAT1 and RACK1, similar to that seen in cells infected with the Rubulavirus MuV, whereas a similar experiment showed that MeV V and C do not form a complex with the IFN-γ receptor[88]. It has been suggested that the interaction of MeV V and C with the IFN-α/β receptor "freezes" the receptor complex and prevents the normal phosphorylation and activation of STAT1 in response to IFN-α. Similarly, the apparent lack of interaction between MeV V and C and the IFN-γ receptor may explain the inability of wild MeV to block IFN-γ signaling.

Surprisingly, it thus appears that wild MeV strains and the vaccine strain MeVEd differ in their ability to block IFN-α/β and IFN-γ signaling and in the mechanisms by which they achieve their antagonism of the IFN response. MeVEd blocks both IFN-α/β and IFN-γ signaling via its V and C proteins as well as suppressing IL-6 and v-src signaling, and the mechanism by which this occurs may involve complexes that include the V protein and STAT1, STAT2, STAT3 and IRF-9, and a block of signaling downstream of STAT phosphorylation. In contrast, wild MeV seems to antagonize only IFN-α/β signaling via a block of STAT1 and STAT2 phosphorylation as a result of interactions between the V protein and the IFN-α/β receptor.

4.4 Henipavirus: Binding and sequestration of STAT proteins

Both members of the Henipavirus genus, NiV and HeV, express multiple proteins from their P genes (Figure 1), most of which have the potential to antagonize IFN signaling. Initial studies observed that the expression of NiV and HeV V proteins blocked both IFN-α/β and IFN-γ signaling and that in such cells the IFN-stimulated nuclear translocation of STAT1 and STAT2 was prevented[90,91]. Both V proteins are predominantly cytoplasmic and interact with STAT1 and STAT2, forming high molecular weight complexes resulting in the cytoplasmic sequestration of STAT1 and STAT2. Furthermore, V-STAT1-STAT2 complexes have been isolated from V-expressing cells using gel filtration techniques[90,91]. The regions of NiV V required for STAT interactions have been mapped to the amino-terminus of the protein, also present in the W and P proteins, with the STAT1 binding site lying between residues 100 and 150[92,93] and the region for optimal STAT2 binding encompassing a larger region of residues 100 to 300[92]. However, deletion of seven residues from position 230 to 237 abrogated STAT2 binding, suggesting that this region, highly conserved between NiV V and HeV V, forms part of a STAT2 interaction site.

The region of STAT1 interacting with NiV V falls between residues 509 and 712, an area that includes both the SH2 and linker domains. Experiments in U3A cells lacking STAT1 demonstrated that the interaction between NiV and HeV V proteins and STAT2 is dependent on the presence of STAT1, and complementation of these cells with STAT1 restored the interaction of V with STAT2. IFN-α/β signaling assays illustrate that the presence of the STAT1 binding domain in NiV V is essential for an efficient block of IFN signaling, suggesting that the interaction with STAT1 is the main factor in the antagonism of IFN signaling by NiV V. However, a truncated version of V that did not interact with STAT1 but weakly interacted with STAT2 had the ability to block IFN signaling by around 20%[92].

The other products of the NiV P gene, namely the W, C and P proteins, have also been demonstrated to antagonize IFN signaling and to prevent the establishment of an IFN-induced antiviral state to varying degrees[93,94]. As well as binding STAT1 and STAT2, Nipah V was shown to shuttle between the nucleus and cytoplasm, leading to the suggestion that Nipah V can enter the nucleus, bind to STAT1 and carry it back to the cytoplasm[90]. However, more recent studies have show that IFN signaling is also blocked in the presence of export-deficient NiV V proteins showing that IFN antagonism and the shuttling of V are separable functions[92]. Furthermore, although NiV W blocks IFN signaling, STAT1 is seen in the nucleus of cells expressing NiV W, which itself has a predominantly nuclear distribution, suggesting that it is the interaction of STAT1 with V or W, rather than its sequestration in a particular cellular compartment that is required for the antagonism of IFN signaling[92,93]. Indeed, even the P protein of Nipah blocks IFN signaling, presumably through the common amino terminal domain. It also appears that Nipah C protein partially blocks IFN signaling, but it is not yet clear how this is achieved[94]. Although the V, W, P and C proteins of Henipaviruses all appear to be able to block the IFN response, it seems likely that some or all of these proteins have other, as yet undefined properties that may eventually help to explain the highly pathogenic nature of these viruses, so different to that of most other members of the *Paramyxoviridae.*

4.5 Pneumovirus: IFN resistance via the NS1 and NS2 proteins

The replication of HRSV is relatively unaffected by both the pre-treatment of cells with IFN-α/β and the addition of IFN-α/β after infection and is also resistant to the effects of endogenously produced IFN stimulated by treatment with poly(I):poly(C) dsRNA[16]. However, the observation that HRSV infection does not prevent the expression of the IFN-stimulated gene for MxA and does not allow the replication of IFN-sensitive viruses has led to the suggestion that, rather than preventing the establishment of an antiviral state via a blockade of IFN signaling as seen for other paramyxoviruses, HRSV is relatively resistant to the effects of IFN. Indeed, Young *et al*[39] showed that neither IFN-α/β nor IFN-γ signaling is blocked in cells infected with HRSV and that the phosphorylation of STAT1 and the subsequent formation of ISGF3 and GAF complexes proceeds as in uninfected cells. Similarly, Bossert *et al*[95] reported that whilst bovine RSV (BRSV) blocked IFN production (below) the virus did not specifically inhibit IFN signaling. However, there has been a recent claim that HRSV infection of respiratory epithelial cells leads to the antagonism of IFN-α/β signaling via the proteasome-mediated degradation of STAT2[96], but the reason(s) for these apparently contradictory results currently remains unclear.

4.5.1 Role of NS1 and NS2 in IFN resistance

As already detailed, HRSV and BRSV use a number of additional genes to encode their accessory gene products rather than encoding multiple products in the single P gene. Two of these accessory genes encode the non-structural NS1 and NS2 proteins, abundantly transcribed during infections due to their position at the extreme 3' end of the genome. Recombinant HRSV and BRSV lacking one or both NS proteins are attenuated in chimpanzees and calves respectively, and replicate particularly poorly in IFN-competent as compared to IFN-deficient cells. Furthermore, the growth of NS-deleted rBRSV in IFN-competent cells was improved by the addition of antibodies against the IFN-α/β receptor, suggesting that the NS proteins have a role in resistance to IFN[26,97-103]. Expression of both NS1 and NS2 of BRSV by a recombinant rabies virus (rRV) protected the normally IFN-sensitive rabies virus from the IFN response, as did expression of HRSV

NS1 and NS2 and combinations of the human and bovine NS proteins, again indicating that both NS1 and NS2 are involved in IFN antagonism[21,100].

Although there is some conflict in the literature with regard to the induction of IFN by HRSV[104-112], the general consensus is that HRSV is a poor inducer of IFN and the picture which is emerging suggests that NS1 and NS2 act synergistically to inhibit the induction of IFN[21,95,103,113]. Thus infection of bovine cells with wild-type BRSV, or rBRSV ΔNS1, does not produce significant amounts of IFN-α/β, whereas infection with rBRSV ΔNS2 and rBRSV ΔNS1/NS2 leads to the stimulation of IFN-β mRNA and production of large amounts of IFN-α/β, suggesting a specific role for NS2 in blocking the transcriptional activation of the IFN-β promoter[103]. This was confirmed by a study showing that infection with wild-type BRSV blocks the activation of the IFN-β promoter by both viral infection and poly(I):poly(C) whilst rBRSV ΔNS1/NS2 does not[95]. Similarly, the NS proteins of HRSV reduce the transcription of IFN-α and IFN-β mRNA and the production of IFN-α/β, and although rHSRV ΔNS1/NS2 gives the largest increase in IFN levels compared to wild-type HRSV, both proteins were found to have a role in blocking IFN production, with NS1 contributing more to this function, in contrast to BRSV where NS2 appears to be more important[113].

Infection of cells with wild-type BRSV blocks the phosphorylation and thus the activation of IRF-3, but not NF-κB or AP1, whereas IRF-3 is phosphorylated in rBRSV ΔNS1/NS2-infected cells[95]. This blockade of IRF-3 phosphorylation requires both NS1 and NS2, although NS2 alone was found to have some ability to block IFN production, as observed by Valarcher *et al*[103]. It is not currently clear whether HRSV NS1 and NS2 proteins block IRF-3 phosphorylation, but the evidence that BRSV and HRSV NS proteins are to some extent interchangeable suggests that their IFN evasion mechanisms will be similar.

Despite the homology of the NS proteins of HRSV and BRSV (69% and 84% sequence identity for NS1 and NS2 respectively), they are best adapted to counteract the IFN responses of their natural hosts and thus the ability of these proteins to resist the effects of IFN may determine the viral host range of RSV. Thus, rBRSV expressing the NS1 and NS2 proteins of HRSV replicates efficiently in human and simian cells but is attenuated in bovine cells fully permissive for wild-type BRSV[21]. Furthermore, it was shown that wild-type BRSV only forms plaques in human MRC5 and Hep2 cells engineered to be non-responsive to IFN[26]. HRSV is also known to be IFN sensitive in murine cells, perhaps due to the inability of its NS proteins to inhibit the IFN response in such cells[114].

5. CONCLUSIONS

Although viruses from all genera within the *Paramyxoviridae* family antagonize the IFN response, there are a wide variety of mechanisms by which this antagonism occurs. It appears that viruses from both sub-families suppress the production of IFN via the V protein for members of the *Paramyxovirinae* and the NS proteins for members of the *Pneumovirinae*. Members of the *Paramyxovirinae* also use the V protein and/or other products of the viral P gene, such as C and W, to antagonize IFN signaling. Some of these viruses block both IFN-α/β and IFN-γ signaling pathways, whilst others block only IFN-α/β signaling by various mechanisms including interactions with the IFN receptor complex, targeted degradation of components of the signaling pathway and sequestration of signaling molecules in high molecular weight complexes.

Several viruses encode multiple proteins that act as IFN antagonists and it is likely that these proteins are multifunctional and have other, as of yet undefined roles in the viral life cycle that may or may not be concerned with immune evasion. For example, it remains to be ascertained why the V proteins of *Henipaviruses* are significantly bigger than the V proteins of other paramyxoviruses, and what if any are the additional functions associated with these larger proteins. As discussed, studies designed to tell us more about how paramyxoviruses evade cellular antiviral responses may shed further light in such diverse areas as the molecular pathogenesis of virus infections, virus host range and virus persistence, as well as potentially having implications for the development of novel approaches to creating attenuated viruses or antiviral drugs.

REFERENCES

1. Wang, L.-F. & Eaton, B. Emerging Paramyxoviruses. *The Infectious Disease Review - microbes of man, animals and the environment* **3:**52-69
2. Lamb, R. A. & Kolakofsky, D. in *Fields' Virology* (eds. Fields, B. N., Knipe, D. M., Howley, P. M. & Griffin, D. E.) 1305-1340 (Lippincott Williams and Wilkins, Philadelphia, 2001)
3. Stark, G. R., Kerr, I. M., Williams, B. R., Silverman, R. H. & Schreiber, R. D. How cells respond to interferons. *Annu Rev Biochem* **67:** 227-64
4. Goodbourn, S., Didcock, L. & Randall, R. E. Interferons: cell signaling, immune modulation, antiviral response and virus countermeasures. *J Gen Virol* **81:**2341-64
5. Sen, G. C. Viruses and interferons. *Annu Rev Microbiol* **55:**255-81
6. Levy, D. E. & Garcia-Sastre, A. The virus battles: IFN induction of the antiviral state and mechanisms of viral evasion. *Cytokine Growth Factor Rev* **12:**143-56
7. Guidotti, L. G. & Chisari, F. V. Noncytolytic control of viral infections by the innate and adaptive immune response. *Annu Rev Immunol* **19:**65-91

8. Grandvaux, N., tenOever, B. R., Servant, M. J. & Hiscott, J. The interferon antiviral response: from viral invasion to evasion. *Curr Opin Infect Dis* **15:**259-67
9. Garcia-Sastre, A. Mechanisms of inhibition of the host interferon alpha/beta-mediated antiviral responses by viruses. *Microbes Infect* **4:**647-55
10. Katze, M. G., He, Y. & Gale, M., Jr. Viruses and interferon: a fight for supremacy. *Nat Rev Immunol* **2:**675-87
11. Biron, C. A. & Sen, G. C. in *Fields' Virology* (eds. Fields, B. N., Knipe, D. M., Howley, P. M. & Griffin, D. E.) 321-349 (Lippincott Williams and Wilkins, Philadelphia, 2001)
12. Gotoh, B., Komatsu, T., Takeuchi, K. & Yokoo, J. Paramyxovirus accessory proteins as interferon antagonists. *Microbiol Immunol* **45:**787-800
13. Horvath, C. M. Silencing STATs: lessons from paramyxovirus interferon evasion. *Cytokine Growth Factor Rev* **15:**117-27
14. Andrejeva, J., Young, D. F., Goodbourn, S. & Randall, R. E. Degradation of STAT1 and STAT2 by the V proteins of simian virus 5 and human parainfluenza virus type 2, respectively: consequences for virus replication in the presence of alpha/beta and gamma interferons. *J Virol* **76:**2159-67
15. Poole, E., He, B., Lamb, R. A., Randall, R. E. & Goodbourn, S. The V proteins of simian virus 5 and other paramyxoviruses inhibit induction of interferon-beta. *Virology* **303:**33-46
16. Atreya, P. L. & Kulkarni, S. Respiratory syncytial virus strain A2 is resistant to the antiviral effects of type I interferons and human MxA. *Virology* **261:**227-41 (1999).
17. Bender, I. B. Paget's disease. *J Endod* **29:**720-3
18. Rall, G. F. Measles virus 1998-2002: progress and controversy. *Annu Rev Microbiol* **57:**343-67
19. Rima, B. K., Gassen, U., Helfrich, M. H. & Ralston, S. H. The pro and con of measles virus in Paget's disease: con. *J Bone Miner Res* **17:**2290-2; author reply 2293
20. Takayanagi, H., Kim, S. & Taniguchi, T. Signaling crosstalk between RANKL and interferons in osteoclast differentiation. *Arthritis Res* **4 Suppl 3:**S227-32
21. Bossert, B. & Conzelmann, K. K. Respiratory syncytial virus (RSV) nonstructural (NS) proteins as host range determinants: a chimeric bovine RSV with NS genes from human RSV is attenuated in interferon-competent bovine cells. *J Virol* **76:**4287-93
22. Parisien, J. P., Lau, J. F. & Horvath, C. M. STAT2 acts as a host range determinant for species-specific paramyxovirus interferon antagonism and simian virus 5 replication. *J Virol* **76:**6435-41
23. Park, M. S., Garcia-Sastre, A., Cros, J. F., Basler, C. F. & Palese, P. Newcastle disease virus V protein is a determinant of host range restriction. *J Virol* **77:**9522-32
24. Young, D. F. et al. Single amino acid substitution in the V protein of simian virus 5 differentiates its ability to block interferon signaling in human and murine cells. *J Virol* **75:**3363-70
25. Didcock, L., Young, D. F., Goodbourn, S. & Randall, R. E. Sendai virus and simian virus 5 block activation of interferon- responsive genes: importance for virus pathogenesis. *J Virol* **73:**3125-33
26. Young, D. F. et al. Virus replication in engineered human cells that do not respond to interferons. *J Virol* **77:**2174-81
27. Garcia-Sastre, A. et al. Influenza A virus lacking the NS1 gene replicates in interferon-deficient systems. *Virology* **252:**324-30
28. Levy, D. E. & Darnell, J. E., Jr. Stats: transcriptional control and biological impact. *Nat Rev Mol Cell Biol* **3:**651-62
29. Ulane, C. M., Rodriguez, J. J., Parisien, J. P. & Horvath, C. M. STAT3 ubiquitylation and degradation by mumps virus suppress cytokine and oncogene signaling. *J Virol* **77:**6385-93

30. Didcock, L., Young, D. F., Goodbourn, S. & Randall, R. E. The V protein of simian virus 5 inhibits interferon signaling by targeting STAT1 for proteasome-mediated degradation. *J Virol* **73:**9928-33
31. He, B. et al. Recovery of paramyxovirus simian virus 5 with a V protein lacking the conserved cysteine-rich domain: the multifunctional V protein blocks both interferon-beta induction and interferon signaling. *Virology* **303:**15-32
32. Wansley, E. K. & Parks, G. D. Naturally occurring substitutions in the P/V gene convert the noncytopathic paramyxovirus simian virus 5 into a virus that induces alpha/beta interferon synthesis and cell death. *J Virol* **76:**10109-21
33. Weissman, A. M. Themes and variations on ubiquitylation. *Nat Rev Mol Cell Biol* **2:**169-78
34. Jackson, P. K. et al. The lore of the RINGs: substrate recognition and catalysis by ubiquitin ligases. *Trends Cell Biol* **10:**429-39
35. Ulane, C. M. & Horvath, C. M. Paramyxoviruses SV5 and HPIV2 Assemble STAT Protein Ubiquitin Ligase Complexes from Cellular Components. *Virology* **304:**160-6
36. Hariya, Y., Yokosawa, N., Yonekura, N., Kohama, G. & Fuji, N. Mumps virus can suppress the effective augmentation of HPC-induced apoptosis by IFN-gamma through disruption of IFN signaling in U937 cells. *Microbiol Immunol* **44:**537-41
37. Kubota, T., Yokosawa, N., Yokota, S. & Fujii, N. C terminal CYS-RICH region of mumps virus structural V protein correlates with block of interferon alpha and gamma signal transduction pathway through decrease of STAT 1-alpha. *Biochem Biophys Res Commun* **283:**255-9
38. Nishio, M. et al. High resistance of human parainfluenza type 2 virus protein-expressing cells to the antiviral and anti-cell proliferative activities of alpha/beta interferons: cysteine-rich V-specific domain is required for high resistance to the interferons. *J Virol* **75:**9165-76
39. Young, D. F., Didcock, L., Goodbourn, S. & Randall, R. E. Paramyxoviridae use distinct virus-specific mechanisms to circumvent the interferon response. *Virology* **269:**383-90
40. Parisien, J. P. et al. The V protein of human parainfluenza virus 2 antagonizes type I interferon responses by destabilizing signal transducer and activator of transcription 2. *Virology* **283:**230-9
41. Huang, Z., Krishnamurthy, S., Panda, A. & Samal, S. K. Newcastle disease virus V protein is associated with viral pathogenesis and functions as an alpha interferon antagonist. *J Virol* **77:**8676-85
42. Young, V. A. & Parks, G. D. Simian virus 5 is a poor inducer of chemokine secretion from human lung epithelial cells: identification of viral mutants that activate interleukin-8 secretion by distinct mechanisms. *J Virol* **77:**7124-30
43. Parisien, J. P., Lau, J. F., Rodriguez, J. J., Ulane, C. M. & Horvath, C. M. Selective STAT protein degradation induced by paramyxoviruses requires both STAT1 and STAT2 but is independent of alpha/beta interferon signal transduction. *J Virol* **76:**4190-8
44. Andrejeva, J., Poole, E., Young, D. F., Goodbourn, S. & Randall, R. E. The p127 subunit (DDB1) of the UV-DNA damage repair binding protein is essential for the targeted degradation of STAT1 by the V protein of the paramyxovirus simian virus 5. *J Virol* **76:**11379-86
45. Young, D. F., Didcock, L. & Randall, R. E. Isolation of highly fusogenic variants of simian virus 5 from persistently infected cells that produce and respond to interferon. *J Virol* **71:**9333-42
46. Nishio, M., Garcin, D., Simonet, V. & Kolakofsky, D. The carboxyl segment of the mumps virus V protein associates with Stat proteins in vitro via a tryptophan-rich motif. *Virology* **300:**92-9
47. Yokosawa, N., Yokota, S., Kubota, T. & Fujii, N. C-terminal region of STAT-1alpha is not necessary for its ubiquitination and degradation caused by mumps virus V protein. *J Virol* **76:**12683-90

48. Randall, R. E. & Bermingham, A. NP:P and NP:V interactions of the paramyxovirus simian virus 5 examined using a novel protein:protein capture assay. *Virology* **224:**121-9
49. Paterson, R. G., Leser, G. P., Shaughnessy, M. A. & Lamb, R. A. The paramyxovirus SV5 V protein binds two atoms of zinc and is a structural component of virions. *Virology* **208:**121-31
50. Lin, G. Y., Paterson, R. G. & Lamb, R. A. The RNA binding region of the paramyxovirus SV5 V and P proteins. *Virology* **238:**460-9
51. Lin, G. Y., Paterson, R. G., Richardson, C. D. & Lamb, R. A. The V protein of the paramyxovirus SV5 interacts with damage-specific DNA binding protein. *Virology* **249:**189-200
52. Lin, G. Y. & Lamb, R. A. The paramyxovirus simian virus 5 V protein slows progression of the cell cycle. *J Virol* **74:**9152-66
53. Leupin, O., Bontron, S. & Strubin, M. Hepatitis B virus X protein and simian virus 5 V protein exhibit similar UV-DDB1 binding properties to mediate distinct activities. *J Virol* **77:**6274-83
54. Pintard, L., Willems, A. & Peter, M. Cullin-based ubiquitin ligases: Cul3-BTB complexes join the family. *Embo J* **23:**1681-7
55. Wertz, I. E. et al. Human De-etiolated-1 regulates c-Jun by assembling a CUL4A ubiquitin ligase. *Science* **303:**1371-4
56. Croze, E. et al. Receptor for activated C-kinase (RACK-1), a WD motif-containing protein, specifically associates with the human type I IFN receptor. *J Immunol* **165:**5127-32
57. Usacheva, A. et al. The WD motif-containing protein receptor for activated protein kinase C (RACK1) is required for recruitment and activation of signal transducer and activator of transcription 1 through the type I interferon receptor. *J Biol Chem* **276:**22948-53
58. Kubota, T., Yokosawa, N., Yokota, S. & Fujii, N. Association of mumps virus V protein with RACK1 results in dissociation of STAT-1 from the alpha interferon receptor complex. *J Virol* **76:**12676-82
59. Kato, A., Kiyotani, K., Sakai, Y., Yoshida, T. & Nagai, Y. The paramyxovirus, Sendai virus, V protein encodes a luxury function required for viral pathogenesis. *Embo J* **16:**578-87
60. Kato, A. et al. Importance of the cysteine-rich carboxyl-terminal half of V protein for Sendai virus pathogenesis. *J Virol* **71:**7266-72
61. Delenda, C., Hausmann, S., Garcin, D. & Kolakofsky, D. Normal cellular replication of Sendai virus without the trans-frame, nonstructural V protein. *Virology* **228:**55-62
62. Delenda, C., Taylor, G., Hausmann, S., Garcin, D. & Kolakofsky, D. Sendai viruses with altered P, V, and W protein expression. *Virology* **242:**327-37
63. Kurotani, A. et al. Sendai virus C proteins are categorically nonessential gene products but silencing their expression severely impairs viral replication and pathogenesis. *Genes Cells* **3:**111-24
64. Hasan, M. K. et al. Versatility of the accessory C proteins of Sendai virus: contribution to virus assembly as an additional role. *J Virol* **74:**5619-28
65. Komatsu, T., Takeuchi, K., Yokoo, J., Tanaka, Y. & Gotoh, B. Sendai virus blocks alpha interferon signaling to signal transducers and activators of transcription. *J Virol* **74:**2477-80
66. Garcin, D., Curran, J., Itoh, M. & Kolakofsky, D. Longer and shorter forms of Sendai virus C proteins play different roles in modulating the cellular antiviral response. *J Virol* **75:**6800-7
67. Takeuchi, K. et al. Sendai virus C protein physically associates with Stat1. *Genes Cells* **6:**545-57
68. Komatsu, T., Takeuchi, K., Yokoo, J. & Gotoh, B. Sendai virus C protein impairs both phosphorylation and dephosphorylation processes of Stat1. *FEBS Lett* **511:**139-44

69. Saito, S., Ogino, T., Miyajima, N., Kato, A. & Kohase, M. Dephosphorylation failure of tyrosine-phosphorylated STAT1 in IFN-stimulated Sendai virus C protein-expressing cells. *Virology* **293:**205-9
70. Gotoh, B., Komatsu, T., Takeuchi, K. & Yokoo, J. The C-terminal half-fragment of the Sendai virus C protein prevents the gamma-activated factor from binding to a gamma-activated sequence site. *Virology* **316:**29-40
71. Garcin, D., Marq, J. B., Goodbourn, S. & Kolakofsky, D. The amino-terminal extensions of the longer Sendai virus C proteins modulate pY701-Stat1 and bulk Stat1 levels independently of interferon signaling. *J Virol* **77:**2321-9
72. Garcin, D., Marq, J. B., Strahle, L., le Mercier, P. & Kolakofsky, D. All four Sendai Virus C proteins bind Stat1, but only the larger forms also induce its mono-ubiquitination and degradation. *Virology* **295:**256-65
73. Garcin, D., Latorre, P. & Kolakofsky, D. Sendai virus C proteins counteract the interferon-mediated induction of an antiviral state. *J Virol* **73:**6559-65
74. Gotoh, B. et al. Knockout of the Sendai virus C gene eliminates the viral ability to prevent the interferon-alpha/beta-mediated responses. *FEBS Lett* **459:**205-10
75. Kato, A. et al. Y2, the smallest of the Sendai virus C proteins, is fully capable of both counteracting the antiviral action of interferons and inhibiting viral RNA synthesis. *J Virol* **75:**3802-10
76. Garcin, D., Curran, J. & Kolakofsky, D. Sendai virus C proteins must interact directly with cellular components to interfere with interferon action. *J Virol* **74:**8823-30
77. Strahle, L., Garcin, D., Le Mercier, P., Schlaak, J. F. & Kolakofsky, D. Sendai virus targets inflammatory responses, as well as the interferon-induced antiviral state, in a multifaceted manner. *J Virol* **77:**7903-13
78. Durbin, A. P. et al. Recovery of infectious human parainfluenza virus type 3 from cDNA. *Virology* **235:**323-32
79. Durbin, A. P., McAuliffe, J. M., Collins, P. L. & Murphy, B. R. Mutations in the C, D, and V open reading frames of human parainfluenza virus type 3 attenuate replication in rodents and primates. *Virology* **261:**319-30
80. Smallwood, S. & Moyer, S. A. The L polymerase protein of parainfluenza virus 3 forms an oligomer and can interact with the heterologous Sendai virus L, P and C proteins. *Virology* **318:**439-50
81. Malur, A. G., Hoffman, M. A. & Banerjee, A. K. The human parainfluenza virus type 3 (HPIV 3) C protein inhibits viral transcription. *Virus Res* **99:**199-204
82. Valsamakis, A. et al. Recombinant measles viruses with mutations in the C, V, or F gene have altered growth phenotypes in vivo. *J Virol* **72:**7754-61
83. Tober, C. et al. Expression of measles virus V protein is associated with pathogenicity and control of viral RNA synthesis. *J Virol* **72:**8124-32
84. Patterson, J. B., Thomas, D., Lewicki, H., Billeter, M. A. & Oldstone, M. B. V and C proteins of measles virus function as virulence factors in vivo. *Virology* **267:**80-9
85. Radecke, F. & Billeter, M. A. The nonstructural C protein is not essential for multiplication of Edmonston B strain measles virus in cultured cells. *Virology* **217:**418-21
86. Schneider, H., Kaelin, K. & Billeter, M. A. Recombinant measles viruses defective for RNA editing and V protein synthesis are viable in cultured cells. *Virology* **227:**314-22
87. Fujii, N. et al. Oligo-2',5'-adenylate synthetase activity in K562 cell lines persistently infected with measles or mumps virus. *J Gen Virol* **69 (Pt 8):**085-91
88. Yokota, S. et al. Measles virus suppresses interferon-alpha signaling pathway: suppression of Jak1 phosphorylation and association of viral accessory proteins, C and V, with interferon-alpha receptor complex. *Virology* **306:**135-46
89. Takeuchi, K., Kadota, S. I., Takeda, M., Miyajima, N. & Nagata, K. Measles virus V protein blocks interferon (IFN)-alpha/beta but not IFN-gamma signaling by inhibiting STAT1 and STAT2 phosphorylation. *FEBS Lett* **545:**177-82

90. Rodriguez, J. J., Parisien, J. P. & Horvath, C. M. Nipah virus V protein evades alpha and gamma interferons by preventing STAT1 and STAT2 activation and nuclear accumulation. *J Virol* **76:**11476-83
91. Rodriguez, J. J., Wang, L. F. & Horvath, C. M. Hendra virus V protein inhibits interferon signaling by preventing STAT1 and STAT2 nuclear accumulation. *J Virol* **77:**11842-5
92. Rodriguez, J. J., Cruz, C. D. & Horvath, C. M. Identification of the nuclear export signal and STAT-binding domains of the Nipah virus V protein reveals mechanisms underlying interferon evasion. *J Virol* **78:**5358-67
93. Shaw, M. L., Garcia-Sastre, A., Palese, P. & Basler, C. F. Nipah virus V and W proteins have a common STAT1-binding domain yet inhibit STAT1 activation from the cytoplasmic and nuclear compartments, respectively. *J Virol* **78:**5633-41
94. Park, M. S. et al. Newcastle disease virus (NDV)-based assay demonstrates interferon antagonist activity for the NDV V protein and the Nipah virus V, W, and C proteins. *J Virol* **77:**1501-11
95. Bossert, B., Marozin, S. & Conzelmann, K. K. Nonstructural proteins NS1 and NS2 of bovine respiratory syncytial virus block activation of interferon regulatory factor 3. *J Virol* **77:**8661-8
96. Ramaswamy, M., Shi, L., Monick, M. M., Hunninghake, G. W. & Look, D. C. Specific Inhibition of Type I Interferon Signal Transduction by Respiratory Syncytial Virus. *Am J Respir Cell Mol Biol* **30:**893-900
97. Teng, M. N. & Collins, P. L. Altered growth characteristics of recombinant respiratory syncytial viruses which do not produce NS2 protein. *J Virol* **73:**466-73
98. Teng, M. N. et al. Recombinant respiratory syncytial virus that does not express the NS1 or M2-2 protein is highly attenuated and immunogenic in chimpanzees. *J Virol* **74:**9317-21
99. Whitehead, S. S. et al. Recombinant respiratory syncytial virus bearing a deletion of either the NS2 or SH gene is attenuated in chimpanzees. *J Virol* **73:**3438-42
100. Schlender, J., Bossert, B., Buchholz, U. & Conzelmann, K. K. Bovine respiratory syncytial virus nonstructural proteins NS1 and NS2 cooperatively antagonize alpha/beta interferon-induced antiviral response. *J Virol* **74:**8234-42
101. Jin, H. et al. Evaluation of recombinant respiratory syncytial virus gene deletion mutants in African green monkeys for their potential as live attenuated vaccine candidates. *Vaccine* **21:**3647-52
102. Jin, H. et al. Recombinant respiratory syncytial viruses with deletions in the NS1, NS2, SH, and M2-2 genes are attenuated in vitro and in vivo. *Virology* **273:**210-8
103. Valarcher, J. F. et al. Role of alpha/beta interferons in the attenuation and immunogenicity of recombinant bovine respiratory syncytial viruses lacking NS proteins. *J Virol* **77:**8426-39
104. Hall, C. B., Douglas, R. G., Jr., Simons, R. L. & Geiman, J. M. Interferon production in children with respiratory syncytial, influenza, and parainfluenza virus infections. *J Pediatr* **93:**28-32
105. Krilov, L. R., Hendry, R. M., Godfrey, E. & McIntosh, K. Respiratory virus infection of peripheral blood monocytes: correlation with ageing of cells and interferon production in vitro. *J Gen Virol* **68 (Pt 6):**1749-53
106. Loveys, D. A., Kulkarni, S. & Atreya, P. L. Role of type I IFNs in the in vitro attenuation of live, temperature-sensitive vaccine strains of human respiratory syncytial virus. *Virology* **271:**390-400
107. McIntosh, K. Interferon in nasal secretions from infants with viral respiratory tract infections. *J Pediatr* **93:**33-6
108. Garofalo, R. et al. Respiratory syncytial virus infection of human respiratory epithelial cells up-regulates class I MHC expression through the induction of IFN-beta and IL-1 alpha. *J Immunol* **157:**2506-13

109. Chonmaitree, T., Roberts, N. J., Jr., Douglas, R. G., Jr., Hall, C. B. & Simons, R. L. Interferon production by human mononuclear leukocytes: differences between respiratory syncytial virus and influenza viruses. *Infect Immun* **32:**300-3
110. Roberts, N. J., Jr., Hiscott, J. & Signs, D. J. The limited role of the human interferon system response to respiratory syncytial virus challenge: analysis and comparison to influenza virus challenge. *Microb Pathog* **12:**409-14
111. Hall, C. B., Douglas, R. G., Jr. & Simons, R. L. Interferon production in adults with respiratory syncytial viral infection. *Ann Intern Med* **94:**53-5
112. Bell, D. M., Roberts, N. J., Jr. & Hall, C. B. Different antiviral spectra of human macrophage interferon activities. *Nature* **305:**319-21
113. Spann, K. M., Tran, K. C., Chi, B., Rabin, R. L. & Collins, P. L. Suppression of the Induction of Alpha, Beta, and Gamma Interferons by the NS1 and NS2 Proteins of Human Respiratory Syncytial Virus in Human Epithelial Cells and Macrophages. *J Virol* **78:**4363-9
114. Hanada, N., Morishima, T., Nishikawa, K., Isomura, S. & Nagai, Y. Interferon-mediated self-limiting growth of respiratory syncytial virus in mouse embryo cells. *J Med Virol* **20:**363-70

Chapter 7

THE STRATEGY OF CONQUEST

The interaction of herpes simplex virus with its host

SUNIL J. ADVANI and BERNARD ROIZMAN
The Marjorie B. Kovler Viral Oncology Laboratories, The University of Chicago, 910 East 58th Street, Chicago, IL, USA

1. INTRODUCTION

Herpes simplex virus 1 (HSV-1) is a member of the family *herpesviridae*. Members of this family include viruses endogenous to virtually every species of animal life. In addition to HSV-1 and HSV-2, the list of herpesviruses infecting humans includes varicella-zoster virus, cytomegalovirus, Epstein-Barr virus, human herpesviruses 6A, 6B and 7, and Kaposi's sarcoma associated herpes virus[1]. Members of this family are relatively large enveloped double-stranded DNA viruses and the source of much human disease. HSV-1 establishes infections at the mucosal surfaces of the oral cavity and genitalia. Upon replication in the epithelial cells, the virus is transported retrograde to the nuclei of sensory neurons of dorsal root ganglia where it establishes latent infection. Periodically, HSV-1 reactivates, initiates the lytic cycle, and is transported anterograde to a site at or near the portal of entry of the virus. The spread of the virus from this site is responsible for the recurrent herpetic sores experienced by a fraction of individuals infected with this virus[2].

The HSV-1 particle consists of viral DNA encased by a capsid. Surrounding the capsid is the tegument consisting of >10 proteins including the virion host shutoff protein (*vhs*) encoded by the U_L41 gene and the α trans-inducing factor (α-TIF) encoded by the U_L48 gene. The tegument is enclosed by an envelope studded with the viral glycoproteins. HSV-1 DNA is approximately 152 kb in size and encodes at least 84 unique proteins. The

P. Palese (ed.), Modulation of Host Gene Expression and Innate Immunity by Viruses, 141-161.

genome consists of two unique sequences, unique long (U_L) and unique short (U_S) separated by inverted repeat sequences. Open reading frames (ORFs) within the unique regions are numbered sequentially, U_L1-U_L56 and U_S1-U_S12. ORFs located within the inverted repeats are present in two copies per genome and include α0, α4, $\gamma_1$34.5, ORF P, ORF O, and the coding sequence for the latency associated transcript (LAT). Viral genes are transcribed sequentially in a coordinate fashion. The α (immediate-early) genes are transcribed first followed by β (early) and γ (late) genes. The six α genes are α0, α4, α27, α22, α47 and U_S1.5, and encode the infected cell proteins (ICP) 0, 4, 27, 22, 47 and U_S1.5, respectively. These proteins express multiple functions that include regulation of gene expression as well as prevention of host responses to infection. The β genes encode primarily proteins involved in viral DNA synthesis. The onset of DNA synthesis enhances or initiates the transcription of a large family of late or γ viral genes. The γ genes are further differentiated into γ_1 (partially dependent on viral DNA synthesis) and γ_2 (totally dependent on viral synthesis). Viral proteins express multiple functions; the 84 unique proteins most likely express several hundred functions[2].

The ability of HSV-1 to successfully replicate within the cell is entirely dependent on its ability to interact with cellular constituents. In essence, HSV-1 blocks antiviral responses and simultaneously redirects cellular processes for it own use. In counteracting antiviral responses, HSV-1 evades immune surveillance and blocks innate antiviral responses, such as apoptosis and activation of protein kinase R (PKR). Moreover, the virus redirects cellular proteins to assist in its replicative cycle. The virus overtakes cellular transcription and translation machinery and modulates cell cycle regulated proteins. This process begins from the onset of viral entry into the cell. Thus *vhs* mediates the degradation of mRNA, which in itself is beneficial from multiple vantages to the virus. Antiviral responses that are transcription dependent are muted, and there is increased availability of the cellular translation machinery for viral transcripts. Another tegument protein, α-TIF, complexes with Oct-1 and host cell factor (HCF) to enhance the transcription of viral α genes.

This article focuses on a few key examples of the interaction of HSV-1 with several cellular proteins. The objective of the interactions is to enhance viral replication and to block host responses to infection. A key feature of these interactions is the degradation of the mRNAs at the onset of viral replication and the capture, degradation or posttranslational modification of host proteins to serve the needs of the virus.

2. THE DEGRADATION OF PML BY THE ICP0 – AN E3 UBIQUITIN LIGASE

Promyelocytic leukemia protein (PML) is a component and nucleation factor for PML nuclear bodies, PML-NB (also known as ND10, Kremer bodies, or PML oncogenic domains)[3]. The PML-NB is composed of multiple proteins, including Sp100, Sp140 Daxx, CBP, pRB, and p53. PML is found diffusely in the nucleoplasm, however upon sumoylation, PML nucleates and recruits other members of the PML-NB. The functions of PML-NB are multiple and include immune surveillance, regulation of apoptosis, tumor suppression and transcriptional co-activation. Interferons (IFNs) upregulate the transcription of components of PML-NB (i.e. PML, Sp100, and Sp140) and increase the formation of PML-NB[4]. Importantly, overexpression of PML has been shown to inhibit VSV and influenza viral replication[5].

ICP0, the product of the α0 gene, is a multifunctional protein, and at high multiplicities of infection, it is not essential for viral growth in cell culture. The gene is located in the inverted repeats flanking the unique long sequence of viral DNA and is therefore present in 2 copies per genome. The major function of ICP0 is transactivation of genes introduced into cells by infection or transfection. ICP0 also acts as an E3 ubiquitin ligase.[6] α0 is one of the few viral genes containing introns. Exon 1 encodes just 19 amino acids. Exon 2 encodes amino acids 20-241, contains a RING finger domain, and is separated by a small intron from exon 3 which encodes amino acids 242-775. The transactivating functions have been mapped to the first 100 amino acids of the ICP0[7]. The ring finger in exon 2 encodes a E3 ubiquitin ligase activity. Curiously, ICP0 encodes a second, independent E3 ubiquitin ligase activity within exon 3[8]. Here we will focus on the role of ICP0 in the destruction of PML and dispersal of PML-NB via its E3 ubiquitin ligase activity.

ICP0 localizes at or near PML-NB early in infection[9,10]. As viral infection proceeds, PML-NB are dispersed, and there is a loss of proteins associated with PML-NB. Cells transfected with ICP0 also exhibit dispersal of PML-NB associated proteins, indicating that ICP0 is sufficient for the degradation of PML-NB. The dispersion and loss of PML-NB by ICP0 is associated with the loss of PML as early as 2 hrs. after infection[11,12]. Treatment of cells with proteasome inhibitors (i.e. MG132 or lactacystin lactone) prevents HSV-1-mediated loss of PML. Also, dispersal of PML-NB and loss of PML are dependent on the RING finger of ICP0. PML is not the only protein of PML-NB that is targeted for destruction by ICP0. The interferon stimulated protein Sp100 is also decreased in HSV-1 infected cells[12]. There appears to be specificity in the destruction of sumoylated proteins in HSV-1 infected cells inasmuch as the sumoylated protein RanGAP1 is not decreased in HSV-1 infected cells.

The ubiquitin-proteasome – dependent loss of PML in HSV-1 infected cells suggests that ICP0 manipulates the ubiquitin-proteasomal pathway. Initially, a GST fusion protein encoding a portion of ICP0 exon 3 (aa 568-773) was shown to behave as an E3 ubiquitin ligase when reacted with the E2 ubiquitin conjugating-enzyme cdc34[8]. However, the degradation of PML mapped to the RING finger domain of exon 2. Curiously, ICP0 has a second, distinct E3 ubiquitin ligase activity mapping to the RING finger[13,14]. The ring finger domain of ICP0 acts as an E3 ligase in conjunction with the E2 ubiquitin conjugating enzymes UbcH5a and UbcH6, and cells transfected with a dominant negative form of UbcH5a do not exhibit HSV-1 induced degradation of PML or of Sp100[15].

The functions attributed to PML-NB are numerous and include antiviral responses mediated by interferons (IFN). Overexpression of PML precludes the HSV-1 induced dispersal of PML-NB but has no effect on viral replication[16]. The absence of PML as in PML-/-murine fibroblasts also has not effect on viral replication[17]. However HSV-1 mutant viruses lacking ICP0 have increased sensitivity to IFN treatment[18]. One explanation for this observation is that IFNs act through PML-NB to elicit their antiviral effects. In wild-type murine fibroblasts (PML+/+) IFN treatment reduced viral replication greater than 1,000 fold. In contrast, viral replication is only minimally affected in PML-/- murine fibroblasts treated with either IFNα or IFNγ. Taken together, the evidence indicates that HSV-1 ICP0 acts as an E3 ubiquitin ligase in conjunction with the E2 ubiquitin conjugating-enzyme UbcH5a to degrade PML and Sp100 to blunt the antiviral effects of IFN[8].

While we have focused on the interaction of ICP0 and PML, ICP0 also mediates the degradation of other cellular proteins. The centromeric proteins CENP-A and CENP-C in addition to DNA-dependent protein kinase are degraded in an ICP0 dependent manner[19-21]. It remains to be determined as to which E2 is utilized by ICP0 for the degradation of these proteins.

It is noteworthy that for a variety of reasons reported elsewhere, the ubiquitin ligase functions of ICP0 do not account for its transactivating functions[8]. The latter remain to be elucidated.

3. MODIFICATION OF CELL CYCLE PROTEINS BY HSV-1

In the course of the cell cycle, the cyclin-dependent kinases (cdk) remain stable, but their activity depends on availability and phosphorylation state of cyclins, their natural partners[22]. HSV-1 stabilizes D-type cyclins, cyclins D1 and D3. D-type cyclins activate the G_1 cdks (i.e. cdk4 and cdk6). Importantly, ICP0 binds cyclin D3, and a viral mutant deleted of ICP0 fails to stabilize cyclins D3 and D1[23,24]. To achieve cell cycle specific activation of cdks, cyclins are transcribed during the appropriate phase and then degraded in an ubiquitin- proteasome dependent manner. As noted earlier, ICP0 encodes a second E3 ubiquitin ligase within exon 3[24]. This domain of ICP0 acts as an E3 ubiquitin ligase in conjunction with the E2 ubiquitin conjugating enzyme, cdc34. Cdc34, also known as UbcH3, is the E2 ubiquitin conjugating enzyme responsible for the normal turnover of D type cyclins. By enabling the degradation of cdc34, ICP0 blocks the turnover of cyclin D3 and cyclin D1 although all of the studies to date indicate that it uses only cyclin D3[25]. The stability of D-type cyclins results in activation of cdk4 but not of cdk2 in wild-type virus-infected cells[24]. The target of activated cdk4 are the pocket proteins (i.e. pRB) that bind to and sequester E2F proteins. The absence of activated cdk2 and the apparent modifications of E2F proteins in the course of viral infection argue strongly against activation of S phase cellular genes[26,27]. The function of activated cdk4 remains unknown. It is noteworthy that the substitution D199A in ICP0 abolishes the interaction of ICP0 with cyclin D3 and precludes the localization of ICP0 in the ND10 structures or the translocation of ICP0 from the nucleus to the cytoplasm[8,24]. Consistent with this role is the observation that overexpression of cyclin D3 by insertion of the gene into HSV-1 under an α promoter accelerates the translocation of ICP0 to the cytoplasm.

Activation of cdc2 and its diversion to perform novel functions is another example of viral control of cell cycle proteins. In uninfected cells the G_2/M cdk, cdc2 (cdk1)[28] is inhibited by the kinases Wee-1 and Myt-1 that target Thr-14 and Tyr-15 of cdc2. These inhibitory phosphates on cdc2 are removed by cdc25C phosphatase. In addition, cyclins A and B are transcribed during G_2/M and bind and activate cdc2. cdk2 is not essential for progress through the cell cycle[29]. The activation of cdk4 but not of cdk2 suggests selectivity and a specific diversion to perform, possibly in conjunction with the stabilized, active cyclin D3, specific functions required by the virus. In the case of cdc2, the objective is this protein inasmuch as its natural partners, cyclins A and B are degraded. In this instance, the objective is to marry cdc2 to a new viral partner. To appreciate fully the function of cdc2 in infected cells, it is necessary to describe the function of several viral proteins.

Genetically engineered mutants lacking ICP22 or the U_L13 protein kinase share properties and are severely attenuated in murine models of HSV-1 replication. ICP22 is an α protein, while U_L13 is a viral serine/threonine protein kinase. The congruence of phenotypes may be due at least in part to the U_L13 mediated post-translational modifications of ICP22[30,31]. The phenotype of deletion mutants is cell-type dependent. Whereas in transformed human cell lines (i.e. HEp-2 and Vero cells), ICP22 and U_L13 are not essential; they are required for optimal expression of a subset of γ_2 genes exemplified by U_S11, U_L38, and U_L41 ORFs in primary human cells, rabbit skin cells, or in rodent cell lines[32,33]. The U_L42 gene encodes the viral DNA polymerase accessory factor, which acts as a processivity factor for the viral DNA polymerase encoded by the U_L30 ORF. U_L42 is an essential viral gene expressed with β kinetics. While U_L42 protein plays an important role in viral DNA synthesis, it is made in excessive amounts in relation to the abundance of its partner, the DNA polymerase. One implication of this observation is that U_L42 may have other functions than those associated with the DNA polymerase. As described below, one function of U_L42 that emerged in recent studies is the cdc2-mediated recruitment of topoisomerase IIα for viral gene transcription[34].

Cdc2 kinase is activated in HSV-1 infected cells between 4 and 8 hrs. after infection. HSV-1 infection results in the disappearance of the inhibitory isoforms of cdc2 resulting from phosphorylation of Thr-14 and Tyr-15 concurrently with a reduction in the level of the inhibitory kinase, Wee1 and hyperphosphorylation (i.e. activation) of cdc25C. In this manner, cdc2 is "primed" to be activated by HSV-1. However, the cognate partners of cdc2, cyclins A and B, do not accumulate in HSV-1 infected cells. In fact, a dramatic reduction in the levels of both cyclin A and B is seen in HSV-1 infected cells compared to those of mock-infected cells. Activation of cdc2 requires the presence of viral proteins ICP22 and U_L13. In consequence, infection with either mutant virus does not result in the loss of cyclins A or B, cdc25C hyperphosphorylation, or cdc2 activation. Interestingly, a cdc2-domininant negative construct decreases the accumulation of the γ_2 protein U_S11, but not those of representative α, β, or γ_1 proteins in wild-type HSV-1 infected cells[35]. While HSV-1 γ_2 gene transcription is dependent on viral DNA synthesis, mutants of ICP22 and U_L13 are not defective with respect to viral DNA synthesis in restricted cells. These mutants, therefore, discriminate viral DNA synthesis from the transcription of a subset of γ_2 genes. These results suggest HSV-1 activates cdc2 via ICP22 and U_L13 for efficient γ_2 gene expression.

While cdc2 is active in HSV-1 infected cells, its cyclin partners are degraded. Cyclins share a sequence termed the cyclin box that mediates binding to cdks. Of all viral proteins only U_L42 has a degenerate cyclin box and physically interacts with cdc2. cdc2 complexed with U_L42 protein is active[36]. A model consistent with available evidence is that ICP22 and U_L13 "prime" cdc2 for binding to U_L42.

Cellular substrates for cdc2 are numerous, and include topoisomerase IIα. Topoisomerase IIα becomes activated by phosphorylation during the G_2/M phase of the cell cycle, and regulates gene transcription. Topoisomerase IIα is phosphorylated in HSV-1 infected cells in a manner similar to the topoisomerase IIα accumulating in cells arrested in G_2/M by nocodazole[37]. Cells transfected with U_L42 also exhibit elevated levels of topoisomerase II activity as would be expected if cdc2 were active. Of particular significance is the observation that U_L42 binds to topoisomerase IIα in a cdc2-phosphorylation dependent manner and the two proteins colocalize. Consistent with this hypothesis, U_L42 associates with topoisomerase IIα from wild-type HSV-1 infected cells but not with that of cells infected with an ICP22 mutant virus in which cdc2 is not active. What is the functional consequence of U_L42 associating with topoisomerase II? HSV-1 DNA synthesis requires topoisomerase II activity and HSV-1 infected cells treated with inhibitors of topoisomerase II are blocked at the stage of viral DNA synthesis[38]. Thus, a role for topoisomerase II downstream of viral DNA synthesis may be difficult to uncover in the context of infected cells. As noted above, ICP22 HSV-1 mutants are able to synthesize viral DNA but are also defective in the expression of a subset of γ_2 viral genes. This suggests that while topoisomerase II is involved in the synthesis of progeny viral DNA, ICP22 recruits topoisomerase II to U_L42 in a cdc2 dependent manner for γ_2 gene transcription.

4. PREVENTION OF SHUTOFF OF PROTEIN SYNTHESIS BY ACTIVATED PROTEIN KINASE R (PKR)

Earlier studies have shown that HSV-1 infected cells accumulate large amounts of complementary RNAs homologous to at least 50% of the genome[39,40]. This is not surprising inasmuch as viral genes are encoded on both strands of HSV-1 DNA, and in several instances, ORFs overlap[2]. The presence of large amounts of complementary RNAs is likely to generate dsRNA, which in turn activates dsRNA-dependent PKR. To become active, dsRNA-bound PKR forms dimers and transphosphorylates its dimer partner. Activated PKR has numerous effects that include phorpshorylation of the α subunit of translation initiation factor 2 (eIF-2α) at ser 51 and the activation of the nuclear factor κB (NF-κB)[41,42]. Phosphorylation of eIF-2α results in the total shutoff of protein synthesis, both viral and cellular, and as would be expected, significantly diminishes yields of progeny virus. Since many viruses generate complementary RNAs in the course of their replication, they have evolved a variety of strategies to block the shutoff of protein synthesis by PKR. HSV-1 has evolved two proteins encoded by $\gamma_1$34.5 and U_S11 genes that are capable of blocking the shutoff of protein synthesis by PKR. Curiously, it uses predominantly only the $\gamma_1$34.5 protein for this purpose.

The $\gamma_1$34.5 gene is located within the inverted repeats flanking the U_L sequence of the viral genome and is therefore present in two copies per genome. The product of the $\gamma_1$34.5 gene, ICP34.5 is a γ_1 protein of approximately 260 residues. It consists of a long amino-terminal domain linked by a variable number of repeats of 3 amino acids to a carboxyl-terminal domain of 73 amino acids. The carboxyl-terminal domain is nearly identical in sequence to the C-terminal domain of a much larger protein known as growth arrest and DNA damage (GADD) 34 protein[43]. Mutants lacking the $\gamma_1$34.5 gene yield lower titers in most cells in culture and are severely attenuated in murine models. The LD_{50} of wild-type HSV-1 is approximately 10^2 PFU upon direct intracranial inoculation. For a virus deleted in both copies of $\gamma_1$34.5 (R3616), the LD_{50} is greater than 10^6 PFU[44]. Due to the severe attenuation of R3616 compared to wild-type HSV-1 upon mouse intracranial inoculation, ICP34.5 is frequently described as a neurovirulence factor.

A characteristic of the $\gamma_1$34.5 deletion mutants is the shutoff of protein synthesis that begins approximately 5 hrs. after infection concurrent with the onset of viral DNA synthesis and is virtually complete by 13 hrs. after infection. A surprising finding is that PKR is activated in both wild-type- and mutant virus-infected cells but that eIF-2α is phosphorylated only in mutant virus-infected cells[45,46]. The solution to this puzzle came from the observation that $\gamma_1$34.5 interacts strongly in a yeast-2 hybrid system with protein phosphatase 2α[47]. In lysates of wild-type virus infected cells but not in those of mutant virus-infected cells, the activity of the phosphatase IIα is redirected to dephosphorylate eIF2α at the expense of dephosphorylation of natural substrates of the phosphatase. While the domain homologous to GADD34 is essential for the dephosphorylation of eIF2a, a short sequence immediately adjacent to the triplet repeat in the carboxyl-terminal domain contains the highly conserved motif of a phosphatase IIα accessory factor[48].

It is noteworthy that of the herpesvirus genomes sequenced to date, only HSV-1, HSV-2 and the simian B virus genomes contain homologs of the $\gamma_1$34.5 gene. The $\gamma_1$34.5 gene appears to be an evolutionarily recent acquisition. It represents the only HSV-1 gene other than enzymes involved in DNA synthesis and repair, with extensive homology to a portion of a cellular gene. Other herpesviruses must have an alternative mechanism for evading the consequences of accumulation of complementary RNAs.

The potential role of U_S11 in blocking the effects of accumulation of complementary RNAs emerged from serial passages of a $\gamma_1$34.5 deletion mutant in cultured cells. In the viral genome, the γ_2 U_S11 gene is located immediately 3' to the α47 gene. The selective pressure exerted in this system resulted in the selection of a second site mutant with better growth characteristics and sustained protein synthesis[49]. Analyses of this mutant revealed the spontaneous deletion of the coding sequence of the α47 gene and the promoter of the U_S11 gene. As a consequence, the α47 promoter was juxtaposed to the coding sequence of the U_S11 gene converting its expression from very late to early in the course of viral replication[50].

The role of the modified U_S11 in blocking the effects of accumulation of complementary RNAs emerged in several studies. U_S11 binds PKR and appears to block its activation since in mutant virus-infected cells, eIF-2α is not phosphorylated and the phosphatase activity is not redirected to phosphorylate eIF2α. In *in vitro* assays, addition of U_S11 prior to the activation of PKR by dsRNA blocked phosphorylation of eIF-2α, whereas addition of U_S11 after activation of PKR had no effect on the phosphorylation of eIF2α[51-53].

In light of the observation that the $\gamma_1 34.5$ gene is a relatively recent entrant into the herpesvirus family, U_S11 could be viewed as a more ancient defense mechanism against PKR. The problem is that the U_S11 gene is not conserved across the three subfamilies of *herpesviridae.* The possibility that a more ancient mechanism for evading the effects of dsRNA persists at least in a cryptic form emerged from serial passages of a mutant lacking both U_S11 and $\gamma_1 34.5$ genes. These studies led to the isolation of a mutant exhibiting better growth characteristics and increased virulence in the mouse model but the site of the new mutation could not be mapped. A characteristic of this mutant is increased phosphatase activity with no apparent specificity in the infected cells[54].

A key question is that given the availability of two genes capable of blocking the effects of activated PKR, why did HSV chose to use the $\gamma_1 34.5$ gene in preference to the U_S11 gene? The question is particularly appropriate since the $\gamma_1 34.5$ protein dephosphorylates eIF-2α but totally ignores the activated PKR, whereas U_S11 binds to and prevents activation of PKR. Among the possible answers, at least two merit further study. First, the possibility exists that additional effects of activated PKR unrelated to phosphorylation of eIF-2α are blocked by other viral genes or that HSV has found a specific use for the activated PKR. At least the latter hypothesis is supported by tangible evidence inasmuch as, in HSV-1 infected cells, activated PKR is required to activate NF-κB[55]. However, while HSV-1 yields higher titers in cells in which PKR and NF-κB are not activated than in cells in which both PKR and NF-κB are active, the opposite is true in cells from which p50, p65 or both NF-κB genes had been deleted[56]. While the activated NF-κB has been associated with a large number of transcripts expressed and accumulating after infection, the ones examined in detail (IEX-1, c-fos, IκBα) are rapidly degraded in a *vhs* dependent manner and their translation products do not accumulate. A possible role of NF-κB in blocking apoptosis is examined in detail below.

5. HSV BLOCKS APOPTOSIS INDUCED BY EXOGENOUS AGENTS OR VIRAL GENE PRODUCTS

Apoptosis or programmed cell death is a response to injury to the cell either by interaction of specific proteins with "death" receptors on the cell surface or as a consequence of a stress response caused by exogenous agents (hyperthermia, sorbitol, TNFα), or by the accumulation and function of viral gene products[2,57,58]. Cell death is caused by the degradation of key proteins as a consequence of proteolytic activation of pro-caspases. Cells contain numerous pro- and anti apoptotic proteins (e.g. the Bcl-2 family of proteins that are either activated to trigger apoptosis or titer out activated pro-apoptotic proteins to block apoptosis). HSV-1 has been shown to block apoptosis induced by either exogenous agents or by viral mutants. To date, mutants lacking one of 6 genes have been reported to induce apoptosis in a cell-type dependent manner; these are mutants lacking the regulatory genes α4 or α27 or the genes encoding the major subunit of ribonucleotide reductase (U_L 39), glycoproteins D (gD, U_S 6) and J (gJ, U_S 7), or the latency associated transcript (LAT)[57-62]. α4 and α27 are regulatory genes; most likely required for the synthesis of key anti-apoptotic viral proteins expressed at later times during viral replication. The involvement of the major component of the ribonucleotide reductase in blocking apoptosis is poorly understood. Most of the studies done in recent years focused on the protein kinase encoded by U_S3, gC, gJ and LAT.

The anti-apoptotic activity of the U_S3, protein kinase was identified in studies of a mutant lacking the a4 gene[59]. This mutant induces apoptosis in a variety of cell lines. Apoptosis induced by this mutant is blocked by overexpression of Bcl-2 or by the U_S3 protein kinase[60]. The U_S3 protein kinase also blocks apoptosis induced by activated pro-apoptotic members of the Bcl-2 family[63-65]. Recent studies led to the realization that U_S3 and cAMP-dependent protein kinase A (PKA) share similar substrate specificities and that activation of PKA blocked apoptosis induced by the α4 deletion mutant or by activation of the pro-apoptotic protein BAD[66]. The activation of PKA appears to be the result of phosphorylation of the regulatory subunit RIIα by the U_S3 protein kinase. The anti-apoptotic activity of U_S3 thus reflects the concerted activity of U_S3 and PKA activated by the viral enzyme[66]. The key substrates of U_S3/PKA kinases remain to be identified.

HSV-1 encodes 12 glycosylated proteins of which only 5 play an essential role in viral entry[2]. Thus gC and gB attach the virus to the heparin sulfate moiety on cell surface proteins, gD interacts with one of at least 3 known cellular receptors and gD, gH and gL fuse the viral envelope to the plasma membrane. gD deletion mutants come in two flavors. Those produced in cells ectopically expressing gD but infected with the gD- virus yield virus particles (gD-/+) that lack the gene but contain gD in the envelope and hence can infect cells expressing the gD receptors. Cells lacking gD but infected with gD-/+ mutants also yield virus particles but these lack both the gD gene and the gD in the envelope. These particles can attach to the cell's surface but fusion of the envelope with the plasma membrane does not ensue. Instead, the attached particles are endocytosed and degraded. Curiously, both gD-/- and gD-/+ mutants induce apoptosis, but based on the requirements for blocking apoptosis, the mechanisms by which apoptosis is induced appears to differ[67,68]. gD-/+ particles are blocked by overexpression of gJ, intact gD, or as little as the ectodomain of gD. In contrast gD-/- particles are blocked by gJ and either the intact gD or a mixture of plasmids encoding the transmembrane domain plus the cytoplasmic domain (TM-C) and transmembrane domain plus the ectodomain (E-TM). The possibility that the E-TM and TM-C form a "heterodimer" is supported by the observation that gD has an unpaired cysteine located in the transmembrane domain. TM-C or E-TM carrying a substitution in the unpaired cysteine no longer block apoptosis. Two other observations support the hypothesis that gD-/- and gD-/+ differ in the mechanisms by which they induce apoptosis. First, chloroquine blocks apoptosis induced by gD-/- virus but not by gD-/+ virus. In contrast, overexpression of mannose 6 phosphate receptor blocks apoptosis induced by both types of particles. Second, while gD-/+ particles induce apoptosis at relatively low multiplicities of infection, the gD-/- particles induce apoptosis only at high multiplicities of infection. Current models suggests that gD-/- is taken up by endocytosis and that a massive discharge of lysosomal enzymes causes cell death. gD present in endocytic vesicles blocks apoptosis by interacting with the cation independent mannose 6 phosphate, a regulator of lysosomal enzymes. Ectopically expressed gD, intact or consisting of the postulated heterodimer described above, is transported to the plasma membrane and ultimately becomes a component of lysosomal vesicles. In contrast, gD-/+ mutant most likely triggers apoptosis late in infection, during the transport of particles lacking gD into the extracellular space. The same vesicle are likely to carry soluble gD (ectodomain) which could interact with the mannose 6 phosphate receptor and block lysosomal enzyme discharge[67,68].

In recent studies gJ appears to play a prominent role in immune evasion. The LAT gene maps in the inverted repeats, 3' and partially antisense to the α0 gene. There is no convincing evidence that LAT encodes one or more proteins and the functions of LATs are largely unknown[2]. A current model is that LAT consists of 2 and 1.5 kb introns of an 8.3 kb transcript extending almost the entire length of the inverted repeat flanking the long unique sequences. Recent evidence suggests LAT can protect neurons from apoptotic insults induced by the resident latent virus. Thus, in animal models, virus inoculated into the cornea is translocated to the trigeminal ganglion where it establishes a latent infection. In rabbits infected with a LAT-/α0+ mutant the trigeminal ganglia contained apoptotic neurons in greater quantity than ganglia of rabbits infected with wild-type and mutant repaired viruses[69]. Furthermore, cells transfected with a plasmid encoding the stable 2 kb LAT are relatively resistant to cell death induced by pro-apoptotic stimuli[69,70]. It has been reported that LAT also protects cells from apoptosis induced by the Bcl-2 family member Bax[7].

Recently it has been reported that LAT affects the splicing of Bcl-x mRNA[72]. Specifically the long form Bcl-x_L blocks apoptosis, whereas an alternative splice product results in a shorter protein, Bcl-x_S, which is pro-apoptotic. Infection of cells with a virus mutated in LAT results in the accumulation of Bcl-x_S, whereas wild-type infected cells express only Bcl-x_L. The effects appear to be specific inasmuch as no alternative splice products of Bcl-2 or Bak were identified. It has been suggested that LAT affects the accumulation of alternative splice products of Bcl-x by interacting with splicing factors[73]. Confirmation of this model may explain the evidence that LAT+ viruses are present in larger mounts in ganglia and hence reactivate more frequently than LAT- viruses[71]. If resident virus induces apoptosis and LAT blocks this event, the number of neurons harboring virus could be expected to be greater after an extended time in animals infected with LAT+ than with LAT- virus. An unresolved question is the mechanism by which apoptosis is induced.

The number of viral gene functions dedicated to prevention of apoptosis suggests that the latter is a major mechanism by which multicellular organisms attempt to prevent the spread of infection. The evolution of viral countermeasures designed to defeat this cellular defense mechanism ensured the survival of viruses in existence today. Recently it has been suggested that HSV induces NF-κB to block apoptosis and that apoptosis is induced by HSV mutants only in transformed but not in untransformed cells[74]. The available data do not support this conclusion. Specifically, a virus mutant that replicates but fails to activate PKR and consequently does not induce

NF-κB did not induce apoptosis in the human neuroblastoma cell line SK-N-SH. Unlike the wild-type sibling cells, murine PKR-/- fibroblasts are resistant to apoptosis by α4 mutant virus. The same was found in the case of murine fibroblasts from mice lacking p50, or p65 or both the p50 and p65 components of NF-κB[55,56]. However, wild-type (sibling) cells and the PKR-/-, or NF-κB-/- cell were sensitive to apoptosis induced by sorbitol or TNFα. It would seem that in the case of HSV-1, as in the case of other viruses, activation of NF-κB tends to predispose rather than block the cell from undergoing apoptosis[75,76].

6. IMMUNE EVASION: HSV BLOCKS MHC CLASS I AND II PATHWAYS

Major histocomapatability complex (MHC) class I antigen presentation is a component of the host adaptive immunity to exogenous pathogens (i.e. bacteria and viruses). Unlike the specialized antigen presenting cells of the MHC class II pathway, MHC class I molecules are expressed by most nucleated cells. Endogenously synthesized proteins are degraded by ubiquitin-proteasomal pathway into peptide fragments. These peptide fragments are then translocated through the ER and loaded onto MHC class I molecules. Lack of peptide loading in the ER results in unstable MHC class I molecules. The transporter associated with antigen processing (TAP) mediates peptide translocation through the ER. TAP itself is composed of a heterodimer, TAP1 and TAP2. Following peptide loading, MHC class I molecules dissociate from TAP and make their way to the cell surface via the Golgi. During transport to the cell surface, MHC class I molecules are progressively post-translationally modified (e.g. sialation). Sialation of MHC class I molecules results in endoglycosidase H resistance. Once on the cell surface, MHC class I molecules are recognized by CD8+ cytotoxic T cell lymphocytes, and the infected cell is destroyed. While MHC class I molecules are expressed at low steady state levels, many of the proteins involved in this pathway are upregulated by interferons during viral infection.

The U_S12 gene encodes an 88 residue α protein designated ICP47. This protein is not essential for viral replication in cell culture and has no known regulatory function. The sole function identified to date is to block presentation of antigenic peptides. In HSV-1 infected cells, there is a reduction in MHC class I molecules at the cell surface and reduced CD8+ mediated cell killing[77,78]. One explanation is that *vhs*, contained within the tegument, degrades mRNA of interferon-stimulated genes including MHC class I components. However in viral mutants lacking the gene encoding *vhs*, a block in MHC class I presentation is also observed. Therefore a second mechanism must exist to deal with antigen presentation. MHC class I molecules are not sialated in HSV-1 infected cells compared to those of uninfected cells[79,80]. Thus, MHC class I molecules remain endoglycosidase H sensitive 2 hrs. after infection in contrast to MHC class I in mock infected cells that became endoglycosidase H resistant. The block in cell surface transport is specific in as much as the transferrin receptor is sialated in HSV-1 infected cells. The early block in MHC class I presentation is mediated by ICP47, the product of the α47 gene.

To block MHC class I peptide loading, ICP47 interacts with the cytoplasmic face of both proteins of the TAP heterodimer, TAP1 and TAP2[81,82]. ICP47 specifically blocks peptide transport by interacting with the peptide-binding domain of TAP[83,84]. Thus, ICP47 acts as a competitive inhibitor of antigenic peptides by binding to TAP. Curiously, HSV-1 ICP47 binds human TAP more efficiently than murine TAP. This species-specific association suggests that the HSV-1 α47 gene has coevolved with its host to tightly bind human TAP.

The MHC class II pathway is responsible for sampling exogenous antigens by antigen presenting cells (APC). Such antigens (i.e. viral proteins) are internalized by endocytosis and loaded onto MHC class II molecules within endosomal or lysosomal compartments. Initially, MHC class II molecules associate with invariant chain (Ii) in the endoplasmic reticulum. Ii prevents premature loading of MHC class II molecules with endogenous peptides (i.e. self antigens). Within the endosomal compartments, the acidic environment results in Ii degradation and allows for MHC class II to bind internalized antigens. MHC class II activates CD4+ T lymphocytes, which secrete cytokines that in turn activate both CD8+ T cells and B cells, with subsequent antibody production.

Infection of APC with HSV-1 inhibits their ability to activate CD4+ T cells[85]. In the presence of HSV-1 infected APC, CD4+ T cell clones do not proliferate and fail to secrete cytokines (i.e. IFN-γ and IL-2) upon antigen stimulation. This phenotype can be seen in APC transfected with the gene encoding ICP22. Unlike ICP47 however, ICP22 does not cause a reduction in the surface expression of MHC class II or costimulatory molecules in lymphoblastoid APC. However in a glioblastoma cell line stably expressing MHC class II transcriptional activator, HSV-1 infection results in MHC class II cell surface downregulation[86]. Curiously, infection with HSV-1 mutated in either *vhs* or ICP34.5 resulted in elevated cell surface expression of MHC class II molecules. This suggests that upon HSV-1 infection, the cell responds by elevating MHC class II surface expression, and that both *vhs* and ICP34.5 blunt this response to allow the infected cell to avoid recognition by the immune response. The effects on MHC class II surface expression with HSV-1 and mutant infection correlated with the level of endocytosis. Thus, wild-type HSV-1 infected cells had reduced levels of endocytosis compared to cells infected with a mutant lacking the $\gamma_1 34.5$ gene. Lowering the basal rate of endocytosis by APC again allows the virus to escape detection by CD4+ cells.

As noted above, the pathway of MHC class II maturation is dependent on Ii. Curiously, gB shares sequence homology to Ii, and the homology is within the domain of Ii that binds a specific class of MHC class II molecules, HLA-DR1[87]. In cells infected with HSV-1, Ii expression is decreased[86,88]. Moreover, gB associates with MHC class II molecules. Unlike MHC class I molecules that remain endoglycosidae H sensitive, MHC class II molecules show endoglycosidae H resistance, indicating transport from the ER to the Golgi. However, MHC class II molecules associated with gB do not reach the cell surface and instead remain within cytoplasmic vesicles.

The ability of HSV-1 to evade detection and activation of immune responses is crucial for the virus to replicate and spread from cell to cell in infected individuals. To this end, HSV-1 has developed multiple means to deal with both MHC class I and class II immune responses.

7. CONCLUSIONS

HSV has developed a wide repertoire of functions designed to thwart the host cell from synthesizing new proteins designed to block infection or to enable resident proteins from activating a metabolic path inimical to viral replication or to signal to the environment of the cells that it is infected. The repertoire of functions encoded by the virus includes specific degradation of cellular proteins and RNA and posttranslational modification of cellular proteins to block their function. Two strategic evolutionary "decisions" standout. First, in contrast to other herpes viruses, HSV has purloined a single short stretch of amino acids from GADD34 but no other host genes. The strategy is not to evade the host by mimicry but rather by modification of cellular functions. The other aspect of the fundamental strategy employed by the virus is to target cellular proteins and divert them to perform functions other than those normally performed in the uninfected cell.

The analyses of viral gene functions carried over the past decade have revealed one important fact: cells have evolved a wide range of defensive measures. Viruses and especially HSV have evolved countermeasures designed to thwart these responses. The selective pressures generated by cellular responses to infection are evident from the numerous genes evolved by HSV to thwart just one cellular defense mechanism. For example the *vhs* protein among other functions blocks the repopulation of the short lived Jak1 protein. ICP0 blocks exogenous interferon from activating the interferon dependent antiviral response by degrading PML and dispersing the PML-NB. ICP34.5 insures that activation of PKR does not result in total shutoff of protein synthesis. Studies in progress suggest that we have not exhausted either the repertoire of cellular responses to infection or the viral countermeasures.

ACKNOWLEDGEMENTS

We thank Ralph Weichselbaum for his support. These studies were aided by grants from the National Cancer Institute (CA78766, CA71933, CA83939, CA87661, and CA88860), the United States Public Health Service.

REFERENCES

1. Roizman, B., and Pellett, P. E. The Family Herpesviridae. In: Fields' Virology 4rd Edition, D.M. Knipe, P. Howley, D.E Griffin, R. A Lamb, M. A. Martin, B. Roizman, and Stephen E. Straus, Editors, Lippincott-Williams and Wilkins, New York, N.Y, 2001, pp2381-2397.
2. Roizman, B. and Knipe, D.M. 2001. The replication of herpes simplex viruses. In Fields Virology 4th Ed. (D.M. Knipe, P. Howley, D.E. Griffin, R.A. Lamb, M.A. Martin, B. Roizman, and S.E. Straus, eds.), Lippincott/The Williams & Wilkins Co., New York, pp.2399-2459.
3. Zhong S., Salomoni P., and Pandolfi, P.P. 2000. The transcriptional role of PML and the nuclear body. *Nat. Cell Biol.* **2:**85-90
4. Regad, T. and Chelbi-Alix, M.K. 2001. Role and fate of PML nuclear bodies in response to interderon and viral infections. *Oncogene* **20:**7274-7286
5. Chelbi-Alix, M.K., Quignon, F., Pelicano, L, Koken M.H.M., and de The H. 1998. Resistance to virus infection conferred by the interferon-induced promyelocytic leukemia protein. *J. Virol.* **72:**1043-1051
6. Hagglund, R. and Roizman, B. 2004. Role of ICP0 in the strategy of conquest of the host cell by herpes simplex virus 1. *J. Virol.* **78:**2169-2178
7. Lium, E.K., Panagiotidis, C.A., Wen, X., and Silverstein, S.J. 1998. The NH2 terminus og the herpes simplex virus type 1 regulatory protein ICP0 contains promoter-specific transcription activation domain. *J. Virol.* **72:**7785-7795
8. Van Sant, C, Hagglund, R, Lopez, P, and Roizman, B. 2001. The infected cell protein 0 of herpes simplex virus 1 dynamically interacts with proteasomes, binds and activates the cdc34 E2 ubiquitin-conjugating enzyme, and possesses in vitro E3 ubiquitin ligase activity. *Proc. Natl. Acad. Sci. USA* **98:**8815-8820
9. Maul, G.G. and Everett, R.D. 1994. The nuclear location of PML, a cellular member of the C3CH4 zinc-binding domain protein family, is rearranged during herpes simplex virus infection by the C3CH4 viral protein ICP0. *J. Gen. Virol.* **75:**1222-1233

10. Everett, R.D. and Maul, G.G. 1994. HSV-1 IE protein vmw110 causes redistribution of *PML. EMBO J.* **13:**5062-5059
11. Everett, R.D., Freemont, P., Saitoh, H., Dasso, M., Orr, A., Kathoria, M., and Parkinson, J. 1998. The disruption of ND10 during herpes simplex virus infection correlates with the vmw110- and proteasome-dependent loss of several PML isoforms. *J. Virol.* **72:**6581-6591
12. Chelbi-Alix, M.K. and de The, H. 1999. Herpes virus induced proteasome dependent degradation of the nuclear bodies-associated PML and sp100. *Oncogene* **18:**935-941
13. Hagglund, R., Van Sant, C., Lopez, P., and Roizman, B. 2002. Herpes simplex virus 1-infected cell protein 0 contains two E3 ubiquitin ligase sites specific for different E2 ubiquitin-conjugating enzymes. *Proc. Natl. Acad. Sci. USA* **99:**631-636
14. Boutell, C., Sadis, S., and Everett, R.D. 2002. Herpes simplex virus type 1 immediate-early protein ICP0 and its isolated ring finger domain act as ubiquitin E3 ligases in vitro. *J. Virol.* **76:**841-850
15. Gu, H., and Roizman, B. 2003. The degradation of promyelocytic leukemia and sp100 proteins by herpes simplex virus 1 is mediated by the ubiquitin-conjugating enzyme UbcH5a. *Proc. Natl. Acad. Sci. USA* **100:**8963-8968
16. Lopez, P., Jacob, R.J., and Roizman, B. 2002. Overexpression of promyelocytic leukemia protein precludes dispersal of ND10 structures and has no effect on accumulation of infectious herpes simplex virus 1 or its proteins. *J. Virol.* **76:**9355-9367.
17. Chee, A.V., Lopez, P., Pandolfi, P.P., and Roizman, B. 2003. Promyelocytic leukemia protein mediates interferon-based anti-herpes simplex virus 1 effects. *J. Virol.* **77:**7101-7105
18. Harle, P., Sainz, Jr. B., Carr, D.J., and Halford, D.J. 2002. The immediate early protein, ICP0, is essential for the resistance of herpes simplex virus to interferon-α/β. *Virology* **293:**295-304
19. Everett, R.D., Earnshaw, W.C., Findlay, J., and Lomonte, P. 1999. Specific destruction of kinetochore protein CENP-C and disruption of cell division by herpes simplex virus immediate-early protein vmw110. *EMBO J.* **18:**1526-1538
20. Parkinson, J., Lees-Miller, S.P., and Everett, R.D. 1999. Herpes simplex virus type 1 immediate-early protein vmw110 induces the proteasome-dependent degradation of the catalytic subunit of DNA-dependent protein kinase. *J. Virol.* **73:**650-657
21. Lomonte, P, Sullivan, K.F., and Everett, R.D. 2001. Degradation of nucleosome-associated centromeric histone H3-like protein CENP-A induced by herpes simplex virus type 1 protein ICP0. *J. Biol. Chem.* **276:**5829-5835
22. Heichman, K.A. and Roberts, J.M. 1994. Rules to replicate by. *Cell* **79:**557-562
23. Kawaguchi, Y., Van Sant, C., and Roizman, B. 1997. Herpes simplex virus 1 a regulatory protein ICP0 interacts with and stabilizes the cell cycle regulator cyclin D3. *J. Virol.* **71:**7328-7336
24. Van Sant, C., Lopez P., Advani, S.J., and Roizman, B. 2001. Role of cyclin D3 in the biology of herpes simplex virus ICP0. *J. Virol.* **75:**1888-1898
25. Hagglund, R., and Roizman, B. 2003. Herpes simplex virus mutant in which the ICP0 HUL-1 E3 ubiquitin ligase site is disrupted stabilizes cdc34 but degrades D-type cyclins and exhibits diminished neurotoxicity. *J. Virol.* **77:**13194-13202
26. Advani, S.J., Weichselbaum, R.R., and Roizman, B. 2000. E2F proteins are posttranslationally modified concomitantly with a reduction in nuclear binding activity in cells infected with herpes simplex virus 1. *J. Virol.* **74:**7842-7850
27. Ehmann, G.L., McLean, T.I., and Bachenheimer, S.L. 2000. Herpes implex virus type 1 infection imposes a G1/S block in asynchronously growing cells and prevents G1 entry in quiescent cells. *Virology* **267:**335-349
28. Nurse, P. 1994. Ordering S phase and M phase in the cell cycle. Cell 79:547-550.
29. Tetsu, O. and McCormick, F. 2003. Proliferation of cancer cells despite cdk2 inhibition. *Cancer Cell* **3:**233-245

30. Poon, A.P.W, Ogle, W.O. and Roizman, B. 2000. The posttranslational processing of infected cell protein 22 mediated by viral protein kinases is sensitive to amino acid substitutions at distant sites and can be cell-type specific. *J. Virol.* **74:**11210-11214
31. Purves, F,C., Ogle, W.O., and Roizman, B. 1993. Processing of the herpes simplex virus regulatory protein α22 mediated by the UL13 protein kinase determines the accumulation of a subset of alpha and gamma mRNAs and proteins in infected cells. *Proc. Natl. Acad. Sci. USA* **90:**6701-6705
32. Sears, A.E., Halliburton ,I.W., Meignier, B., Silver, S., and Roizman. 1985. Herpes simplex virus 1 mutant defective in the α22 gene: growth and gene expression in permissive and restrictive cells and establishment of latency in mice. *J. Virol.* **55:**338-356
33. Purves, F.C. and Roizman, B. 1992. The U_L13 gene of herpes simplex virus 1 encodes the function for posttranslational processing associated with the phosphorylation of the regulatory protein ?22. *Proc. Natl. Acad. Sci. USA* **89:**7310-7314
34. Advani, S.J., Brandimarti, R., Weichselbaum, R.R., and Roizman, B. 2000. The disappearance of cyclins A and B and the increase of the G_2/M-phase cellular kinase cdc2 in herpes simplex virus 1-infected cells requires the expression of the α22/U_S1.5 and U_L13 viral genes. *J. Virol.* **74:**8-15
35. Advani, S.J., Weichselbaum, R.R., and Roizman, B. 2000. The role of cdc2 in the expression of herpes simplex virus genes. *Proc. Natl. Acad. Sci. USA* **97:**10996-11001
36. Advani, S.J, Weichselbaum, R.R., and Roizman, B. 2001. cdc2 cyclin-dependent kinase binds and phosphorylates herpes simplex virus 1 U_L42 DNA synthesis processivity factor. *J. Virol.* **75:**10326-10333
37. Advani, S.J., Weichselbaum, R.R., and Roizman, B. 2003. Herpes simplex virus 1 activates cdc2 to recruit topoisomerase IIα for post-DNA synthesis expression of late genes. *Proc. Natl. Acad. Sci. USA* **100:**4825-4830
38. Ebert, S.N., Subramanian, D., Shtrom, S.S., Chung, I.K., Parris, D.S., and Muller, M.T. 1994. Association between the p170 form of human topoisomerase II and progeny viral DNA in cells infected with herpes simplex virus type 1. *J. Virol.* **68:**1010-1020
39. Kozak, M. and Roizman, B. RNA synthesis in cells infected with herpes simplex virus. IX. Evidence for accumulation of abundant symmetric transcripts in nuclei. *J. Virol.* **15:**36-40, 1975
40. Jacquemont, B. and Roizman, B. Ribonucleic acid synthesis in cells infected with herpes simplex virus. X. Properties of viral symmetric transcripts and double-stranded RNA prepared from them. *J. Virol.* **15:**707-713, 1975
41. Schneider RJ and Mohr I. 2003. Translation initiation and viral tricks. *Trends in Biochemical Sciences* **28:**130-136
42. Zamanian-Daryoush, M., Mogensen, T.H., DiDonato, J.A., and Williams, B.R.G. NF-kappaB activation by double-stranded-RNA-activated protein kinase (PKR) is mediated through NF-kappaB-inducing kinase and IkappaB kinase. *Mol. Cell. Biol.* **20:**1278-1290, 2000
43. Chou, J. and Roizman B. 1994. Herpes simplex virus 1 $\gamma_1$34.5 gene function, which blocks the host response to infection maps in the homologous domain of the genes expressed during growth arrest and DNA damage. *Proc. Natl. Acad. Sci. USA* **91:**5247-5251
44. Chou, J., Kern E.R., Whitley, R.J., and Roizman, B. 1990. Mapping of herpes simplex virus-1 neurovirulence to $\gamma_1$34.5, a gene nonessential for growth in cell culture. *Science* **250:**1262-1266
45. Chou, J. and Roizman, B. 1992. The $\gamma_1$34.5 gene of herpes simplex virus 1 precludes neuroblastoma cells from triggering total shutoff of protein synthesis characteristic of programmed cell death in neuronal cells. *Proc. Natl. Acad. Sci. USA* **89:**3266-3270
46. Chou, J, Chen, J.J., Gross, M., and Roizman, B. 1995. Association of a Mr 90,000 phosphoprotein with protein kinase PKR in cells exhibiting enhanced phosphorylation of translation initiation factor eIF-2α and premature shutoff of protein synthesis after

infection with $\gamma_1 34.5^-$ mutants of herpes simplex virus 1. *Proc. Natl. Acad. Sci. USA* **92:**10516-10520

47. He, B., Gross, M., and Roizman, B. 1997. The $\gamma_1 34.5$ protein of herpes simplex virus 1 complexes with protein phosphatase 1α to dephosphorylate the α subunit of the eukaryotic translation initiation factor 2 and prelude the shutoff of protein synthesis by double-stranded RNA-activated protein kinase. *Proc. Natl. Acad. Sci. USA* **94:**843-848
48. He, B., Gross, M., and Roizman, B. The $\gamma_1 34.5$ protein of herpes simplex virus 1 has the structural and functional attributes of a protein phosphatase 1 regulatory subunit and is present in a high molecular weight complex with the enzyme in infected cells. *J. Biol. Chem.* **273:**20,737-20,743, 1998
49. Mohr, I. and Gluzman, Y. 1996. A herpesvirus genetic element which affects translation in the absence of the viral GADD34 function. *EMBO J.* **15:**4759-4766
50. He, B., Chou, J., Brandimarti, R., Mohr, I., Gluzman, Y., and Roizman, B. 1997. Suppresion of the phenotype of $\gamma_1 34.5^-$ herpes simplex virus 1: failure of activated RNA-dependent protein kinase to shut off protein synthesis is associated with a deletion in the domain of the α47 gene. *J. Virol.* **71:**6049-6054
51. Cassady, K.A., Gross, M., and Roizman, B. 1998. The second-site mutation in the herpes simplex virus recombinants lacking the $\gamma_1 34.5$ gene precludes shutoff of protein synthesis by blocking the phosphorylation of eIF-2α. *J. Virol.* **72:**7005-7011
52. Cassady, K.A., Gross, M., and Roizman, B. 1998. The herpes simplex virus US11 protein effectively compensates for the $\gamma_1 34.5$ gene if present before activation of protein kinase R by precluding its phosphorylation and that of the α subunit of eukaryotic translation initiation factor 2. *J. Virol.* **72:**8620-8626
53. Poppers, J., Mulvey, M., Khoo, D., and Mohr, I. 2000. Inhibition of PKR activation by the proline-rich RNA binding domain of the herpes simplex virus type 1 $U_S 11$ protein. *J. Virol.* **74:**11215-11221
54. Cassady, K.A., Gross, M., G. Gillespie, G. Y., and Roizman, B. 2002. Second-site mutation outside of the US10-12 domain of Δγ134.5 herpes simplex virus 1 recombinant blocks the shutoff of protein synthesis induced by activated protein kinase R and partially restores neurovirulence. *J. Virol.* **76:**942-949
55. Taddeo, B., Luo T.R., Zhang, W., and Roizman, B. 2003. The activation of NF-κB in cells productively infected with herpes simplex virus 1 is dependent on activated protein kinase R and plays no apparent role in blocking apoptosis. *Proc. Nat. Acad. Sci. USA* **100:**12408-12413
56. Taddeo, B., Zhang, W., Laakemnan, F., and Roizman, B. Cells lacking NF-kB or in which NF-kB is not activated vary with respect to ability to sustain herpes simplex virus 1 replication and are not susceptible to apoptosis induced by a replication incompetent virus. *J. Virol.* in press, 2004
57. Leopardi, R. and Roizman, B. 1996. The herpes simplex virus major regulatory protein ICP4 blocks apoptosis induced by the virus or by hyperthermia. *Proc. Natl. Acad. Sci. USA* **93:**9583-9587
58. Galvan, V. and Roizman, B. 1998. Herpes simplex virus 1 induces and blocks apoptosis at multiple steps during infection and protects cells from exogenous inducers in a cell-type-dependent manner. *Proc. Natl. Acad. Sci. USA* **95:**3931-3936
59. Leopardi, R., Van Sant, C., and Roizman, B. 1997. The herpes simplex virus protein kinase US3 is required for protection from apoptosis induced by the virus. *Proc. Natl. Acad. Sci. USA* **94:**7891-7896
60. Aubert, M. and Blaho, J. 1999. The herpes simplex virus type 1 regulatory protein ICP27 is required for the prevention of apoptosis of infected human cells. *J. Virol.* **73:**2803-2813
61. Galvan, V., Brandimarti, R., Munger, J, and Roizman, B. 2000. Bcl-2 blocks a caspase-dependent pathway of apoptosis activated by herpes simplex virus 1 infection in HEp-2 cells. *J. Virol.* **74:**1931-1938

62. Zhou, G., Galvan, V., Campadelli-Fiume, G., and Roizman, B. 2000. Glycoprotein D or J delivered in trans blocks apoptosis in SK-N-SH cells induced by a herpes simplex virus 1 mutant lacking intact genes expressing both proteins. *J. Virol.* **74:**11782-11791
63. Jerome, K.R., Chen, Z., Lang, R., Torres, M.R., Hofmeister, J., Smith, S., Fox, R., Froelich, C.J., and Corey, L. 2001. HSV-1 and glycoprotein J inhibit caspase activation and apoptosis induced by granzyme B or Fas. *J. Immunol.* **167:**3928-3935
64. Munger, J. and Roizman, B. 2001. The US3 protein kinase of herpes simplex virus 1 mediates the posttranslational modification of BAD and prevents BAD-induced programmed cell death in the absence of other viral proteins. *Proc. Natl. Acad. Sci. USA* **98:**10410-10415
65. Ogg, P.D., McDonell, P.J., Ryckman, B.J., Knudson, C.M., and Roller RJ. 2004. The HSV-1 US3 protein kinase is sufficient to block apoptosis induced by overexpression of a variety of Bcl-2 family members. *Virology* **319:**212-224
66. Benetti, L., and Roizman, B. 2004. The herpes simplex virus protein kinase US3 activates and functionally overlaps the protein kinase S to block apoptosis. *Proc. Nat. Acad. Sci USA* **101:**9411-9416
67. Zhou, G, Avitabile, E., Campadelli-Fiume, G., and Roizman B. 2003. The domains of glycoprotein D required to block apoptosis induced by herpes simplex virus 1 are largely distinct from those involved in cell-cell fusion and binding to nectin1. *J. Virol.* **77:**3759-3767
68. Zhou, G. and Roizman, B. 2002. Cation-independent mannose 6-phosphate receptor blocks apoptosis induced by herpes simplex virus 1 mutants lacking glycoprotein D and is likely the target of antiapoptotic activity of the glycoprotein. *J. Virol.* **76:**6197-6204
69. Perng, G.C., Jones, C, Ciacci-Zanella, J., Stone, M., Henderson, G., Yukht, A., Slanina, S,M., Hofman, F.M., Ghiasi, H., Nesburn, A.B., and Wechsler, S.L. 2000. Virus-induced neuronal apoptosis blocked by the herpes simplex virus latency-associated transcript. *Science* **287:**1500-1503
70. Ahmed, M., Lock, M., Miller, C.G., and Fraser, N.W. 2002. Regions of the herpes simplex virus type 1 latency-associated transcript that protect cells from apoptosis in vitro and protect neuronal cells in vivo. *J. Virol.* **76:**717-729
71. Inman M, Perng, G.C., Henderson, G., Ghiasi, H., Nesburn, A.B., Wechsler, S.L., and Jones C. 2001. Region of herpes simplex virus type 1 latency-associated transcript sufficient for wild-type spontaneous reactivation promotes cell survival in tissue culture. *J. Virol.* **75:**3636-3646
72. Peng, W., Henderson, G., Perng, G.C., Nesburn, A.B., Wechsler, S.L., and Jones, C. 2003. The gene that encodes the herpes simplex virus type 1 latency-associated transcript influences the accumulation of transcripts (Bcl-xL and Bcl-xS) that encode apoptotic regulatory proteins. *J. Virol.* **77:**10714-10718
73. Ahmed, M. and Fraser, N.W. 2001. Herpes simplex virus type 1 2-kilobase latency-associated transcript intron associates with ribosomal proteins and splicing factors. *J. Virol.* **75:**12070-12080
74. Goodkin, M.L., Ting, A.T., and Blaho, J.A. NF-kB is required for apoptosis prevention during herpes simplex virus type 1 infection. *J. Virol.* **77:**7261-7280, 2003
75. Marianneau, P., Cordona, A., Edelman, L., Deubel, V., and Despres, P. Dengue virus replication in human hepatoma cells activates NF-kB which in turn induces apoptotic cell death . *J. Virol.* **71:**3244-3249, 1997
76. Lin, K-I., Lee, S-H., Narayanan, R., Baraban, J.M., Hardwick, J.M., and Ratan, R.R. Thiol agents and Bcl-2 identify an alphavirus-induced apoptotic pathway that requires activation of the transcription factor NF-kappa B. *J. Cell Biol.* **131:**1149-1161, 1995
77. Posavad, C.M. and Rosenthal, K.L. 1992. Herpes simplex virus-infected human fibroblasts are resistant and inhibit cytotoxic T lymphocytic activity. *J. Virol.* **66:**6264-6272

78. Koelle, D.M., Tigges, M.A., Burke, R.L., Symington, F.W., Riddell, S.R., Abbo, H., and Corey, L. 1993. Herpes simplex virus infection of human fibroblasts and keratinocyes inhibits recognition by cloned CD8+ cytotoxic T lymphocyes. *J. Clin Invest.* **91**:961-968
79. Hill, A., Barnett, B.C., McMichael, A.J., and McGeoch, D.J. 1994. HLA Class I molecules are not transported to the cell surface in cells infected with herpes simplex virus types 1 and 2. *J. Immunol.* **152**:2736-2741
80. York, I.A., Roop, C., Andrews, D.W., Riddell, S.R., Graham, F.L, and Johnson, D.C. 1994. A cytosolic herpes simplex virus protein inhibits antigen presentation to CD8+ T lymphocytes. *Cell* **77**:525-535
81. Hill, A., Jugovic, P., York, I., Russ, G., Bennink, J., Yewdell, J., Ploegh, H., and Johnson, D. 1995. Herpes simplex virus turns off the TAP to evade host immunity. *Nature* **375**:411-415
82. Fruh, K., Ahn, K., Djaballah, H., Sempe, P., van Endert, P.M., Tampe, R., Peterson, P.A., and Yang, Y. 1995. A viral inhibitor of peptide transporters for antigen presentation. *Nature* **375**:415-418
83. Tomazin, R, Hill, A.B., Jugovic, P, York, I., van Endert, P., Ploegh, H.L., Andrews, D.W., and Johnson, D.C. 1996. Stable binding of the herpes simplex virus ICP47 protein to the peptide binding site of TAP. *EMBO J.* **15**:3256-3266
84. Ahn, K, Meyer, T.H., Uebel, S., Sempe, P., Djaballah, H., Yang, Y., Peterson, P.A., Fruh, K., and Tampe, R. 1996. Molecular mechanism and species specificity of TAP inhibition by herpes simplex virus protein ICP47. *EMBO J.* **15**:3247-3255
85. Barcy, S. and Corey, L. 2001. Herpes simplex inhibits the capacity of lymphoblastoid B cell lines to stimulate CD4+ T cells. *J. Immunol.* **166**:6242-6249
86. Trgovcich, J., Johnson, D., and Roizman, B. 2002. Cell surface histocompatability complex class II proteins are regulated by the products of the g134.5 and UL41 genes of herpes simplex virus 1. *J. Virol.* **76**:6974-6986
87. Sievers, E., Neumann, J., Raftery, M., Sconrich, G., Eis-Hubinger, A.M, and Koch, N. 2002. Glyocprotein B from strain 17 of herpes simplex virus type I contains an invariant chain homologous sequence that binds to MHC class II molecules. *Immunology* **107**:129-135
88. Neumann, J, Eis-Hubinger, A.M., and Koch, N. 2003. Herpes simplex virus type I targets the MHC class II processing pathway for immune evasion. *J. Immunol.* **171**:3075-3083

Chapter 8

IMMUNOMODULATION BY POXVIRUSES

Insights into virus-host interactions by selectively deleting poxvirus genes

JAMES B. JOHNSTON[*] and GRANT McFADDEN[*,#]
[*]*Biotherapeutics Research Group, Robarts Research Institute, London, ON, CAN;* [#]*Department of Microbiology and Immunology, University of Western Ontario, London, ON, CAN*

1. INTRODUCTION

The *Poxviridae* encompass a nearly ubiquitous family of DNA viruses capable of infecting a broad spectrum of vertebrates (*Chordopoxvirinae*) and insects (*Entomopoxvirinae*)[1]. The Chordopoxviruses in particular include several viruses of economic and social importance to humans, and thus are the most extensively studied and best characterized (Table 1). Poxviruses are notable among DNA viruses for their large virion size and the ability to replicate within the cytoplasm of infected cells autonomous of the host nuclear machinery. Poxviruses also possess one of the largest viral genomes, ranging in size from 135 kb to 290 kb and encoding as many as 260 open reading frames (ORFs). The poxvirus genome consists of linear double-stranded DNA characterized by termini that form covalently closed hairpin loops and flanking terminal inverted repeat (TIR) regions that contain varying numbers of genes whose positions and orientations are mirrored at the opposing ends of the genome (Figure 1)[1]. In general, genes that are centrally located in the genome are conserved among poxviruses and have common essential molecular functions, such as replication and virion assembly. Genes located closer to the termini tend to be more variable and are considered non-essential because they are commonly dispensable for

P. Palese (ed.), Modulation of Host Gene Expression and Innate Immunity by Viruses, 163-195.

replication in culture[1]. However, increasing interest has been engendered by the products of these terminally located genes, which include a diverse array of proteins that function in host-range restriction and modulation or inhibition of the host responses to infection.

Table 1. The *Chordopoxvirinae* genera and representative species.

Genus	Species	Reservoir Host	Abbreviation
Avipoxvirus	Fowlpox virus	Birds	FPV
Capripoxvirus	Sheep pox virus	Sheep	ShPV
	Lumpy skin disease virus	Buffalo	LSDV
Leporipoxvirus	Myxoma virus	Rabbit	MV
	Shope fibroma virus	Rabbit	SFV
	Malignant fibroma virus	Rabbit	MRV
Molluscipoxvirus	Molluscum contagiosum virus	Human	MCV
Orthopoxvirus	Variola virus	Human	VaV
	Vaccinia virus	Unknown	VV
	Cowpox virus	Rodent	CPV
	Ectromelia virus	Rodent	EV
	Monkeypox virus	Rodent	MPV
	Rabbitpox virus	Unknown	RPV
Parapoxvirus	Orf virus	Sheep	ORFV
	Pseudocowpox virus	Cattle	PCPV
Suipoxvirus	Swinepox virus	Swine	SPV
Yatapoxvirus	Yaba monkey tumor virus	Primate	YMTV
	Tanapox	Primate	TPV

The immunomodulatory strategies of viruses are highly dependent on their inherent coding capacity. Viruses with smaller genomes and limited capacity to encode nonessential proteins, such as human immunodeficiency virus, survive against the pressures of the host immune response through such mechanisms as targeting and destroying immune cells or replicating and mutating at a rate that exceeds the ability of the host to compensate[2]. In contrast, larger DNA viruses like poxviruses can encode a range of accessory proteins with the sole function of evading the immune response. The sequences of over two dozen poxvirus genomes have been determined and immunomodulatory proteins targeting all facets of innate and adaptive immune responses have been reported[3-5]. In fact, so diverse are these genes that a single immunomodulatory protein that is common to all poxviruses has yet to be identified. The obvious sequence similarity between some of these genes and the cDNA versions of related cellular counterparts often provides insight into their function. However, the evolutionary origins of other poxvirus ORFs are more obscure, and many immunomodulatory proteins have no known cellular homolog or have putative functions that cannot be predicted based on sequence similarity. Thus, genetic analyses employing recombinant DNA technologies are essential tools for studying the contribution of specific viral genes and proteins to virus-host interactions and viral pathogenesis.

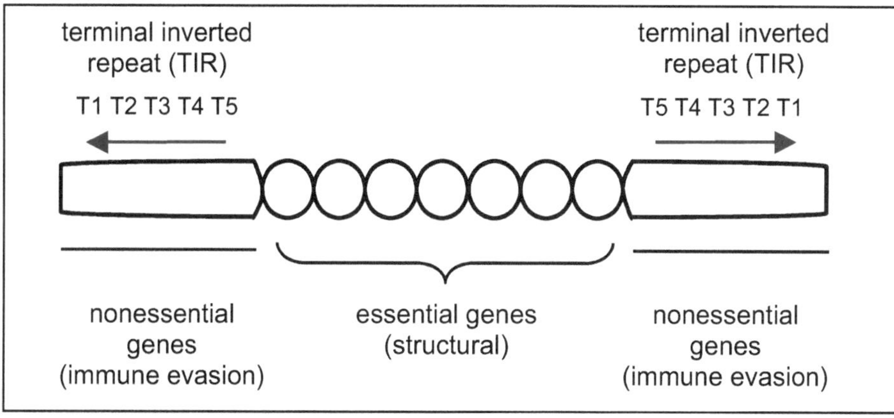

Figure 1. Structural organization of a consensus poxvirus genome.

2. SELECTIVELY DELETING POXVIRUS GENES

2.1 Manipulating the poxvirus genome

Much of what is known about poxviral pathogenesis can be traced back to discoveries made with knock-out (KO) viruses, mutant viruses in which the targeted disruption of a specific viral gene produces phenotypic changes reflective of the normal biological function of its protein product[6]. Advances in molecular cloning techniques continue to dramatically broaden the scope of genetic engineering that may be achieved; however, manipulating the poxvirus genome presents certain challenges unique among DNA viruses. During the course of a normal infection, poxvirus ORFs are transcribed by viral enzymes in the cytoplasm and never encounter the nuclear splicing machinery[1]. In contrast, transfected poxvirus genomes or DNA-based plasmid constructs containing poxvirus ORFs are transported to the nucleus and inadvertent activation of cryptic splice sites can result in the expression of aberrant products. As a result, the genomes of poxviruses cannot initiate a replicative cycle when transfected alone into a permissive cell line and some poxviral genes cannot be successfully expressed from mammalian expression plasmids[1]. The shear size and complexity of the poxvirus genome is also a technical disadvantage. Unlike smaller DNA viruses, poxvirus genomes far exceed the coding capacity of conventional cloning plasmids, while their unique structural features and complex replication scheme preclude conventional cellular mechanisms of DNA synthesis and gene expression[1].

Given these hurdles, KO virus strategies are particularly well suited as experimental tools in the study of immune modulation by poxviruses. Because poxvirus immunomodulators are seldom essential for replication in cultured cells, deletion of these genes produces a virus that is generally viable and can be propagated in permissive cells. Many poxviral immunomodulatory proteins also have related cellular counterparts, whose function and mechanism of action are well characterized, thereby providing a yardstick against which to compare phenotypic changes associated with a KO virus. The use of KO viruses is not without its disadvantages, however. The redundancy inherent to poxvirus immune evasion strategies means that the effect of a deletion may be masked by the activities of intact immunomodulatory proteins with complementary functions. Similarly, it is more difficult to determine the underlying cause of a particular phenotype when the deleted viral gene encodes a multi-functional immunomodulator. As discussed below, analysis of the effect of deleting immunomodulatory genes also requires a suitable animal model. Thus, poxviruses like MCV that lack either an animal model or a tissue culture system cannot be effectively studied using KO strategies[7].

2.2 Constructing knock-out poxviruses: Current and future strategies

The typical method for generating KO mutants of poxviruses employs a step-wise approach reminiscent of early marker-transfer mutagenesis strategies[8]. Of interest, this process closely resembles the ancestral retrotranscription/recombination events by which many immunomodulatory proteins were likely originally acquired by poxviruses from their vertebrate hosts. Several variations of this protocol have been reported, the most common of which is shown in Figure 2. In this strategy, a segment of the poxvirus genome encoding the gene to be deleted and flanking poxvirus DNA is first cloned into a plasmid transfer vector. The gene of interest is then inactivated by deleting a central segment of the ORF and inserting a selectable marker under the control of a poxviral promoter. Deletion of these nucleotides can be achieved using unique restriction endonuclease sites contained within the gene, but more commonly requires that novel restriction sites be introduced by PCR mutagenesis. In theory, gene expression can also be interrupted using PCR to introduce stop codons or frameshift mutations within the ORF in question. A wide variety of selection

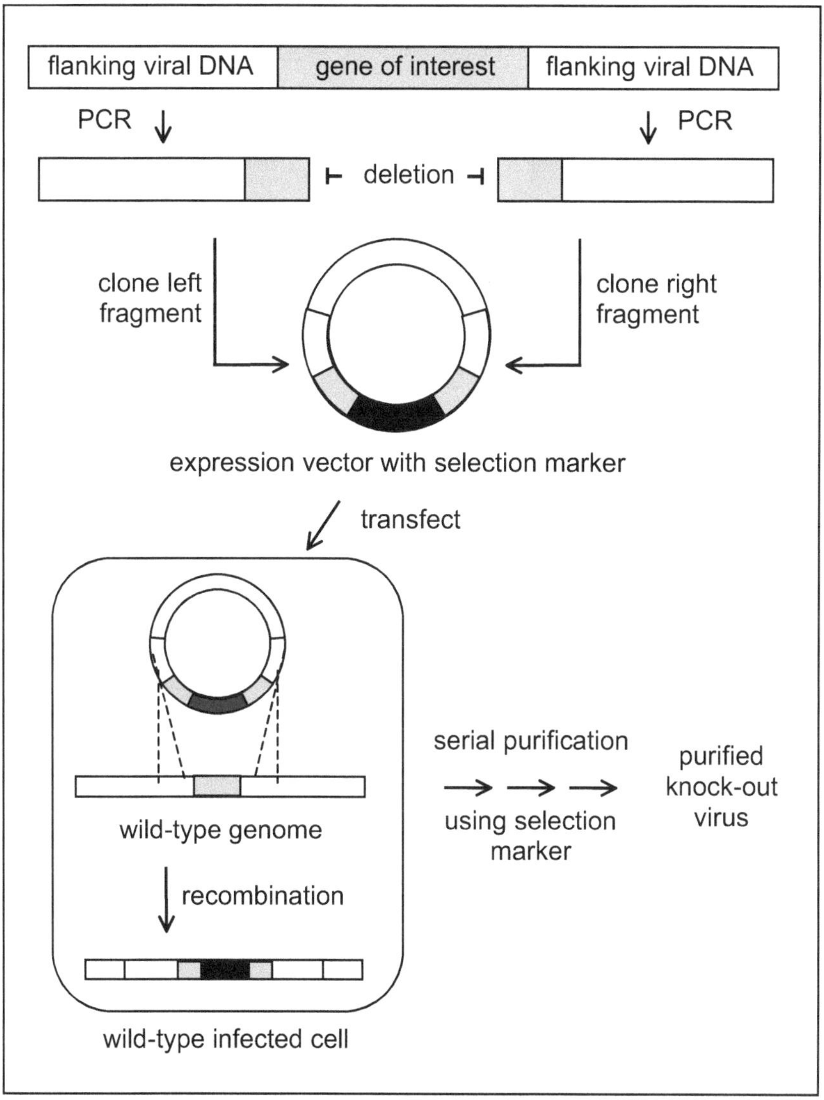

Figure 2. General strategy for constructing knock-out poxviruses.

markers are also routinely used, including sequences encoding dominant-selectable chemical resistance or sequences that encode products that can be used to visually distinguish mutant viruses, such as green fluorescence protein. The vector containing the disrupted gene is then transfected into a permissive cell line that is subsequently infected at a low multiplicity of infection with the wild-type strain of that poxvirus. Within infected cells, two-site homologous recombination between the wild-type genome and the mutated sequence encoded by the vector produces viral genomes containing the desired mutation. KO virus is subsequently isolated from wild-type by multiple rounds of serial purification using the selection criteria determined by the inserted marker gene.

The need for more sophisticated strategies to functionally analyze larger viral genomes has stimulated several technological advances, most notably cloning vehicles that accept larger inserts such as bacterial artificial chromosomes (BACs). BACs can accommodate viral genomes up to 300 kb in size, exhibit a high degree of fidelity and stability even at low copy number, and can be manipulated in bacteria and then reconstituted as infectious virus by transfection into eukaryotic cells[9]. Despite successes with other large DNA viruses, namely baculovirus[10] and herpesviruses[11], the inherent difficulties associated with the hairpin termini of poxvirus genomes made exploitation of the BAC system more challenging. Recently, Domi and Moss reported the construction of a stable VV BAC that was based on circularization of head-to-tail concatemers of VV DNA[12]. Moreover, expression of the construct in the presence of helper poxvirus resulted in the rescue of infectious virus. Although the applicability of this strategy to other poxviruses remains to be shown, BAC technology has the potential to greatly simplify the manipulation of poxvirus genomes and the production of mutant viruses.

2.3 Animal models of poxvirus pathogenesis

The focus of this chapter is the use of KO viruses to study the role of virus-encoded immune modulators in poxvirus pathogenesis. With that intent, we necessarily concentrate on how deletion or mutation of a given gene influences poxvirus virulence in a relevant animal model. Investigation of poxvirus immunomodulatory proteins using KO viruses has largely employed murine models in the study of orthopoxviruses, such as VV, and rabbit models in the study of leporipoxviruses, such as MV (Table 2). VV exhibits a broad host range that includes several mammalian species, but the natural reservoir for the virus is unknown[1]. The route of inoculation also significantly impacts upon disease course and VV infection is achieved under experimental conditions using several different routes. In contrast, MV is an obligate rabbit pathogen that is benign in its natural host species, the S. American *Sylvilagus* rabbit, but causes a lethal disseminated disease (myxomatosis) in European *Oryctolagus* rabbits[13]. MV is transmitted under natural conditions by arthropod vectors and intradermal injection is the most common mechanism for introducing virus. Despite these differences, both

VV and MV cause generalized disseminated infections characterized by the formation of a primary lesion at the initial site of inoculation and a viremia that spreads the infection through the host lymphoreticular system to establish internal and external lesions in secondary organs and tissues. The immunosuppressive capacities of MV in particular, manifests in the development of supervening bacterial infections that ultimately lead to the death of host. Thus, targeted disruption of poxviral immunomodulatory genes can impact greatly on disease progression in these models.

Table 2. Animal models for the study of poxvirus pathogenesis.

Model Species	Poxviruses Studied	Inoculation Routes	Advantages	Disadvantages
mouse	VV EV CPV	Nasal Dermal Cerebral Peritoneal	Low cost Genetically defined Specific reagents Low input titer Rapid disease course	Variable outcomes Not reservoir host
rabbit	MV SFV MRV	Dermal	Reservoir host Low input titer Rapid disease course Defined pathology	Higher cost Few reagents Not genetically defined

2.4 A caveat

Although a useful tool, the information obtained from KO analyses alone is not sufficient to definitively assign a particular function to a viral gene product. Extensive research into the biological activity of purified proteins in relevant *in vitro* assays is necessary to confirm the function predicted by bioinformatic analyses. Similarly, more advanced biochemical studies are required to provide insight into the nature of the interaction between poxviral immunomodulatory proteins and their targets, such as species-specificity, interaction kinetics and critical residues within binding domains. Several recent reviews have provided in depth accounts of the advances made using these techniques for the study of poxviruses and the reader is referred to them for more information[3,5,14-19].

3. OVERVIEW OF POXVIRUS IMMUNOMODULATORY STRATEGIES

The coevolution of poxviruses and their hosts is reflected not only in the complexity and versatility of the mammalian immune system, but also in the diversity and efficacy of the strategies employed by poxviruses to overcome those host antiviral responses. Poxvirus immunomodulatory proteins can be divided by function into three strategic classes: virostealth proteins, virotransducers and viromimetics (virokines and viroceptors) (Figure 3)[13]. Virostealth encompasses a general strategy in which the visible signals of infection are masked in order to reduce the ability of cell-mediated immune responses to recognize and eliminate infected cells. Virotransducers are viral proteins that act intracellularly to inhibit innate antiviral pathways and the signal transduction cascades that mediate host range. Virokines and viroceptors represent virus-encoded proteins that mimic host cytokines or their receptors, respectively, thereby blocking extracellular communication signals and promoting a protected microenvironment for the virus within immuno-exposed tissues. An overview of the most common poxvirus immunomodulatory strategies and representative viral proteins is provided in Table 3. Specific examples of these strategies emphasizing gene knock-out analysis are explored in greater detail below.

4. VIROSTEALTH - PLAYING HIDE AND SEEK

Innate and educated cytolytic cells, such as cytotoxic T-lymphocytes (CTL) and natural killer (NK) cells, are critical for the rapid identification and clearance of virus-infected cells[20]. To evade this component of the acquired immune system, viruses attempt to downregulate recognition receptors and shift the balance of stimulatory signals available to immune effector cells below the threshold required to initiate an antiviral response[21, 22]. Two common targets of viruses are CD4, a co-receptor in exogenous antigen-induced T-cell activation, and the major histocompatibility complex (MHC), the principal antigen presenting receptors. The impact of regulating the expression of these receptors is exemplified in the pathogenesis of MV and VV. Infection with MV causes severe systemic immunosuppression in rabbits that correlates with the rapid downregulation of cell surface MHC I[23] and CD4 [24]. In contrast, VV only moderately influences MHC I levels and fails to cause similar immune suppression upon infection[23].

The capacity to down-regulate both MHC I and CD4 expression maps to the product of the MV M153R gene, a protein containing a LAP (leukemia-associated protein) domain found in the other viral proteins that influence MHC levels, such as the K3 and K5 proteins of human herpesvirus (HHV)-8 [25-27]. Through interactions mediated by this domain, M153R localizes to the endoplasmic reticulum (ER) where it targets β2-microglobulin-asociated MHC I molecules for retention and degradation via the late endolysosomal/lysosomal pathway[26, 27]. The importance of M153R in MV pathogenesis is demonstrated by the finding that *in vitro* cells infected with a KO virus lacking this gene do not exhibit the MHC I loss associated with wild-type infection and are more susceptible to CTL-mediated cytolysis[23, 27]. Similarly, M153R KO virus is attenuated *in vivo*, causing an infection in susceptible rabbits characterized by increased mononuclear infiltration at the primary site of infection and rapid clearance of virus-infected cells[27]. Genes predicted to encode LAP proteins have been identified in several other poxviruses (SFV, SPV, YLDV and LSDV), but the function of their products has not been confirmed[28].

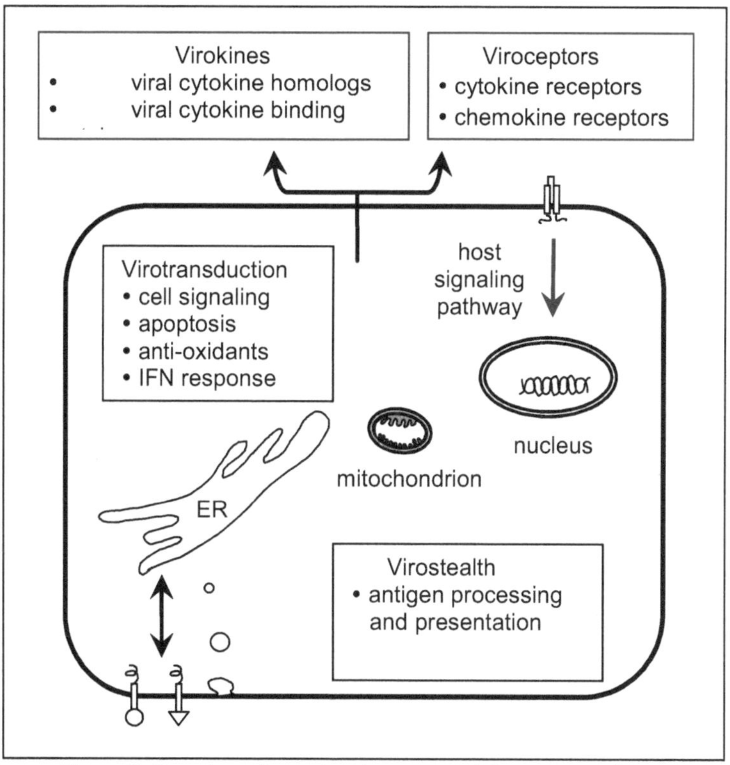

Figure 3. Overview of poxvirus immunomodulatory strategies.

5. VIROTRANSDUCTION - A FAILURE TO COMMUNICATE

5.1 Inhibition of apoptosis

In order to disconnect antiviral signals from their biological effects and create an environment in infected cells conducive to the production of progeny virions, poxviruses target the function of critical elements within the host cell signal transduction machinery. Many of these strategies converge on the concerted activities of diverse apoptotic pathways, thereby preventing the elimination of infected cells and host-induced shutdown of protein synthesis[29]. The extent to which the coding capacity of the poxvirus genome is dedicated to regulating these intracellular events emphasizes the important role in counteracting virus infections attributable to host apoptotic responses.

5.1.1 Modulation of caspase activity

Caspases are pro-apoptotic proteases that represent key regulatory elements in the apoptotic cascade[30]. Following infection, poxviruses rapidly express proteins that modulate caspase function by blocking their activation or acting as suicide substrates. The orthopoxvirus cytokine response modifier A (CrmA)/SPI-2, is a versatile serine protease inhibitor (serpin) that targets both intrinsic and extrinsic apoptotic pathways, as well as host inflammatory responses[31]. As the prototypical poxvirus caspase suicide substrate, CrmA inhibits the activity of caspases-8 and -10 to block apoptotic events initiated by stress- and cytokine-activated signaling pathways, and caspase-1 activity to suppress processing of pro-inflammatory cytokines[32]. Through inhibition of the serine protease, granzyme B, CrmA also protects cells from perforin-dependent apoptosis induced by CTLs and NK cells[33]. SPI-2 orthologs have also been shown to inhibit apoptosis induced by Fas ligand or TNF-α, but these proteins are generally less effective than CrmA[34]. Despite these functions, deletion of CrmA from CPV[35] or SPI-2 from VV[36] only moderately influences virulence in intranasal murine models and has little impact on host inflammatory responses to infection.

MV SERP-2 is a poxviral serpin that belongs to the same functional class as CrmA/SPI-2, a property that is reflected in its ability to weakly inhibit caspase-1 to modulate inflammation and granzyme B to protect lymphoid cells from apoptosis[37,38]. In contrast to CrmA/SPI-2, however, SERP-2 is an important virulence factor in MV pathogenesis. Loss of SERP-2 function markedly decreases the mortality associated with MV infection of susceptible rabbits, concurrent with a more rapid acute inflammatory response and the absence of secondary lesion formation.

The products of the MCV MC159L and MC160L genes are the prototypical poxvirus-encoded inhibitors of caspase activation in response to pro-apoptotic stimuli[39]. Both MC159L and 160L are classified as viral FLICE/caspase-8 inhibitory molecules (vFLIPS), homologs of cellular death effector domain (DED)-containing proteins that bind both caspase-8 and the Fas-associated death domain (FADD) adaptor molecule and prevent

Table 3. Representative poxviral immunomodulatory proteins.

Class	Strategy	Function/Target	Poxviral Protein
Virostealth	MHC homolog	Class I MHC	MCV033L
	Downregulation	Class I MHC	MV M153R
		CD4/Fas	MV M153R
Virotransduction	IFN inhibition	dsRNA binding	VV E3L
		eIF2α homolog	VV K3L
	Caspase inhibition	apoptosis	CPV CrmA,VV SPI-2
	Cytokine signaling	inflammation	VV SPI-3
		viral FLIP	MCV159R
		toll-like receptors	VV A46R, A52R
		NF-κB	MCV159L, MV M150R
		STAT-1	VV H1L
	Apoptosis regulators	mitochondria	MV M11L
		ER	MV M-T4
		ankyrin proteins	MV M-T5
		oxidative stress	MCV066L
Viromimics			
Virokines	Growth factor homolog	EGF	MV MGF, VV VGF
		VEGF	ORFV vVEGF
	Semaphorin homolog	inflammation	VV A39R
	Cytokine homolog	IL-10	ORFV vIL-10
	Chemokine homolog	MIP-1β	MCV MC148R
	IL-18 binding	IL-18	MCV54L, EV p13
	Cytokine binding	TNF	TPV 2L
		GMCSF/IL-2	ORFV GIF
	Chemokine binding	C, CC, CXC	MV M-T7
		CC	MV M-T1, VV CCI
	Complement binding	C3 convertase	VV VCP, CPV IMP
	Serpins	serine proteases	MV Serp-1
Viroceptors	TNF-R homolog	TNF	MV M-T2, EV CD30 CPV CrmB/C/D/E
	IFN-R homolog	Type 1 IFN	VV B19R
		Type 2 IFN	VV B8R, MV M-T7
	IL-R homolog	IL-1β	VV B15R
	Chemokine receptor	CCR8	TPV 7L, 145R

recruitment of pro-caspase-8 to death receptor complexes at the cell membrane[40]. Recent evidence has shown that the DEDs of MC159L cannot be functionally interchanged with those from other proteins and that mutations within regions of MC159L, other than the DED motifs, abrogate its anti-apoptotic activity[41,42]. This suggests that MC159L may interact with cellular factors other than those within the death receptor complex to exert its effect. For example, MCV MC159L inhibits the activation of NF-κB, a critical intermediate of the signaling pathways of cytokines like TNF and IFN[43].

5.1.2 Other anti-apoptotic strategies

Poxviral proteins have also been shown to specifically target key regulatory factors at subcellular sites involved in the coordination of apoptotic signals, such as mitochondria and the ER. Of particular interest are the apoptotic regulators encoded by the MV M11L, M-T4 and M-T5 genes. The M11L product lacks sequence homology to known cellular proteins, yet it is targeted to mitochondria where it inhibits pro-apoptotic changes in mitochondrial integrity that include the loss of inner mitochondrial membrane potential (MMP)[44]. Recently, this activity was shown to involve the interaction of M11L with a component of the mitochondrial permeability transition (PT) pore, the peripheral benzodiazepine receptor (PBR), thereby preventing cytochrome C release[45]. In comparison, M-T4 is an ER-resident protein that may inhibit stress responses to infection[46], while M-T5 is a cytosolic anti-apoptotic ankyrin repeat protein homologous to host range proteins identified in other poxviruses[47]. Despite overt differences in cellular localization and the absence of common structural features, KO virus analyses involving these proteins have revealed strikingly similar profiles *in vitro* and *in vivo*. All three proteins are important virulence factors and determinants of host range, deletion of which is associated with the rapid induction of apoptosis following infection of cultured rabbit leukocytes. Consistent with this property, infection of susceptible rabbits with viruses lacking M11L, M-T4 or M-T5 induces a markedly elevated acute inflammatory response at sites of infection and is associated with the absence of secondary lesion formation[46,47]. These findings are consistent with poor virus dissemination due to enhanced host antiviral responses and the loss of the cellular vectors required for spread of the virus *in vivo*.

Poxviruses also encode proteins that target other elements of the apoptotic process, such as the MCV MC066 protein. MC066 possesses glutathione peroxidase activity, catalyzing the conversion to water of reactive oxygen species such as hydrogen peroxide[48]. In this manner, MC066 protects cells from apoptosis arising from the oxidative stress caused by ultra-violet (UV) irradiation or activated macrophages and neutrophils.

This capacity to block UV-induced apoptosis has also been demonstrated for other poxvirus proteins, including EV p28 and SFV N1R[49, 50].

5.2 Modulation of the interferon response

The activities of poxvirus anti-apoptotic proteins are closely intertwined with strategies that target intracellular elements in the interferon (IFN) response pathway. Consequently, all poxviruses have evolved strategies to directly inhibit these responses[5]. Two IFN-dependent enzymatic cascades, the double-stranded RNA (dsRNA)-dependent protein kinase R (PKR) and the 2',5'-oligoadenylate synthetase (OAS) pathways, largely mediate the antiviral events that promote cell cycle arrest and apoptosis following infection[51]. Both of these enzymes are activated by the dsRNA produced during poxviral transcription and initiate cascades that inhibit viral protein synthesis and often lead to apoptosis by activating caspase-8[52]. Thus, both MC159L and CrmA/SPI-2 may be considered inhibitors of IFN activity.

The VV E3L and K3L are the prototypical examples of poxvirus proteins that target these enzymes, but proteins with similar functions have been identified in numerous species, including MV (M029L, M156R), YLDV (34L, 12L), VaV (E3L, C3L), SFV (S029L, 008.2L/R), SPV (C8L), EV (E3L, nonfunctional K3L) and ORFV (ORF20L)[53-58]. Mechanistically, E3L is a dsRNA-binding protein that sequesters dsRNA and prevents activation of PKR and OAS[59], although it can also bind directly to PKR to inhibit its activity and prevent phosphorylation of diverse host proteins associated with cell-cycle arrest[60,61]. The K3L gene product, a homolog of the eukaryotic initiation factor (eIF)-2α subunit, functions as a nonphosphorylated pseudosubstrate of PKR to competitively inhibit eIF2α phosphorylation[62].

The importance of modulating the IFN pathway to successful virus replication is exemplified by the fact that both E3L and K3L have been identified as determinants of the broad host range exhibited by VV[63]. Deletion of either gene alters the cell tropism of VV in culture, a property believed to reflect inherent variability in the expression and activity of IFN response elements between cell types and across species. *In vivo*, loss of E3L or K3L function renders VV apathogenic in murine models, such that infection with KO viruses causes decreased morbidity, weight loss and lethality compared to wild-type VV[64].

5.3 Uncoupling cell signaling events

Less direct strategies to modulate host antiviral responses employed by poxviruses include viral proteins that disrupt the signaling events induced by cytokines. For example, poxviral genes whose products are aimed at the transcription factor nuclear factor (NF)-κB, include MCV MC159L, VV A46R and A52R, and MV M150R. In addition, the VV H1L gene encodes a phosphatase that blocks IFN-induced activation of the signal transducer and activator of transcription (STAT)-1[65]. Of note, MCV, which lacks E3L and K3L homologs, likely uses MC159L to inhibit IFN-mediated PKR-induced NF-κB activation[43]. The products of both the VV A46R and A52R genes contain Toll-like/IL-1 receptor (TIR) domains that enable these proteins to disrupt NF-κB signaling pathways that transduce the biological effects of several host toll-like receptors (TLR) and cytokines, including interleukin (IL)-1 and tumor necrosis factor (TNF)[66,67]. Moreover, deletion of A52R has been shown to decrease VV virulence in murine intranasal models, reducing both the weight loss and clinical symptoms associated with wild-type infections[66]. Loss of M150R function also impacts greatly on MV pathogenesis. Like M-T5, M150R is an ankyrin repeat protein that has been shown to co-localize with NF-κB in the nucleus of infected cells and inhibit pro-inflammatory cytokine responses[68]. Deletion of M150R renders MV apathogenic in rabbits, resulting in an elevated acute inflammatory response at primary sites of infection.

6. VIROMIMICRY - SEEING A FAMILIAR FACE

Poxviruses also encode diverse virokines and viroceptors that target the specific mechanisms by which the host coordinates and regulates early inflammatory responses, including complement control proteins, IFNs, pro-inflammatory cytokines, chemokines and growth factors. Viroceptors, which can be either secreted or localized to the surface of infected cells, are related to cellular receptors and act by competing for ligands that promote antiviral immune or inflammatory processes. In contrast, virokines are secreted viral proteins that mimic host molecules, such as cytokines, complement regulators, or their inhibitors[69].

6.1 Viroceptors

6.1.1 TNF viroceptors

To inhibit the potent pro-apoptotic and pro-inflammatory properties of TNF, many poxviruses encode soluble proteins that resemble secreted versions of the extracellular domains of cellular TNF receptors, termed vTNFRs[70]. The primary function of vTNFRs is that of molecular scavengers that bind to and sequester TNF, thereby blocking the interaction between the ligand and its native receptor. The T2-like vTNFRs found only in MV and

SFV are secreted glycoproteins that bind TNF with high affinity[71]. SFV T2 has been reported to bind both TNF-α and TNF-β from several species[72], but MV T2 (M-T2) exhibits specificity for rabbit TNF-α[73]. The orthopoxvirus Crm-like vTNFRs, of which four major classes have been identified (Crm B, C, D and E) vary widely according to the viral strain in both distribution and biological activity[70, 74]. CrmD is found primarily in poxviruses that lack CrmB and CrmC[75], while functional CrmE homologs have been identified only in CPV and select VV strains[74,76]. Members of a fifth Crm-like vTNFR family that closely resemble CD30 have also been identified in CPV and EV[74,77]. The EV CD30 homolog has been shown to block IFNγ production and inhibit host inflammatory responses[78].

Despite the importance of controlling TNF activity, many VV strains have discontinuous and nonfunctional Crm-like genes[79], and naturally arising MV strains deficient in M-T2 have been reported[80]. However, a role for vTNFRs in poxvirus pathogenesis has been demonstrated using KO viruses. *In vitro*, rabbit lymphocytes infected with MV, disrupted in both copies of the M-T2 gene, rapidly undergo apoptosis and abortive infection[81]. Consistent with poor virus spread in an immunocompetent host, M-T2 KO virus is markedly attenuated in susceptible rabbits and exhibits decreased lethality that is characterized by a pathology in which opportunistic bacterial infections are less frequent, primary lesions are smaller and less pronounced and secondary lesions are largely absent[82]. Because of virus strain-dependent variability in the expression of Crm-like vTNFRs, the role played by these proteins in pathogenesis is more commonly studied by expressing their ORFs in a background in which they are normally absent. For example, recombinant VV expressing CPV CrmB, CrmC or CrmE is more virulent in mice than wild-type virus, causing rapid weight loss and mortality[76]. However, the contribution of the two vTNFRs encoded by VV (strain USSR), CrmE and A53R, has been assessed in both intradermal and intranasal mouse models using KO viruses. Deletion of A53R does not impact on virulence following infection by either route[76], but loss of CrmE results in marked attenuation of the virus when delivered by the intranasal route[76]. Thus Crm-like vTNFRs likely contribute to pathogenesis, but in a manner that reflects the complex regulation of TNF in the host.

6.1.2 IFN viroceptors

All poxviruses employ at least one mechanism to disrupt IFN activity, underscoring the integral role this cytokine family plays in host antiviral responses. As with TNF, many poxviruses encode soluble viral mimics of both Type I (α/β) and Type II (γ) IFN receptors (IFN-R) to sequester these cytokines. For example, the B8R genes of both VV and EV encode proteins that bind IFN-γ from several species[83], although only the B8R homolog of EV inhibits murine IFN-γ despite the ability of both viruses to infect mice[58]. Both viruses also encode a protein (B18R) that closely resembles the IL-1 receptor but actually strongly interacts with Type I IFNs from several species[58,84]. Of note, the VV B18R product has been detected as both a secreted protein and localized to the surface of infected and uninfected cells[85], suggesting that it protects infected cells from the direct action of IFN-α/β and uninfected cells from IFN-induced resistance to infection. In terms of IFN regulation, the activity of the MV M-T7 IFN-R is specific to rabbit IFN-γ[86]. However, purified M-T7 protein also exhibits the surprising ability to bind to diverse families of human chemokines[87].

Deletion of both copies of MV M-T7 attenuates the virus and leads to elevated inflammatory responses in primary lesions[86]. Given the contribution of chemokines to inflammation, however, this finding must be viewed in the context of the capacity for M-T7 to bind both chemokines and IFN-γ. Deletion of the VV B8R gene has been reported to either enhance or not affect virulence in mice[88, 89], although B8R-deletants are attenuated in other rodent species such as rabbits[89]. The increased virulence reported in some mice infected with B8R KO virus is particularly surprising since the protein does not bind murine IFN, possibly indicating that like M-T7, this protein has additional activities[83]. The VV B18R gene product also exhibits low affinity for murine IFN-α/β *in vitro*, but deletion of this gene significantly attenuates the virus. B18R KO virus produces few disease symptoms and limited mortality following infection of mice and exhibits a lack of the neuroinvasive phenotype that characterizes VV[84].

6.1.3 IL-1β viroceptors

Several orthopoxviruses are predicted to express secreted IL-1β receptor homologs, but the pathogenic contribution of these viroceptors has only been determined for VV. The IL-1β receptor encoded by the VV B15R gene has been shown to bind to and block the activity of murine IL-1β in functional bioassays *in vitro*[90]. Although deletion of B15R influences VV pathogenesis in mice, this property varies greatly with the route of inoculation. Intracranial injection of VV deleted for B15R results in significant attenuation compared to wild-type virus[90], but delivery of B15R KO virus intranasally leads to the rapid emergence of clinical symptoms and lethality comparable to wild-type VV[91]. In fact, the febrile response to B15R-deleted virus delivered intranasally is enhanced and the virus is marginally more pathogenic. Thus, the effects of deleting a specific virus gene can vary according to non-genotypic factors such as the inoculation route, likely reflecting regional differences in host immune responses.

6.2 Virokines

6.2.1 Viral IL-18 binding proteins

The activity of mammalian IL-18, a pleiotrophic pro-inflammatory cytokine that induces IFNγ production, is tightly regulated by its natural antagonist, the IL-18 binding protein (IL-18BP). To indirectly regulate IFN responses to infection, many poxviruses have been shown (MCV, EV, VV and CPV) or are predicted (SPV, YLDV, MPV and LSDV) to encode soluble IL-18BP homologs[92-95]. The human poxvirus MCV encodes three putative IL-18BPs, MC051L, MC053L and MC054L, but only MC054L binds human and murine IL-18 with high affinity[95]. For the reasons described above, the contribution of MC054L cannot be assessed in the context of MCV infection. The IL-18BPs encoded by orthopoxviruses that infect rodents, VV, EV and CPV, exhibit much greater affinity for murine IL-18 than human[93]. Moreover, deletion of the IL-18BP ORF (C12L) from VV has been shown to attenuate virulence following intranasal inoculation of mice, producing less weight loss and fewer disease symptoms[96]. In contrast, disruption of the EV IL-18BP gene (p13) by insertional mutagenesis leads to an increase in NK cell activity relative to wild-type virus that is similar to that observed with VV C12L KO virus, but loss of p13 function only minimally attenuates EV in mice[92]. Although this finding correlated with increased clearance of infected cells, it had little effect on the pathogenic outcome of the infection. It is conceivable that the effect of deleting the EV IL-18BP was compensated for by the activity of other immunomodulatory factors, such as the intact EV IFN-γR.

6.2.2 Viral IL-10

IL-10 is a multifunctional cytokine with both immunostimulatory and immunosuppressive effects. Homologs of IL-10 have been identified in the genomes of ORFV, YLDV and LSDV, but only the ORFV IL-10 has been characterized to date and shown to have biological activity similar to that of ovine IL-10[97]. *In vitro*, ORFV IL-10 promotes thymocyte proliferation, co-stimulates mast cell growth and suppresses macrophages activation[98], suggesting a role in immune evasion that involves mimicking the suppressive effects of host IL-10 on immune responses. More recently, ORFV IL-10 has also been implicated in the impairment of acquired immunity by inhibiting the maturation of and antigen presentation by dendritic cells[99], possibly explaining why ORFV can repeatedly infect the same host. In a sheep model, deletion of ORFV IL-10 was found to result in elevated levels of IFN-γ in infected tissue compared to wild-type virus[100].

6.2.3 Viral growth factors

Homologs of mammalian epidermal growth factor (EGF) have been detected in members of virtually all poxvirus genera. The EGF homologs of VV and MV, termed VGF and MGF respectively, are secreted proteins produced early in infection that compete with cellular EGF for receptors (EGFR) expressed in the epithelial cell layers overlying poxviral lesions, the conjunctiva and the respiratory tract[101, 102]. The functional consequences of these interactions, as demonstrated *in vitro* for native VGF[102] and for synthetic peptides comprising the carboxyl portions of MGF[103], include both the generation of mitotic responses and the induction of EGFR autophosphorylation to inhibit receptor downregulation and extend the duration of proliferative signals. Deletion of VGF influences VV pathogenesis in both mice and rabbits. KO virus delivered by an intracranial route exhibits decreased neurovirulence in mice, whereas virus introduced intradermally produces less localized cellular proliferation in lesions[104]. Similarly, infection of rabbits with MV KO virus lacking MGF results in decreased mortality, less hyperplastic lesions and fewer opportunistic bacterial infections[101], characteristics of wild-type infections that can be restored through the addition of VGF and host growth factors[105].

The parapoxviruses, ORFV and PCPV, encode biologically active homologs of mammalian vascular endothelial growth factor (VEGF)-A that possess a receptor binding profile that is unique among VEGF family members in its apparent specificity for VEGF receptor (VEGFR)-2 and neuropilin-1[106]. Purified VEGF from ORFV and PCPV has been shown to stimulate proliferation of vascular endothelial cells and promote vascular permeability[107], suggesting that these proteins function in pathogenesis by contributing to the proliferative and highly vascularized nature of parapoxvirus lesions. Consistent with this hypothesis, disruption of the VEGF-like gene in ORFV results in the loss of the three VEGF activities associated with the parent virus: mitogenesis of vascular endothelial cells, induction of vascular permeability and activation of VEGFR-2[108]. *In vivo* loss of functional VEGF does not impact greatly on virus replication, but it does result in lesions with markedly reduced clinical indications of infection and vascularization, including decreased proliferation of blood vessels in the dermis underlying the site of infection, reduced inflammatory cell influx and abrogation of the distinctive pattern of epidermal proliferation associated with ORFV infections[108].

6.2.4 Viral semaphorins

Semaphorins represent a family of cellular regulatory proteins implicated in both neuronal development and activation of B- and T-lymphocytes. Included in this family are several poxviral proteins, most notably the products of the VV and EV A39R genes, which have been shown to possess the defining 500-amino acid 'sema' domain[109]. Moreover, these proteins interact with a novel virus-encoded semaphorin protein receptor, termed VESPR, a plexin family cell surface receptor for which the natural ligand is unknown. Surprisingly, studies into the function of the EV A39R protein have suggested pro-inflammatory properties that manifest as increased recruitment of immune cells to sites of infection due to IL-6 and -8 upregulation[109]. This strategy likely favors virus dissemination within the host by attracting immune cells that can be subsequently infected. The VV (strain Western Reserve) A39R gene product is naturally truncated and insertion of full-length A39R from VV (strain Copenhagen) was shown to have only minimal effect on pathogenesis in mice[110]. Consistent with its pro-inflammatory potential, moderately increased inflammation leading to larger lesions that were slower to resolve was observed following infection with virus containing A39R[110]. However, disease symptoms and viral titers remained unaffected. Similarly, deletion of the intact gene from VV (strain Copenhagen) did not influence pathogenesis, suggesting that poxviral semaphorins may promote, but are not essential to, infection.

6.2.5 Viral anti-inflammatory serpins

In addition to serpins that modulate host apoptotic and inflammatory responses through caspase inhibition, poxviruses also encode serpins that target other elements within inflammatory cascades. MV SERP-1 is the first poxviral serpin shown to be secreted and, therefore, it is technically a virokine[111]. The orthopoxvirus SPI-3 is not secreted from RPV-infected cells and shares limited sequence homology with SERP-1[112], but the presence of a common P1 arginne residue in the active site of the protein confers a similar inhibitory profile *in vitro*[113]. Both proteins inhibit a range of trypsin-like serine proteinases *in vitro*, including tissue plasminogen activator, urokinase, plasmin, thrombin and factor Xa[113-115], suggesting that the primary function of these serpins is to modulate host inflammatory responses to infection. However, it has been shown using CPV that SERP-1 and SPI-3 are not functionally interchangeable despite these similarities[116]. The precise targets and receptors through which SERP-1 acts *in vivo* are not yet defined, although purified SERP-1 has recently been shown to interact with native vascular urokinase-type plasminogen activator receptors to inhibit inflammatory cell responses in a mouse model[117].

Deleting the SPI-3 gene from the genome of orthopoxviruses such as CPV and VV has limited effect on pathogenesis *in vivo* and KO viruses exhibit only moderate reductions in virulence in murine models[35]. In contrast, SERP-1 is an important virulence factor for MV in its rabbit host. Deletion of SERP-1 markedly reduces the lethality of MV compared to wild-type virus and produces a pathology characterized by rapid induction of host inflammatory responses, reduced leukocyte infiltration and the development of fewer secondary lesions[111]. These observations suggest that SERP-1 is important to the dissemination of MV *in vivo* as well as the control of host inflammatory responses to infection.

6.2.6 Modulation of chemokine function

Chemokines (chemoattractant cytokines) are small, secreted cytokines that contribute to the host efforts to limit virus infections by coordinating the activation and mobilization of leukocytes that mediate inflammatory responses in areas of infection. Consequently, all poxviruses attempt to modulate chemokine activity by encoding chemokine receptor homologs or secreted ligand mimics and chemokine binding proteins (CBPs)[118]. Bioinformatic analyses of the genomes of several poxviruses have identified putative chemokine G-protein-coupled receptor (GPCR) homologs (SPV and YMTV) and ligand mimics (MCV and FPV). With the exception of the product of the MCV MC148R gene, which has been shown to function as a selective antagonist of human CCR8[119], functional studies on these proteins are limited and their role in pathogenesis *in vivo* remains speculative. Greater insight has been gained into the role of poxvirus CBPs, however. Classified as either Type I (low affinity) or Type II (high affinity), their roles in poxvirus virulence have been examined extensively using KO viruses.

The dual-function IFNγ-R homolog of MV, M-T7, is the sole Type I poxvirus CBP identified to date. The capacity for M-T7 to inhibit the activity of a broad spectrum of chemokines, in addition to IFN-γ, likely arises from its ability to interact with the heparin binding domains common to many chemokines[87]. In doing so, M-T7 has the potential to interfere with generalized chemokine binding to glycosaminoglycans and disrupt the localization of a large number of C, CC and CXC chemokines in tissues. Deletion of the M-T7 gene produces marked attenuation that prevents dissemination of the virus to distal sites of infection[86]. Locally, loss of MT-7 function is associated with leukocyte infiltration into the primary dermal sites of viral replication and activation of leukocytes in secondary immune tissues, such as the lymph nodes and spleen[86]. Naturally, the question remains as to whether this phenotype reflects loss of binding to chemokines, IFN-γ or both classes of molecules.

Type II CBPs, also termed CBP-IIs or vCCIs, have been identified in several poxvirus species and are exemplified by the 35 kDa vCCI encoded by many orthopoxviruses[120] and the product of the M-T1 gene of MV[121]. Despite lacking sequence similarity with known mammalian proteins, type II CBPs target the GPCR binding domain conserved among many CC chemokines to competitively inhibit their ability to interact with diverse cellular receptors[122]. Like the high binding affinities exhibited by these proteins, this property likely reflects the unique structure of type II CPBs that distantly resembled the collagen-binding domain of the *Staphlococcus aureus* adhesin molecule[123]. The proposed function of poxvirus CBPs is based on *in vitro* studies that support the ability of these proteins to bind chemokines and impede leukocyte migration. However, loss of type II CBP activity appears to have limited effect on poxvirus virulence *in vivo*. For example, infection of rabbits with MV lacking M-T1 differs from wild-type infections only in the development of heightened localized cellular inflammation in primary lesions, together with a moderate increase in infiltrating monocytes and macrophages early in infection[124]. Similarly, the virulence of a RPV 35 kDa CBP-II KO in mice was shown to differ little from wild-type virus[125]. These results are perhaps unsurprising when the considerable redundancy in chemokine function and the combined effects of other poxviral immunomodulators that indirectly impact on chemokine function are considered.

6.2.7 Control of the complement system

The complement system is an integrated network of cell-associated effector proteins and secreted regulatory proteins that participate in the identification and destruction of invading pathogens, as well as the initiation and amplification of inflammatory responses. The prototypical poxviral complement regulatory protein is the VV complement control protein (VCP), although genes encoding similar products identified in MPV, VaV and CPV[126]. VCP is a highly stable, monomeric secreted protein that inhibits both the classical and alternative complement activation pathways by directly and indirectly promoting the decay of the C3 convertase[127,128]. Recently, VCP has also been shown to interact with cell-surface glycosaminoglycans, inhibiting both chemokine-mediated leukocyte migration and antibody binding to MHC I[129,130]. The VCP homolog of VaV, termed the smallpox inhibitor of complement enzymes (SPICE), closely resembles VV VCP in both structure and activity[131]. Of particular interest given the highly virulent nature of VaV in humans compared to VV is that

SPICE is nearly one hundred fold more potent at inhibiting human C3 activity than VCP[131].

Various murine models have been used to assess the function of the CPV VCP homolog, the inflammation modulatory protein (IMP). Although results varied extensively with host strain and route of inoculation, these studies suggest that modulation of host complement contributes little to poxvirus virulence. For example, IMP KO virus was not attenuated in BALB/c mice inoculated using either footpad injections[132] or a connective tissue air pouch model[133], exhibiting lethality comparable to the wild-type. However, inflammation and mononuclear cellular infiltration at sites of infection were greater in animals infected with the IMP KO virus compared to wild-type CPV. Since BALB/c mice express only low constitutive levels of C3, mice that were either fully deficient in C3 or expressed high levels of C3 were also studied[133]. Infection with either wild-type or KO virus produced similar disease pathologies in C3-deficient mice, but the differences in inflammatory responses elicited by the viruses were comparable to those observed in BALB/c mice despite greater levels of host C3 expression.

7. CONCLUSIONS

The examples provided above illustrate the important contribution of immunomodulatory genes in the progression and resolution of poxvirus infections and their ability to impact on host range, virulence and pathogenesis. However, the redundancy inherent to many poxvirus immune evasion strategies and the capacity for individual immunomodulatory proteins to have multiple functions means that the phenotype exhibited by a KO virus may not necessarily reflect the full extent to which a gene product contributes to pathogenesis. However, certain trends do manifest when KO viruses are compared on the basis of the function of the gene disrupted. As shown in Table 4, deletion of genes whose products regulate host anti- viral responses that influence survival at the level of the infected cell, such as apoptosis, more profoundly affect virulence than genes encoding proteins that modulate more global host antiviral responses, such as chemokine and complement networks. This property is demonstrated at the level of both intracellular virotransducers, such as MV M11L, that target the signals initiated by pro-apoptotic cytokines, and extracellular viroceptors, such as MV M-T2, that target the cytokines themselves. Although manipulation of the latter strategies are important to the efficient spread of the virus once an infection has been established, failure to block innate defense mechanisms evolved to remove infected cells prevents the infection from ever being established. Thus, it is not surprising that poxvirus proteins that modulate these critical host responses are determinants of host range as well as virulence.

Table 4. Summary of poxviral immunomodulators studied with KO poxviruses[1].

Strategy	Mechanism	Protein	Function or Target	Virus	Host Range Gene
Virostealth	MHC downregulation	M153R	PHD protein	MV	No
Virotransduction	Apoptosis regulators	M-T4	ER	MV	Yes
		M-T5	ankyrin protein	MV	Yes
		M11L	mitochondria	MV	Yes
		Serp-2	serpin	MV	No
		SPI-2	serpin	VV	No
	Cytokine signaling	M150R	NF-κB	MV	No
		A46R	TLR	VV	No
		A52R	TLR	VV	No
		B12R	Ser-Thr kinase	VV	No
	IFN inhibition	E3L	dsRNA binding	VV	Yes
		K3L	eIF2α mimic	VV	Yes
	Other	Serp-3	serpin	MV	No
		SPI-1	serpin	VV	Yes
		SPI-3	serpin	VV	No
		M131R	SOD	MV	No
		A45R	SOD	VV	No
Viromimics					
Virokine	Binding proteins	M-T1	CC chemokines	MV	No
		C21L	complement	VV	No
		p13	IL-18BP	EV	No
		C12L	IL-18BP	VV	No
	Viral orthologs	VGF	EGF ortholog	VV	No
			VEGF orthologs	ORFV	No
			IL-10 ortholog	ORFV	No
		Serp-1	serpin	MV	Yes
		A39R	semaphorin	VV	No
Viroceptor	TNF	M-T2	TNF-R	MV	Yes
		CrmE	TNF-R	VV	No
		A53R	TNF-R	VV	No
	IFN	M-T7	Type 2 IFN-R	MV	No
		B8R	Type 2 IFN-R	VV	No
	IL	B15R	IL1β-R	VV	No

1. VGF, viral growth factor; SOD, superoxide dismutase

8. APPLICATIONS OF KO VIRUS TECHNOLOGY

The functions of many poxviral immunomodulatory proteins can be predicted based on bioinformatic analyses and confirmed using *in vitro* functional assays. However, understanding the precise roles of these proteins during infection of a natural host *in vivo* requires the ability to assess their properties in the context of the virus as a whole. The immunomodulatory strategies of poxviruses are so effective that new avenues of research, collectively known as virotherapeutics, have emerged in the attempt to exploit viral immunomodulatory proteins for the treatment of human diseases[134]. As illustrated in Table 5, several of the poxviral proteins described in this chapter have been proposed for use in the treatment of a spectrum of human conditions associated with adverse immune responses. In these applications, viral proteins are used as purified biotherapeutics outside

Table 5. Potential virotherapies based on poxviral immunomodulatory proteins.

Condition	Etiological Factors	Protein(s)
Arteriosclerotic plaque formation	Balloon angioplasty	MV Serp-1
Arterial hyperplasia and scarring	Injury or surgery	MV Serp-1 MV M-T7
Transplant rejection and graft vasculopathy	Allograft or xenograft	MV Serp-1 MV M-T7 VV VCP MCV 148R
Arthritis	Collagen-induced	MV Serp-1
Asthma	Allergic airway hyper-reactivity	MV Serp-1 CPV IMP

the context of the intact virus, and pathogenesis is not a consideration. However, other virotherapies based on live poxviruses that have been modified to be less virulent or to exhibit a specific phenotype are becoming increasingly prevalent. These include the use of poxviruses in vaccines[135-138] and as therapeutic vectors and oncolytic agents[139-142]. The rational design of such therapies requires detailed information about how modification of the viral genome impacts on the biology of poxviruses.

In addition to the therapeutic benefit afforded by the characterization of poxvirus immunomodulators, the study of immune modulation by viruses contributes greatly to our understanding of how the immune system responds to infection and the selective pressures that drive the co-evolution of virus and host. Although a great deal of research has been carried out on the mechanisms of poxvirus replication, the cellular factors that determine cell tropism and permissiveness to infection are still poorly understood. One factor impeding such research is that so many of the genes conserved across poxvirus species encode products with no known function. The eradication of smallpox as a human health concern resulted in the curtailing of research into the specific virus and host mechanisms required to efficiently target anti-smallpox therapies and overcome the significant limitations inherent to current smallpox vaccines[143]. However, the potential use of variola virus as a bioterrorism agent[143] and the recent outbreak of MPV in North America[144] has lent greater significance to this field. By understanding the role of individual genes in poxvirus pathogenesis, greater insight can be gained into novel targets on which to base antiviral strategies.

ACKNOWLEDGEMENTS

G. McFadden holds a Canada Research Chair in Molecular Virology. J.B. Johnston is a Canadian Institutes of Health Research (CIHR) Fellow. We thank J. Barrett and S. Nazarian for critically reviewing the manuscript. We regret that space constraints prevent the citation of more literature pertinent to this subject and extend our apologies to those whose contributions we were unable to recognize.

REFERENCES

1. Moss, B. 2001. Poxviridae: The virus and their replication. In *Fields Virology*, Vol. 2. D. M. Knipe, and P. M. Howley, eds. Lippincott Williams and Wilkins, Philadelphia, p. 2849.
2. Lucas, M., U. Karrer, A. Lucas, and P. Klenerman. 2001. Viral escape mechanisms-Escapology taught by viruses. *Int J Exp Pathol* **82:**269
3. Moss, B., and J. L. Shisler. 2001. Immunology 101 at poxvirus U: Immune evasion genes. *Semin Immunol* **13:**59
4. Alcami, A. 2003. Viral mimicry of cytokines, chemokines and their receptors. *Nat Rev Immunol* **3:**36
5. Seet, B. T., J. B. Johnston, C. R. Brunetti, J. W. Barrett, H. Everett, C. Cameron, J. Sypula, S. H. Nazarian, A. Lucas, and G. McFadden. 2003. Poxviruses and immune evasion. *Annu Rev Immunol* **21:**377
6. Coen, D. M., and R. F. Ramig. 1996. Viral genetics. In *Fields Virology*. B. N. Fields, D. M. Knipe, P. M. Howley, R. M. Chanock, J. L. Melnick, T. P. Monath, B. Roizman, and S. E. Straus, eds. Lippincott-Raven, New York, N.Y., p. 113
7. Smith, K. J., and H. Skelton. 2002. Molluscum contagiosum: Recent advances in pathogenic mechanisms, and new therapies. *Am J Clin Dermatol* **3:**535

8. Flint, S. J., L. W. Enquist, R. M. Krug, V. R. Racaniello, and A. M. Skalka. 2000. Virus cultivation, detection and genetics. In *Principles of Virology*. ASM Press, Washington, D.C., p. 42
9. Lee, E. C., D. Yu, J. Martinez de Velasco, L. Tessarollo, D. A. Swing, D. L. Court, N. A. Jenkins, and N. G. Copeland. 2001. A highly efficient Escherichia coli-based chromosome engineering system adapted for recombinogenic targeting and subcloning of BAC DNA. *Genomics* **73:**56
10. Luckow, V. A., S. C. Lee, G. F. Barry, and P. O. Olins. 1993. Efficient generation of infectious recombinant baculoviruses by site-specific transposon-mediated insertion of foreign genes into a baculovirus genome propagated in Escherichia coli. *J Virol* **67:**4566.
11. Adler, H., M. Messerle, and U. H. Koszinowski. 2003. Cloning of herpesviral genomes as bacterial artificial chromosomes. *Rev Med Virol* **13:**111
12. Domi, A., and B. Moss. 2002. Cloning the vaccinia virus genome as a bacterial artificial chromosome in escherichia coli and recovery of infectious virus in mammalian cells. *Proc Natl Acad Sci USA* **99:**12415
13. Nash, P., J. Barrett, J. X. Cao, S. Hota-Mitchell, A. S. Lalani, H. Everett, X. M. Xu, J. Robichaud, S. Hnatiuk, C. Ainslie, B. T. Seet, and G. McFadden. 1999. Immunomodulation by viruses: The myxoma virus story. *Immunol Rev* **168:**103
14. Alcami, A., and U. H. Koszinowski. 2000. Viral mechanisms of immune evasion. *Trends Microbiol* **8:**410
15. Johnston, J. B., and G. McFadden. 2003. Poxvirus immunomodulatory strategies: Current perspectives. *J Virol* **77:**6093
16. McFadden, G., and P. M. Murphy. 2000. Host-related immunomodulators encoded by poxviruses and herpesviruses. *Cur Opin Microbiol* **3:**371
17. Smith, G. L. 2000. Secreted poxvirus proteins that interact with the immune system. In *Effects of Microbes on the Immune System.* M. W. Cunningham, and R. S. Fujinami, eds. Lippincott Williams and Wilkins, Philadelphia, p. 491.
18. Turner, P., and R. Moyer. 2002. Poxvirus immune modulators: Functional insights from animal models. *Virus Res* **88:**35
19. Zuniga, M. 2002. A pox on thee! Manipulation of the host immune system by myxoma virus and implications for viral-host co-adaptation. *Virus Res* **88:**17
20. Barry, M., and R. C. Bleackley. 2002. Cytotoxic T lymphocytes: All roads lead to death. *Nat Rev Immunol* **2:**401
21. Orange, J. S., M. S. Fassett, L. A. Koopman, J. E. Boyson, and J. L. Strominger. 2002. Viral evasion of natural killer cells. *Nat Immunol* **3:**1006
22. Yewdell, J. W., and A. B. Hill. 2002. Viral interference with antigen presentation. *Nat Immunol* **3:**1019
23. Boshkov, L. K., J. L. Macen, and G. McFadden. 1992. Virus-induced loss of class I MHC antigens from the surface of cells infected with myxoma virus and malignant rabbit fibroma virus. *J Immunol* **148:**881
24. Barry, M., S. F. Lee, L. Boshkov, and G. Mcfadden. 1995. Myxoma virus induces extensive CD4 downregulation and dissociation of p56(lck) in infected rabbit CD4(+) t lymphocytes. *J Virol* **69:**5243
25. Coscoy, L., and D. Ganem. 2000. Kaposi's sarcoma-associated herpesvirus encodes two proteins that block cell surface display of MHC class I chains by enhancing their endocytosis. *Proc Natl Acad Sci USA* **97:**8051
26. Mansouri, M., E. Bartee, K. Gouveia, B. T. Hovey Nerenberg, J. Barrett, L. Thomas, G. Thomas, G. McFadden, and K. Fruh. 2003. The PHD/LAP-domain protein M153R of myxoma virus is a ubiquitin ligase that induces the rapid internalization and lysosomal destruction of CD4. *J Virol* **77:**1427
27. Guerin, J. L., J. Gelfi, S. Boullier, M. Delverdier, F. A. Bellanger, S. Bertagnoli, I. Drexler, G. Sutter, and F. Messud-Petit. 2002. Myxoma virus leukemia-associated protein is responsible for major histocompatibility complex class I and Fas-CD95 down-

regulation and defines scrapins, a new group of surface cellular receptor abductor proteins. *J Virol* **76:**2912
28. Fruh, K., E. Bartee, K. Gouveia, and M. Mansouri. 2002. Immune evasion by a novel family of viral PHD/LAP-finger proteins of gamma-2 herpesviruses and poxviruses. *Virus Res* **88:**55
29. Benedict, C. A., P. S. Norris, and C. F. Ware. 2002. To kill or be killed: Viral evasion of apoptosis. *Nat Immunol* **3:**1013
30. Hengartner, M. O. 2000. The biochemistry of apoptosis. *Nature* **407:**770
31. Ray, C. A., R. A. Black, S. R. Kronheim, T. A. Greenstreet, P. R. Sleath, G. S. Salvesen, and D. J. Pickup. 1992. Viral inhibition of inflammation: Cowpox virus encodes an inhibitor of the interleukin-1 beta converting enzyme. *Cell* **69:**597
32. Tewari, M., and V. M. Dixit. 1995. Fas- and tumor necrosis factor-induced apoptosis is inhibited by the poxvirus CrmA gene product. *J Biol Chem* **270:**3255
33. Tewari, M., W. G. Telford, R. A. Miller, and V. M. Dixit. 1995. Crma, a poxvirus-encoded serpin, inhibits cytotoxic T-lymphocyte-mediated apoptosis. *J Biol Chem* **270:**22705
34. Dobbelstein, M., and T. Shenk. 1996. Protection against apoptosis by the vaccinia virus SPI-2 (B13R) gene product. *J Virol* **70:**6479
35. Thompson, J. P., P. C. Turner, A. N. Ali, B. C. Crenshaw, and R. W. Moyer. 1993. The effects of serpin gene mutations on the distinctive pathobiology of cowpox and rabbitpox virus following intranasal inoculation of Balb/c mice. *Virology* **197:**328
36. Kettle, S., A. Alcami, A. Khanna, R. Ehret, C. Jassoy, and G. L. Smith. 1997. Vaccinia virus serpin b13r (spi-2) inhibits interleukin-1-beta-converting enzyme and protects virus-infected cells from TNF- and Fas-mediated apoptosis, but does not prevent IL-1-beta-induced fever. *J Gen Virol* **78:**677
37. Messud-Petit, F., J. Gelfi, M. Delverdier, M. F. Amardeilh, R. Py, G. Sutter, and S. Bertagnoli. 1998. Serp2, an inhibitor of the interleukin-1beta-converting enzyme, is critical in the pathobiology of myxoma virus. *J Virol* **72:**7830
38. Petit, F., S. Bertagnoli, J. Gelfi, F. Fassy, C. Boucrautbaralon, and A. Milon. 1996. Characterization of a myxoma virus-encoded serpin-like protein with activity against interleukin-1 beta-converting enzyme. *J Virol* **70:**5860
39. Shisler, J. L., S. N. Isaacs, and B. Moss. 1999. Vaccinia virus serpin-1 deletion mutant exhibits a host range defect characterized by low levels of intermediate and late mRNAs. *Virology* **262:**298
40. Bertin, J., R. C. Armstrong, S. Ottilie, D. A. Martin, Y. Wang, S. Banks, G. H. Wang, T. G. Senkevich, E. S. Alnemri, B. Moss, M. J. Lenardo, K. J. Tomaselli, and J. I. Cohen. 1997. Death effector domain-containing herpesvirus and poxvirus proteins inhibit both Fas- and TNFR1-induced apoptosis. *Proc Natl Acad Sci USA* **94:**1172
41. Garvey, T. L., J. Bertin, R. M. Siegel, G. H. Wang, M. J. Lenardo, and J. I. Cohen. 2002. Binding of FADD and caspase-8 to molluscum contagiosum virus MC159 v-FLIP is not sufficient for its antiapoptotic function. *J Virol* **76:**697
42. Garvey, T., J. Bertin, R. Siegel, M. Lenardo, and J. Cohen. 2002. The death effector domains (DEDs) of the molluscum contagiosum virus MC159 v-FLIP protein are not functionally interchangeable with each other or with the DEDs of caspase-8. *Virology* **300:**217
43. Gil, J., J. Rullas, J. Alcami, and M. Esteban. 2001. Mc159l protein from the poxvirus molluscum contagiosum virus inhibits NF-kappa B activation and apoptosis induced by PKR. *J Gen Virol* **82:**3027
44. Everett, H., M. Barry, S. F. Lee, X. J. Sun, K. Graham, J. Stone, R. C. Bleackley, and G. McFadden. 2000. M11L: A novel mitochondria-localized protein of myxoma virus that blocks apoptosis of infected leukocytes. *J Exp Med* **191:**1487
45. Everett, H., M. Barry, X. Sun, S. F. Lee, C. Frantz, L. G. Berthiaume, G. McFadden, and R. C. Bleackley. 2002. The myxoma poxvirus protein, M11L, prevents apoptosis by

direct interaction with the mitochondrial permeability transition pore. *J Exp Med* **196:**1127
46. Barry, M., S. Hnatiuk, K. Mossman, S. F. Lee, L. Boshkov, and G. McFadden. 1997. The myxoma virus M-T4 gene encodes a novel RDEL-containing protein that is retained within the endoplasmic reticulum and is important for the productive infection of lymphocytes. *Virology* **239:**360
47. Mossman, K., S. F. Lee, M. Barry, L. Boshkov, and G. Mcfadden. 1996. Disruption of M-T5, a novel myxoma virus gene member of the poxvirus host range superfamily, results in dramatic attenuation of myxomatosis in infected European rabbits. *J Virol* **70:**4394
48. Shisler, J. L., T. G. Senkevich, M. J. Berry, and B. Moss. 1998. Ultraviolet-induced cell death blocked by a selenoprotein from a human dermatotropic poxvirus. *Science* **279:**102
49. Brick, D. J., R. D. Burke, L. Schiff, and C. Upton. 1998. Shope fibroma virus ring finger protein N1R binds DNA and inhibits apoptosis. *Virology* **249:**42
50. Brick, D. J., R. D. Burke, A. A. Minkley, and C. Upton. 2000. Ectromelia virus virulence factor p28 acts upstream of caspase-3 in response to UV light-induced apoptosis. *J Gen Virol* **81:**1087
51. Sen, G. C. 2001. Viruses and interferons. *Annu Rev Microbiol* **55:**255
52. Barber, G. N. 2001. Host defense, viruses and apoptosis. *Cell Death Differ* **8:**113
53. Cameron, C., S. Hota-Mitchell, L. Chen, J. Barrett, J. X. Cao, C. Macaulay, D. Willer, D. Evans, and G. McFadden. 1999. The complete DNA sequence of myxoma virus. *Virology* **264:**298
54. Willer, D. O., G. McFadden, and D. H. Evans. 1999. The complete genome sequence of shope (rabbit) fibroma virus. *Virology* **264:**319
55. Lee, H. J., K. Essani, and G. L. Smith. 2001. The genome sequence of yaba-like disease virus, a yatapoxvirus. *Virology* **281:**170
56. McInnes, C. J., A. R. Wood, and A. A. Mercer. 1998. Orf virus encodes a homolog of the vaccinia virus interferon resistance gene E3L. *Virus Genes* **17:**107
57. Kawagishi-Kobayashi, M., C. N. Cao, J. M. Lu, K. Ozato, and T. E. Dever. 2000. Pseudosubstrate inhibition of protein kinase PKR by swine pox virus C8L gene product. *Virology* **276:**424
58. Smith, V. P., and A. Alcami. 2002. Inhibition of interferons by ectromelia virus. *J Virol* **76:**1124
59. Chang, H. W., J. C. Watson, and B. L. Jacobs. 1992. The E3L gene of vaccinia virus encodes an inhibitor of the interferon-induced, double-stranded RNA-dependent protein kinase. *Proc Natl Acad Sci USA* **89:**4825
60. Sharp, T. V., F. Moonan, A. Romashko, B. Joshi, G. N. Barber, and R. Jagus. 1998. The vaccinia virus E3L gene product interacts with both the regulatory and the substrate binding regions of PKR - implications for PKR autoregulation. *Virology* **250:**302
61. Smith, E. J., I. Marie, A. Prakash, A. Garcia-Sastre, and D. E. Levy. 2001. IRF3 and IRF7 phosphorylation in virus-infected cells does not require double-stranded RNA-dependent protein kinase R or I kappa B kinase but is blocked by vaccinia virus E3L protein. *J Biol Chem* **276:**8951
62. Carroll, K., O. Elroystein, B. Moss, and R. Jagus. 1993. Recombinant vaccinia virus K3L gene product prevents activation of double-stranded RNA-dependent, initiation factor-2-alpha-specific protein kinase. *J Biol Chem* **268:**12837
63. Langland, J., and B. Jacobs. 2002. The role of the PKR-inhibitory genes, E3L and K3L, in determining vaccinia virus host range. *Virology* **299:**133
64. Brandt, T. A., and B. L. Jacobs. 2001. Both carboxy- and amino-terminal domains of the vaccinia virus interferon resistance gene, E3L, are required for pathogenesis in a mouse model. *J Virol* **75:**850
65. Najarro, P., P. Traktman, and J. A. Lewis. 2001. Vaccinia virus blocks gamma interferon signal transduction: Viral VH1 phosphatase reverses STAT1 activation. *J Virol* **75:**3185

66. Harte, M. T., I. R. Haga, G. Maloney, P. Gray, P. C. Reading, N. W. Bartlett, G. L. Smith, A. Bowie, and L. A. O'Neill. 2003. The poxvirus protein A52R targets toll-like receptor signaling complexes to suppress host defense. *J Exp Med* **197:**343
67. Bowie, A., E. Kiss-Toth, J. A. Symons, G. L. Smith, S. K. Dower, and L. A. J. O'Neill. 2000. A46R and A52R from vaccinia virus are antagonists of host il-1 and toll-like receptor signaling. *Proc Natl Acad Sci USA* **97:**10162
68. Camus-Bouclainville, C., L. Fiette, S. Bouchiha, B. Pignolet, D. Counor, C. Filipe, J. Gelfi, and F. Messud-Petit. 2004. A virulence factor of myxoma virus colocalizes with NF-kappaB in the nucleus and interferes with inflammation. *J Virol* **78:**2510
69. McFadden, G. 2003. Viroceptors: Virus-encoded receptors for cytokines and chemokines. In *Cytokines and Chemokines in Infectious Diseases Handbook.* M. Kotb, and T. Calandra, eds. Humana Press Inc., Totowa, N.J., p. 285.
70. Cunnion, K. M. 1999. Tumor necrosis factor receptors encoded by poxviruses. *Mol Gen Metab* **67:**278
71. Sedger, L., and G. McFadden. 1996. M-T2: A poxvirus TNF receptor homologue with dual activities. *Immunol Cell Biol* **74:**538
72. Smith, C. A., T. Davis, J. M. Wignall, W. S. Din, T. Farrah, C. Upton, G. McFadden, and R. G. Goodwin. 1991. T2 open reading frame from the shope fibroma virus encodes a soluble form of the TNF receptor. *Biochem Biophys Res Commun* **176:**335
73. Schreiber, M., K. Rajarathnam, and G. Mcfadden. 1996. Myxoma virus T2 protein, a tumor necrosis factor (TNF) receptor homolog, is secreted as a monomer and dimer that each bind rabbit TNF alpha, but the dimer is a more potent TNF inhibitor. *J Biol Chem* **271:**13333
74. Saraiva, M., and A. Alcami. 2001. CrmE, a novel soluble tumor necrosis factor receptor encoded by poxviruses. *J Virol* **75:**226
75. Alcami, A., A. Khanna, N. L. Paul, and G. L. Smith. 1999. Vaccinia virus strains Lister, USSR and Evans express soluble and cell-surface tumour necrosis factor receptors. *J Gen Virol* **80:**949
76. Reading, P. C., A. Khanna, and G. L. Smith. 2002. Vaccinia virus crmE encodes a soluble and cell surface tumor necrosis factor receptor that contributes to virus virulence. *Virology* **292:**285
77. Panus, J. F., C. A. Smith, C. A. Ray, T. D. Smith, D. D. Patel, and D. J. Pickup. 2002. Cowpox virus encodes a fifth member of the tumor necrosis factor receptor family: A soluble, secreted CD30 homologue. *Proc Natl Acad Sci USA* **99:**8348
78. Saraiva, M., P. Smith, P. G. Fallon, and A. Alcami. 2002. Inhibition of type 1 cytokine-mediated inflammation by a soluble CD30 homologue encoded by ectromelia (mousepox) virus. *J Exp Med* **196:**829
79. Howard, S. T., Y. S. Chan, and G. L. Smith. 1991. Vaccinia virus homologues of the shope fibroma virus inverted terminal repeat proteins and a discontinuous ORF related to the tumor necrosis factor receptor family. *Virology* **180:**633
80. Saint, K. M., N. French, and P. Kerr. 2001. Genetic variation in Australian isolates of myxoma virus: An evolutionary and epidemiological study. *Arch Virol* **146:**1105
81. Macen, J. L., K. A. Graham, S. F. Lee, M. Schreiber, L. K. Boshkov, and G. Mcfadden. 1996. Expression of the myxoma virus tumor necrosis factor receptor homologue and M11L genes is required to prevent virus-induced apoptosis in infected rabbit T lymphocytes. *Virology* **218:**232
82. Upton, C., J. L. Macen, M. Schreiber, and G. McFadden. 1991. Myxoma virus expresses a secreted protein with homology to the tumor necrosis factor receptor gene family that contributes to viral virulence. *Virology* **184:**370
83. Alcami, A., and G. L. Smith. 1995. Vaccinia, cowpox, and camelpox viruses encode soluble gamma interferon receptors with novel broad species specificity. *J Virol* **69:**4633
84. Symons, J. A., A. Alcami, and G. L. Smith. 1995. Vaccinia virus encodes a soluble type I interferon receptor of novel structure and broad species specificity. *Cell* **81:**551

85. Alcami, A., J. A. Symons, and G. L. Smith. 2000. The vaccinia virus soluble alpha/beta interferon (IFN) receptor binds to the cell surface and protects cells from the antiviral effects of IFN. *J Virol* **74:**11230
86. Mossman, K., P. Nation, J. Macen, M. Garbutt, A. Lucas, and G. Mcfadden. 1996. Myxoma virus M-T7, a secreted homolog of the interferon-gamma receptor, is a critical virulence factor for the development of myxomatosis in European rabbits. *Virology* **215:**17
87. Lalani, A. S., K. Graham, K. Mossman, K. Rajarathnam, I. Clarklewis, D. Kelvin, and G. McFadden. 1997. The purified myxoma virus gamma interferon receptor homolog M-T7 interacts with the heparin-binding domains of chemokines. *J Virol* **71:**4356
88. Symons, J. A., D. C. Tscharke, N. Price, and G. L. Smith. 2002. A study of the vaccinia virus interferon-gamma receptor and its contribution to virus virulence. *J Gen Virol* **83:**1953
89. Verardi, P. H., L. A. Jones, F. H. Aziz, S. Ahmad, and T. D. Yilma. 2001. Vaccinia virus vectors with an inactivated gamma interferon receptor homolog gene (B8R) are attenuated in vivo without a concomitant reduction in immunogenicity. *J Virol* **75:**11
90. Spriggs, M. K., D. E. Hruby, C. R. Maliszewski, D. J. Pickup, J. E. Sims, R. M. Buller, and J. VanSlyke. 1992. Vaccinia and cowpox viruses encode a novel secreted interleukin-1-binding protein. *Cell* **71:**145
91. Alcami, A., and G. L. Smith. 1992. A soluble receptor for interleukin-1 beta encoded by vaccinia virus: A novel mechanism of virus modulation of the host response to infection. *Cell* **71:**153
92. Born, T. L., L. A. Morrison, D. J. Esteban, T. VandenBos, L. G. Thebeau, N. H. Chen, M. K. Spriggs, J. E. Sims, and R. M. L. Buller. 2000. A poxvirus protein that binds to and inactivates il-18, and inhibits NK cell response. *J Immunol* **164:**3246
93. Calderara, S., Y. Xiang, and B. Moss. 2001. Orthopoxvirus IL-18 binding proteins: Affinities and antagonist activities. *Virology* **279:**22
94. Smith, V. P., N. A. Bryant, and A. Alcami. 2000. Ectromelia, vaccinia and cowpox viruses encode secreted interleukin-18-binding proteins. *J Gen Virol* **81:**1223
95. Xiang, Y., and B. Moss. 1999. Il-18 binding and inhibition of interferon gamma induction by human poxvirus-encoded proteins. *Proc Natl Acad Sci USA* **96:**11537
96. Symons, J. A., E. Adams, D. C. Tscharke, P. C. Reading, H. Waldmann, and G. L. Smith. 2002. The vaccinia virus C12L protein inhibits mouse IL-18 and promotes virus virulence in the murine intranasal model. *J Gen Virol* **83:**2833
97. Fleming, S. B., C. A. McCaughan, A. E. Andrews, A. D. Nash, and A. A. Mercer. 1997. A homolog of interleukin-10 is encoded by the poxvirus Orf virus. *J Virol* **71:**4857
98. Imlach, W., C. A. McCaughan, A. A. Mercer, D. Haig, and S. B. Fleming. 2002. Orf virus-encoded interleukin-10 stimulates the proliferation of murine mast cells and inhibits cytokine synthesis in murine peritoneal macrophages. *J Gen Virol* **83:**1049
99. Lateef, Z., S. Fleming, G. Halliday, L. Faulkner, A. Mercer, and M. Baird. 2003. Orf virus-encoded interleukin-10 inhibits maturation, antigen presentation and migration of murine dendritic cells. *J Gen Virol* **84:**1101
100. Haig, D. M., J. Thomson, C. McInnes, C. McCaughan, W. Imlach, A. Mercer, and S. Fleming. 2002. Orf virus immuno-modulation and the host immune response. *Vet Immunol Immunopathol* **87:**395
101. Opgenorth, A., D. Strayer, C. Upton, and G. McFadden. 1992. Deletion of the growth factor gene related to EGF and TGF alpha reduces virulence of malignant rabbit fibroma virus. *Virology* **186:**175
102. Stroobant, P., A. P. Rice, W. J. Gullick, D. J. Cheng, I. M. Kerr, and M. D. Waterfield. 1985. Purification and characterization of vaccinia virus growth factor. *Cell* **42:**383
103. Lin, Y. Z., X. H. Ke, and J. P. Tam. 1991. Synthesis and structure-activity study of myxoma virus growth factor. *Biochemistry* **30:**3310

104. Buller, R. M., S. Chakrabarti, J. A. Cooper, D. R. Twardzik, and B. Moss. 1988. Deletion of the vaccinia virus growth factor gene reduces virus virulence. *J Virol* **62:**866
105. Opgenorth, A., N. Nation, K. Graham, and G. McFadden. 1993. Transforming growth factor alpha, shope fibroma growth factor, and vaccinia growth factor can replace myxoma growth factor in the induction of myxomatosis in rabbits. *Virology* **192:**701
106. Meyer, M., M. Clauss, A. Lepple-Wienhues, J. Waltenberger, H. G. Augustin, M. Ziche, C. Lanz, M. Buttner, H. J. Rziha, and C. Dehio. 1999. A novel vascular endothelial growth factor encoded by Orf virus, VEGF-E, mediates angiogenesis via signalling through VEGFR-2 (KDR) but not VEGFR-1 (FLT-1) receptor tyrosine kinases. *EMBO J* **18:**363
107. Wise, L. M., T. Veikkola, A. A. Mercer, L. J. Savory, S. B. Fleming, C. Caesar, A. Vitali, T. Makinen, K. Alitalo, and S. A. Stacker. 1999. Vascular endothelial growth factor (VEGF)-like protein from orf virus NZ2 binds to VEGFR2 and neuropilin-1. *Proc Natl Acad Sci USA* **96:**3071
108. Savory, L. J., S. A. Stacker, S. B. Fleming, B. E. Niven, and A. A. Mercer. 2000. Viral vascular endothelial growth factor plays a critical role in orf virus infection. *J Virol* **74:**10699
109. Comeau, M. R., R. Johnson, R. F. DuBose, M. Petersen, P. Gearing, T. VandenBos, L. Park, T. Farrah, R. M. Buller, J. I. Cohen, L. D. Strockbine, C. Rauch, and M. K. Spriggs. 1998. A poxvirus-encoded semaphorin induces cytokine production from monocytes and binds to a novel cellular semaphorin receptor, VESPR. *Immunity* **8:**473
110. Gardner, J. D., D. C. Tscharke, P. C. Reading, and G. L. Smith. 2001. Vaccinia virus semaphorin A39R is a 50-55 kDa secreted glycoprotein that affects the outcome of infection in a murine intradermal model. *J Gen Virol* **82:**2083
111. Upton, C., J. L. Macen, D. S. Wishart, and G. McFadden. 1990. Myxoma virus and malignant rabbit fibroma virus encode a serpin-like protein important for virus virulence. *Virology* **179:**618
112. Turner, P. C., and R. W. Moyer. 1995. Orthopoxvirus fusion inhibitor glycoprotein Spi-3 (open reading frame K2L) contains motifs characteristic of serine proteinase inhibitors that are not required for control of cell fusion. *J Virol* **69:**5978
113. Turner, P. C., M. T. Baquero, S. Yuan, S. R. Thoennes, and R. W. Moyer. 2000. The cowpox virus serpin Spi-3 complexes with and inhibits urokinase-type and tissue-type plasminogen activators and plasmin. *Virology* **272:**267
114. Nash, P., A. Whitty, J. Handwerker, J. Macen, and G. McFadden. 1998. Inhibitory specificity of the anti-inflammatory myxoma virus serpin, Serp-1. *J Biol Chem* **273:**20982
115. Lomas, D. A., D. L. Evans, C. Upton, G. McFadden, and R. W. Carrell. 1993. Inhibition of plasmin, urokinase, tissue plasminogen activator, and C1S by a myxoma virus serine proteinase inhibitor. *J Biol Chem* **268:**516
116. Wang, Y. X., P. C. Turner, T. L. Ness, K. B. Moon, T. R. Schoeb, and R. W. Moyer. 2000. The cowpox virus Spi-3 and myxoma virus Serp1 serpins are not functionally interchangeable despite their similar proteinase inhibition profiles in vitro. *Virology* **272:**281
117. Dai, E., H. Guan, L. Liu, S. Little, G. McFadden, S. Vaziri, H. Cao, I. A. Ivanova, L. Bocksch, and A. Lucas. 2003. Serp-1, a viral anti-inflammatory serpin, regulates cellular serine proteinase and serpin responses to vascular injury. *J Biol Chem* **278:**18563
118. Mahalingam, S., and G. Karupiah. 2000. Modulation of chemokines by poxvirus infections. *Cur Opin Immunol* **12:**409
119. Krathwohl, M. D., R. Hromas, D. R. Brown, H. E. Broxmeyer, and K.H. Fife. 1997. Functional characterization of the C-C chemokine molecules encoded by molluscum contagiosum virus types 1 and 2. *Proc Natl Acad Sci USA* **94:**9875
120. Smith, C. A., T. D. Smith, P. J. Smolak, D. Friend, H. Hagen, M. Gerhart, L. Park, D. J. Pickup, D. Torrance, K. Mohler, K. Schooley, and R. G. Goodwin. 1997. Poxvirus

genomes encode a secreted, soluble protein that preferentially inhibits beta chemokine activity yet lacks sequence homology to known chemokine receptors. *Virology* **236:**316

121. Graham, K. A., A. S. Lalani, J. L. Macen, T. L. Ness, M. Barry, L. Y. Liu, A. Lucas, A. Clarklewis, R. W. Moyer, and G. McFadden. 1997. The T1/35kda family of poxvirus-secreted proteins bind chemokines and modulate leukocyte influx into virus-infected tissues. *Virology* **229:**12
122. Seet, B. T., and G. McFadden. 2002. Viral chemokine-binding proteins. *J Leukoc Biol* **72:**24
123. Carfi, A., C. A. Smith, P. J. Smolak, J. McGrew, and D. C. Wiley. 1999. Structure of a soluble secreted chemokine inhibitor vCCI (p35) from cowpox virus. *Proc Natl Acad Sci USA* **96:**12379
124. Lalani, A. S., J. Masters, K. Graham, L. Y. Liu, A. Lucas, and G. McFadden. 1999. Role of the myxoma virus soluble cc-chemokine inhibitor glycoprotein, M-T1, during myxoma virus pathogenesis. *Virology* **256:**233
125. Martinezpomares, L., J. P. Thompson, and R. W. Moyer. 1995. Mapping and investigation of the role in pathogenesis of the major unique secreted 35-kDa protein of rabbitpox virus. *Virology* **206:**591
126. Kotwal, G. J. 2000. Poxviral mimicry of complement and chemokine system components: What's the end game? *Immunol Today* **21:**242
127. Kotwal, G. J., S. N. Isaacs, R. McKenzie, M. M. Frank, and B. Moss. 1990. Inhibition of the complement cascade by the major secretory protein of vaccinia virus. *Science* **250:**827
128. Sahu, A., S. N. Isaacs, A. M. Soulika, and J. D. Lambris. 1998. Interaction of vaccinia virus complement control protein with human complement proteins - factor I-mediated degradation of C3b to iC3b(1) inactivates the alternative complement pathway. *J Immunol* **160:**5596
129. Murthy, K. H. M., S. A. Smith, V. K. Ganesh, K. W. Judge, N. Mullin, P. N. Barlow, C. M. Ogata, and G. J. Kotwal. 2001. Crystal structure of a complement control protein that regulates both pathways of complement activation and binds heparan sulfate proteoglycans. *Cell* **104:**301
130. Smith, S. A., N. P. Mullin, J. Parkinson, S. N. Shchelkunov, A. V. Totmenin, V. N. Loparev, R. Srisatjaluk, D. N. Reynolds, K. L. Keeling, D. E. Justus, P. N. Barlow, and G. J. Kotwal. 2000. Conserved surface-exposed K/R-X-K/R motifs and net positive charge on poxvirus complement control proteins serve as putative heparin binding sites and contribute to inhibition of molecular interactions with human endothelial cells: A novel mechanism for evasion of host defense. *J Virol* **74:**5659
131. Rosengard, A. M., Y. Liu, Z. Nie, and R. Jimenez. 2002. Variola virus immune evasion design: Expression of a highly efficient inhibitor of human complement. *Proc Natl Acad Sci USA* **99:**8808
132. Miller, C. G., S. N. Shchelkunov, and G. J. Kotwal. 1997. The cowpox virus-encoded homolog of the vaccinia virus complement control protein is an inflammation modulatory protein. *Virology* **229:**126
133. Kotwal, G. J., C. G. Miller, and D. E. Justus. 1998. The inflammation modulatory protein (IMP) of cowpox virus drastically diminishes the tissue damage by down-regulating cellular infiltration resulting from complement activation. *Mol Cell Biochem* **185:**39
134. Smith, S. A., and G. J. Kotwal. 2001. Virokines: Novel immunomodulatory agents. *Expert Opin Biol Ther* **1:**343
135. Mayr, A. 2003. Development of a non-immunizing, paraspecific vaccine from attenuated pox viruses: A new type of vaccine. *New Microbiol* **26:**7
136. Graham, B. S. 2002. Clinical trials of HIV vaccines. *Annu Rev Med* **53:**207
137. Tartaglia, J., M. C. Bonnet, N. Berinstein, B. Barber, M. Klein, and P. Moingeon. 2001. Therapeutic vaccines against melanoma and colorectal cancer. *Vaccine* **19:**2571

138. Zavala, F., M. Rodrigues, D. Rodriguez, J. R. Rodriguez, R. S. Nussenzweig, and M. Esteban. 2001. A striking property of recombinant poxviruses: Efficient inducers of in vivo expansion of primed CD8(+) T cells. *Virology* **280:**155
139. Kaufman, H. L. 2003. The role of poxviruses in tumor immunotherapy. *Surgery* **134:**731
140. Kwak, H., H. Horig, and H. L. Kaufman. 2003. Poxviruses as vectors for cancer immunotherapy. *Curr Opin Drug Discov Devel* **6:**161
141. Humrich, J., and L. Jenne. 2003. Viral vectors for dendritic cell-based immunotherapy. *Curr Top Microbiol Immunol* **276:**241
142. Vanderplasschen, A., and P. P. Pastoret. 2003. The uses of poxviruses as vectors. *Curr Gene Ther* **3:**583
143. Bonilla-Guerrero, R., and G. A. Poland. 2003. Smallpox vaccines: Current and future. *J Lab Clin Med* **142:**252
144. Di Giulio, D. B., and P. B. Eckburg. 2004. Human monkeypox: An emerging zoonosis. *Lancet Infect Dis* **4:**15

Chapter 9

INTERFERON ANTAGONISTS ENCODED BY EMERGING RNA VIRUSES

CHRISTOPHER F. BASLER
Department of Microbiology, Mount Sinai School of Medicine, New York, NY, USA

1. INTRODUCTION

Emerging viruses can be considered those which have recently appeared in the human population or which threaten to reemerge in the human population. This definition encompasses a variety of viruses including human specific pathogens such as human immunodeficiency virus (HIV) and hepatitis C virus (HCV), zoonotic viruses such as some arenaviruses or Nipah virus and arboviruses such as West Nile virus. This chapter will look at the mechanisms by which emerging viruses counteract the host IFNα/β system. Its focus will be on several examples of emerging RNA viruses, which are transmitted to humans via either an animal or arthropod host.

2. THE IFNα/β RESPONSE

The IFNα/β proteins are a family of cytokines that bind to a common receptor, the interferon α/β receptor (IFNAR). Originally described as having antiviral activity, IFNα/βs also play important roles in regulating immune responses[1,2]. As will become apparent, much of the work done in the area of "IFN-antagonism" focuses on the specific molecular mechanisms by which an individual viral gene product blocks some aspect of the host IFN response. In some cases, the in vivo importance of a given viral IFN

P. Palese (ed.), Modulation of Host Gene Expression and Innate Immunity by Viruses, 197-220.

antagonist has also been demonstrated, providing a clear demonstration that these proteins act as virulence factors. In the case of many emerging viruses, a role for viral interferon antagonists in pathogenesis has yet to be fully demonstrated. In future *in vivo* studies, it will be important to not only determine whether viral disease is altered when an IFN antagonist function is eliminated, but also to begin to understand the mechanisms by which such viruses are attenuated. For example, does an interferon antagonist only inhibit local induction of an antiviral state, or does it also have a more global effect, perhaps by modulating the ability of IFNα/β to regulate other immune responses?

The IFNα/β proteins are members of the family of what were originally called type I interferons. This family is encoded by, in humans, a single IFNβ gene and a number of IFNα genes. The proteins produced by these genes are secreted and bind to a common receptor, the interferon alpha/beta receptor (IFNAR), and activate a JAK/STAT signaling pathway that leads primarily to the formation of STAT1:STAT2 homodimers which associate with interferon-regulatory factor 9 (IRF9), forming the transcription factor ISGF-3. ISGF-3 accumulates in the nucleus and activates genes with promoters containing interferon-stimulated response elements (ISREs). Treatment of cells with IFNα/β typically induces an antiviral state in which virus replication is usually impaired. The best studied IFN-induced antiviral genes include the dsRNA-activated protein kinase PKR, 2',5-oligoadenylate synthetase (OAS) and the MxA protein. (MxA is the term used to refer to the human protein, Mx1 refers to its mouse homologue). However, other IFN induced genes likely also contribute to IFN's antiviral effects.

IFNα/β production can be induced by a variety of stimuli including viral infection[3] (Figure 1). Several pathways leading to IFNα/β production have recently been identified[3]. In most cell types, the predominant form of IFN first produced in response to virus infection is IFNβ, although select IFNα genes may also be produced. The processes leading to IFNβ production are well studied and involve the activation of the cellular transcription factors NF-κB, interferon regulatory factor 3 (IRF-3) and ATF-2/c-Jun. Most IFNα genes appear to be activated by interferon regulatory factor 7, itself an interferon induced protein. The interferon inducibility of IRF-7 contributes to an enhanced secondary IFN response to virus infection. Because "primed" cells contain IRF-7, subsequent virus infection leads to the activation of IRF-7 and the activation of many IFNα genes. In a manner similar to IRF-7, IRF-5 also turns on expression of IFNα genes, although its activation is virus-specific[4-6]. It should also be noted that some cell types, including plasmacytoid dendritic cells rapidly produce IFNα in response[3].

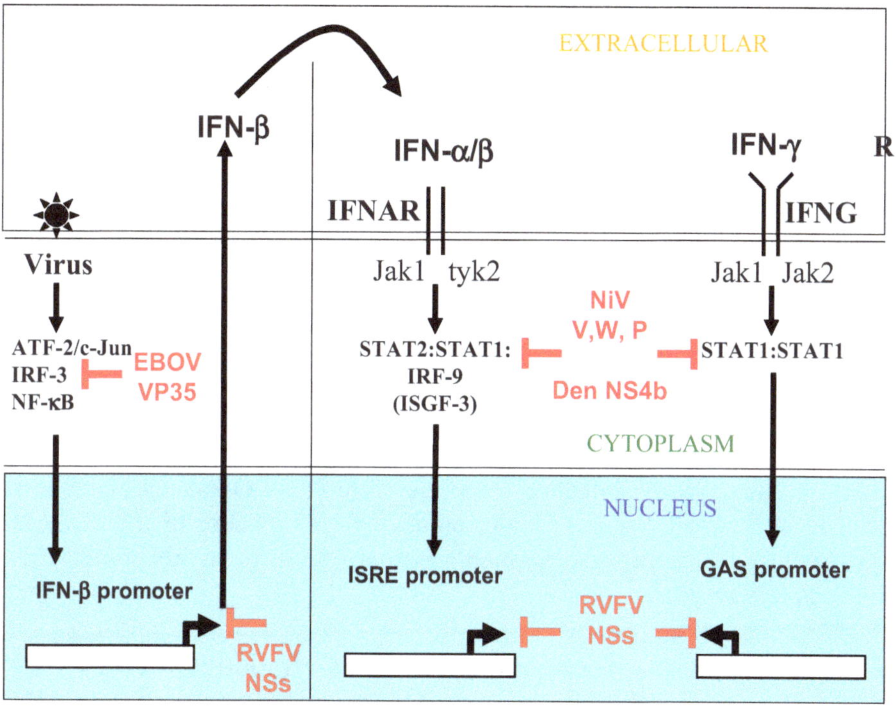

Figure 1. Cellular Targets of Ebola virus, Nipah virus, Rift Valley fever virus and dengue virus interferon antagonists. On the left, the figure depicts in a simplified way, the signaling pathway leading to interferon-beta (IFN-β) gene expression. The cellular transcription factors responsible for activation of the IFN-β promoter include the AP-1 transcription factor complex ATF-2/c-Jun, interferon regulatory factor 3 (IRF-3) and NF-κB. The Ebola virus VP35 protein (EBOV VP35) has been shown to block activation of IRF-3. The right side of the figure depicts the JAK-STAT signaling pathways activated by interaction of either IFNα/β or IFNγ with their respective receptors, IFNAR or IFNGR. The IFNα/β pathway activates the receptor associated kinases Jak1 and Tyk2. These phosphorylate STAT1 and STAT2, which form heterodimers, associate with IRF-9 (forming the transcription factor complex ISGF-3) and activate transcription within the nucleus. The IFNγ pathway activates the receptor associated kinases Jak1 and Jak2 which, in turn, induce STAT1 homodimers which activate IFNγ-responsive genes. The Nipah virus (NiV) P, V and W proteins as well as the dengue virus (Den) NS4b protein all appear to inhibit the activation of STAT1, thus blocking signaling from both IFNAR and IFNGR. The consequence of all of these signaling pathways is the activation of gene expression, including the expression of antiviral genes (not shown). The Rift valley fever virus (RVFV) NSs protein, as well as the NSs of bunyamwera virus (not shown), target cellular transcription preventing the production of IFN gene expression as well as the expression of other genes. See text for details and references.

Several cellular signaling pathways have been described that lead to IFNα/β production. The basic background has been recently reviewed and will be summarized here[3]. These pathways include signaling from specific Toll-like receptors (TLRs) and from TLR-independent pathways. The TLR3 and TLR4 pathways leading to activation of IRF-3 and to IFN production have been fairly well characterized. TLR3, because it is activated by dsRNA, long used as an IFN-inducer and experimentally as a mimic of virus infection, has been suggested to act as a sensor of virus infection. TLR4, which recognizes LPS, is reportedly also activated by certain viral glycoproteins. In addition, TLRs 7 and 8 can recognize some forms of ssRNA and viral RNA resulting in production of IFNα, and TLR9 can recognize CpG DNA and has been implicated in induction of IFN by some DNA viruses. However, many cell types do not express TLRs, yet they still produce IFNα/β in response to virus infection. Other cellular "receptors" that may detect virus infection must therefore exist. One such example is RIG-I, an intracellular protein and putative RNA helicase of the "DEAD-box" family that appears to play an essential role in the dsRNA and virus-induced IFN response[7].

2.1 Viral interferon antagonists as virulence factors

The interferon response of the host provides an early means of eliminating virus infections. In apparent consequence, many viruses have evolved mechanisms to evade the interferon response[8]. For each "arm" of the IFN response described above, examples of viral encoded antagonists exist. Those molecules produced by viruses that counteract any of the various components of the host interferon system are, in this chapter, termed "IFN antagonists." Studies on a variety of viruses indicate that viral IFN antagonists make an important contribution to virulence. The best evidence for this argument are viruses which, due to mutations in genes that counteract the IFN response, show attenuation in hosts able to mount an IFN response. For example, a mutant influenza virus lacking the NS1 protein[9,10], a herpes simplex virus 1 with a mutant ICP34.5 [11,12] or vaccinia viruses with mutant E3Ls[13] lose virulence in mice. However, the same mutant influenza and herpes viruses regain their virulence when inoculated into mice unable to mount a normal IFN response (e.g. STAT1 -/-, IFNAR -/- or PKR -/- mice), demonstrating that the attenuation of these mutants is mediated by IFN[9,14,15]. Based on such observations, it is hypothesized that emerging viruses must also counteract the host interferon response in order to cause disease.

3. FILOVIRUSES

Filoviruses, Ebola and Marburg viruses, are perhaps the most infamous of emerging viruses due to their high lethality in humans and due to reports of dramatic symptoms associated with Ebola hemorrhagic fever. These viruses continue to cause periodic outbreaks of severe viral hemorrhagic fever in humans and are of concern as potential bioweapons[16,17]. Filoviruses were first identified following a 1967 outbreak of hemorrhagic fever among workers in European vaccine production facilities. Those who became ill were working with imported African green monkey kidney cultures, and the outbreak had a 23 percent fatality rate. The causative agent was named Marburg virus after the city in which it was identified. Ebola viruses were first identified in the mid-1970's following outbreaks in Zaire (now the Democratic Republic of Congo) and Sudan. Mortality rates were even higher than with Marburg virus. The Zaire outbreak had a reported mortality rate of 88 percent, while the Sudan outbreak had a 53 percent rate of death[18]. Several additional filovirus outbreaks have occurred in humans, and such outbreaks have been identified with increasing frequency[19]. Examples include outbreaks of Ebola hemorrhagic fever, which occurred in Gabon and the Republic of Congo in 2001 and 2003[20,21].

Filoviruses are cytoplasm-replicating, filament-shaped, enveloped viruses with negative-sense RNA genomes[22]. Four subtypes of Ebola virus, Zaire, Sudan, Ivory Coast (Côte d'Ivoire) and Reston, and one subtype of Marburg virus have been recognized[22]. Ebola Zaire and Sudan viruses have caused highly lethal disease in humans. The Reston Ebola virus has only been found to cause disease in non-human primates, while the few documented cases of human exposure have not caused clinical illness[19]. The Zaire Ebola virus is the most virulent in humans with fatality rates as high as 88 percent during human outbreaks[19]. Despite this information, however, determinants of Ebola virus virulence and of subtype-specific differences in virulence remain unclear[19].

3.1 Immune and cytokine responses during the course of Ebola virus infection

Studies of both human and experimental, non-human primate infections have detected substantial cytokine production and inflammatory responses during the course of Ebola virus infections in vivo[23-26]. It is as yet unclear how these inflammatory/host cytokine responses contribute either directly or indirectly to the clearance of virus in non-fatal cases, or how important a role these host cytokine responses play in the pathology of fatal filovirus infections. However, differences characteristic of either fatal or non-fatal disease have been described[24,27]. It has been suggested that "very early events in Ebola virus infection" determine the ultimate course of the disease, and that survival was characterized by "orderly and well-regulated" immune responses[26,27]. In support of this, examination of 11 asymptomatic individuals infected with Ebola virus identified an early, transient period (2-3 days) in which pro-inflammatory cytokines and chemokines, including IL-6, IL-1β, TNFα, MCP-1, MIP-1α and MIP-1β, were detected[25]. Levels of Ebola virus RNA in asymptomatic patients were very low throughout the course of infection, and it is not clear whether the inflammatory response—reflected by cytokine production—was involved in controlling the infection in these individuals[25]. Cytokine production may also mediate the hemorrhagic manifestations of filovirus infection; it was shown that Marburg virus infection of monocytes/macrophages induces sufficient TNFα production to enhance endothelial permeability[28].

3.2 Ebola virus and IFN

3.2.1 IFNα/β treatment does not prevent Ebola virus disease in non-human primates

IFNα/β has been reported to have at best modest effects against Ebola virus infection. 200 IU/ml of IFN-α2b was found to inhibit Ebola virus replication 100-fold in Vero cells[29] In vivo, intramuscular treatment of cynomolgus monkeys with the high dose of $2x10^7$ IU/kg of IFN-α2b once daily beginning 18 hours post inoculation was tested against Ebola-Zaire virus. This treatment merely delayed viremia and death by about one day[29]. This data suggesting that Ebola Zaire virus is relatively insensitive to IFN treatment fits with the reports of IFN antagonists encoded by the virus. Conceivably, the efficacy of IFN-treatment might be improved if the viral IFN antagonist(s) were to be inhibited.

3.2.2 The role of the IFN response in restricting Ebola virus infection in mice

The ability of the interferon system to prevent Ebola virus disease in mice argues for the importance of IFN antagonists in filovirus pathogenesis. Non-mouse-adapted filoviruses do not cause lethal disease in adult or weanling mice, but can kill newborn mice and SCID mice[30]. However, adaptation by repeated passage in mice yielded a mouse lethal Zaire Ebola virus[31]. This "mouse-adapted" Ebola virus causes fatal illness following intraperitoneal (i.p.) injection, and the pathology of the disease in mice has similarities to the human disease[31]. However, non-adapted Zaire Ebola virus is able to kill mice deleted of either the IFNα/β receptor (IFNAR -/- mice) or STAT1 (STAT1 -/- mice)[32]. In the knock-out mice, the route of virus administration did not matter; i.p. or subcutaneous (s.c.) inoculation led to lethal illness with either mouse-adapted or wild-type Zaire Ebola virus[32]. However, it should be noted that not all filovirus isolates tested were lethal in IFNAR -/- mice. While Sudan Ebola virus and two isolates of guinea pig-passaged Marburg-Musoke virus killed all knock-out mice, and a human isolate of Marburg-Musoke virus killed 1 out of 3 mice, the 1995 Zaire Ebola virus, Reston Ebola virus and Ivory Coast Ebola virus did not kill the knock-out mice[32].

In mice, virulence appears to be influenced in particular by IFN produced early in infection. When mice were infected with the mouse-adapted Ebola virus s.c., a normally non-lethal route, and then injected with anti-IFNα/β antibodies at various times post-infection, early neutralization of IFN (injection with antibody on day 2) led to rapid and complete killing. When anti-IFN antibodies were given at later times post-infection, mice died more slowly and fewer mice died[32]. This data demonstrates that the IFNα/β system plays an important role in restricting filovirus replication in the mouse and suggests that the ability of a filovirus to overcome the host IFN response may be a requirement for lethal infection in a given species. Thus, the ability of Ebola viruses to cause disease in primates may also require virus-encoded mechanisms of host IFN evasion. The variable ability of different filoviruses to cause disease in the IFNAR -/- mice suggests, however, that the IFN system is not the only host-range determinant of filovirus virulence.

Studies on the anti-Ebola virus activity of the drug 3-deazaneoplanocin (c^3-Npc A) also suggest a role for the IFNα/β response in regulating filovirus pathogenesis. c^3-Npc A was found to protect mice from lethal infection with mouse-adapted Ebola virus[32-34]. However, the efficacy of c^3-Npc A could be blocked by co-administration of anti- IFNα/β antibodies and the drug[32]. c^3-Npc A was previously described as an inhibitor of the host cell enzyme S-adenosylhomocysteine hydrolase which exerted an antiviral effect due to its ability to reduce methylation of 5'mRNA caps. Subsequent studies on Ebola virus found that the drug induced a large amount of IFNα/β in virus-infected mice but induced little to no IFN in uninfected mice[34]. There was also a good correlation between the induction of IFN, viral spread and virulence[34]. The authors suggest that the drug somehow blocks the ability of Ebola virus to inhibit host IFN production in the mouse[34].

3.2.3 Evidence for Ebola virus evasion of the IFNα/β system

Several studies suggest that Ebola virus infection counteracts the host cell IFN antiviral response[35-37]. Zaire Ebola virus infection of human umbilical vein endothelial cells (HUVECs) appeared to inhibit signaling through the IFN alpha receptor or the IFN gamma receptor in that induction of gene expression by either IFNα or IFNγ was inhibited[36]. Infection also blocked the IFN-induced activation of IRF-1 and 2',5'-OAS by IFNα or IFNγ[36]. These conclusions were further supported by electrophoretic mobility shift assays (EMSAs) that demonstrated inhibited formation of IFNα/β– or IFNγ-induced transcription factor complexes[36]. In contrast, infection did not block induction of gene expression by IL-1β nor did it block activation of functional NF-κB transcription complexes by IL-1β[36]. Thus, infection does not induce a global inhibition of signaling, but rather targets specific signaling pathways, including the IFN pathways.

Zaire Ebola virus infection also inhibits the ability of HUVECs to produce IFN in response to dsRNA (poly I:polyC) treatment[35]. In uninfected HUVECs, dsRNA treatment enhanced cell surface expression of class I MHC, and induced IRF-1, 2'-5' OAS, PKR, ICAM-1 and IL-6 gene expression. In contrast, in Zaire Ebola virus infected cells, dsRNA-mediated induction of IFN gene expression was impaired[35]. This inhibition was not due to a global block in cellular signaling pathways nor was it due to a general inhibition of host cell gene expression[35]. Also, the effect on IFN gene expression was not due to a global inhibition of dsRNA signaling, as polyI:polyC still activated NF-κB in infected cells[35]. EMSA assays indicated that the dsRNA-treatment of uninfected cells resulted in the inactivation of ISRE and GAS binding complexes, indicating that the dsRNA-treatment induced IFN-production[35].

Studies using human peripheral blood mononuclear cells and macrophages also argue for an Ebola virus imposed block to dsRNA-induced IFNα/β production[37]. When PBMC or human macrophages were infected with Zaire Ebola virus (a 1976 isolate), pro-inflammatory chemokines and cytokines such as MCP-1, MIP-1α, RANTES and TNFα were produced by PBMC, and MIP-1α and TNFα were produced by macrophages. However, IFNα and IFNβ production could not be detected until three days post-infection, and at this time only small amounts of IFN were seen. Evidence for active inhibition of IFNα/β production was also provided. Infected macrophages were treated with dsRNA, but this treatment did not induce IFNα production in infected cells. In contrast, the same treatment did induce IFNα production in uninfected cells[37].

3.2.4 Production of IFN during the course of Ebola virus infection

It should be noted that some authors have reported Zaire Ebola virus infection (a 1995 strain) does induce IFNα in primary human monocyte/macrophage cultures[23]. Why this study disagrees from other studies is not clear, but the use of different isolates or differences in the cell cultures used (for example, different quantities of plasmacytoid dendritic cells may have been present) might account for the different levels of IFN production.

IFNα/β are also produced in Ebola virus-infected patients and in Ebola virus-infected non-human primates[23,26,37]. When sera from Ebola virus-infected patients were examined for cytokines, all had detectable levels of IFNα and IFNβ in addition to other cytokines, whereas pooled serum from uninfected individuals had undetectable levels of IFNα/βs[37]. Of the nine sera examined, the seven sera taken during the acute phase of infection had higher levels of IFNα (as well as MCP-1) than two other sera from infected patients, one taken during the recovery phase and one taken during the convalescent phase of infection[37]. Perhaps Ebola virus blocks production of IFN in a cell type specific manner, such that early targets of infection produce little IFN, but later targets of infection produce more IFN. Characterization of the response of different cell types to Ebola virus infection is therefore of interest. It is also of interest that even the significant levels apparent in infected non-human primates are not sufficient to prevent an invariably fatal outcome.

3.2.5 Identification of VP35 as an Ebola virus-encoded IFNα/β-antagonist

Because Ebola virus reportedly blocks both host cell production of IFNα/β and host cell response to IFN, a search was undertaken for Ebola virus proteins which could act as IFN antagonists. Zaire Ebola virus open-reading frames were therefore cloned into a mammalian expression plasmid and tested for their ability to functionally substitute for the "IFN antagonist" NS1 protein of influenza A virus[38]. Specifically, cells were transfected with expression plasmids and tested for their ability to complement the otherwise impaired growth of a mutant influenza virus lacking the NS1 open reading frame. Of the seven Ebola virus proteins tested, only the VP35 protein significantly enhanced influenza delNS1 virus replication suggesting that VP35 might act as an antagonist of the IFN-response in Ebola virus infected cells[38]. VP35 was also found to be able to inhibit the expression of reporter genes whose expression depends upon the activation by virus or dsRNA of the cellular transcription factor interferon regulatory factor 3 (IRF-3)[38,39]

(Figure 1). These data suggest that VP35 could block virus or dsRNA induced IFN production. The ability of VP35 to block reporter gene activation was sustained even when VP35 was co-expressed with Ebola virus NP, the most abundant member of the Ebola virus RNA replication complex known to interact with VP35[38]. VP35 also retained its IFN antagonist function when it was co-expressed with both NP and VP24 (data not shown), proteins which cooperate with VP35 to form viral nucleocapsids[40]. Thus, it seems likely that VP35 would be able to block the host cell IFN production in Ebola virus infected cells. Further evidence of a VP35 IFN antagonist function was provided by a separate study in which expression of VP35 from replication-defective alphavirus vectors inhibited IFNα production in dendritic cells[41]. To determine whether VP35 can also inhibit signaling from the IFNAR, cells were first transfected with a VP35 expression plasmid and an IFN-responsive-reporter plasmid and subsequently treated with IFNβ. VP35 did not significantly inhibit IFN-induced activation of the ISRE promoter, indicating that VP35 does not efficiently block IFNα/β signaling (data not shown).

3.2.6 VP35 inhibition of IRF-3

More detailed characterization of the effect of VP35 on IRF-3 suggests that the phosphorylation of IRF-3, which serves as the transcription factor's activating signal, is blocked. As a consequence, VP35 prevented the virus-induced, IRF-3-dependent activation of IRF-3 responsive promoters. Further, in a commonly used assay of IRF-3 activation[42,43], VP35 could block the nuclear translocation of and GFP-tagged IRF-3. Finally, Sendai virus induced phosphorylation and dimerization of IRF-3 was inhibited[39]. These results provide strong evidence that VP35 can block the virus-induced activation of IRF-3.

3.2.7 Use of "reverse genetics" to further clarify roles in viral replication

Recently, reverse genetics methods have been established for Ebola virus, permitting the introduction of specific alterations into the viral genome. These techniques should facilitate the evaluation of the IFN antagonist function of VP35 in the context of infectious Ebola virus.

4. HENIPAVIRUSES

Nipah virus (NiV) and Hendra virus (HeV) are zoonotic pathogens which comprise the *Henipavirus* genus of the family *Paramyxoviridae*. As paramyxoviruses, they are enveloped, negative-strand RNA viruses that replicate in the cytoplasm of cells[44]. HeV was identified in Australia in 1994 following an outbreak in horses and has been associated with three human cases, two fatal[45]. NiV was identified following an outbreak in pigs and in humans who had contact with infected pigs. In humans, the virus killed 105 of 276 individuals with clinical disease with the most prominent clinical feature being encephalitis[45]. Nipah virus is comparatively unique among paramyxoviruses in that it readily crosses species barriers to cause disease. This ability may reflect the ability of the virus to counteract the IFN response by multiple mechanisms.

4.1 Henipavirus IFN antagonists

NiV and HeV appear to encode, like many paramyxoviruses, several proteins from within their phosphoprotein (P) gene. Translation from an mRNA transcribed from a faithful copy of the genomic RNA template would yield the phosphoprotein (P protein), a component of the viral RNA polymerase[46]. However, paramyxovirus P genes under go "RNA editing" during their transcription. This process, which involves the addition of non-template encoded G residues to a specific position within the nascent transcript, results in the production of proteins with common amino-termini but different carboxy-termini. In the case of Nipah and Hendra viruses, the +1 G product would be the "V" protein while the +2 G product would be the "W" protein[47]. An additional P gene product, the "C" protein, which would be translated from an alternate, internal AUG start codon, is also predicted for Nipah and Hendra viruses[47].

Evidence suggests that each of the 4 P gene products of Nipah virus are able to counteract the host IFN response at some level[47-51]. Expression of Nipah C, P, V or W was able to counteract the antiviral effects of IFNα/β. This was demonstrated by transfection of cells with expression plasmids encoding the Nipah virus proteins or with control plasmids and subsequent infection of the cells with a Newcastle disease virus (NDV) expressing green fluorescent protein (GFP). Transfection of plasmids resulted in the production of IFNα/β which suppressed viral replication unless a given plasmid also expressed an interferon antagonist[47]. For the V, W and P proteins, this observation correlated with their ability to inhibit IFN signaling[47-49] (Figure 1). For the C protein, reporter gene assays did not suggest a specific mechanism by which it rescues NDV-GFP from the effects of IFN[47].

The manner in which the NiV P gene products inhibit IFN signaling appears to be somewhat unique among paramyxoviruses. The V proteins of other paramyxoviruses also block IFN signaling. However, these other V proteins inhibit IFN signaling by targeting STAT1 or STAT2 for proteasome-mediated degradation[52-57]. In contrast, STAT1 and STAT2 are not degraded by Nipah V, W or P. NiV V instead prevents the phosphorylation of STAT1 and shifts STAT1 and STAT2 into high molecular weight complexes[49]. Similar observations were made for HeV V[51]. When the localization of STAT1 and STAT2 is analyzed by immunofluorescence, STAT1 is retained in the nucleus, even following IFNα/β treatment which otherwise induces STAT1 and STAT2 nuclear accumulation[48,51], and the cytoplasmic localization of NiV V and the cytoplasmic retention of STAT1 and STAT2 reportedly require an Crm1-dependent nuclear export pathway[49]. Interestingly, the NiV P and W proteins also affect STAT protein localization. The P protein, like the V protein, is cytoplasmic and retains STAT1 in the cytoplasm as well[48]. In contrast, the W protein localizes to the nucleus and also relocalizes STAT1 to the nucleus, even in the absence of STAT1 activation[48]. The unique carboxy-terminal region of W has been found to possess a nuclear localization signal which is required for its nuclear accumulation[58]. That this nuclear STAT1 is inactive is evidenced by the ability of W to prevent IFN-induced, STAT1-dependent gene expression[47,48]. Although, a mutant W that localizes to the cytoplasm, and thus is similar to V, still inhibits IFN signaling (unpublished observation).

The ability of NiV V, W and P to inhibit STAT1 and STAT2 function depends upon the presence of a unique amino-terminal STAT1-binding domain. As noted above, the P, V and W share a common amino-terminus and which in NiV and HeV is longer than that of other paramyxoviruses[59]. This amino-terminal extension lacks significant homology to other paramyxoviruses (data not shown). Deletions within this region impair the IFN antagonist function of the proteins, and IFN antagonist function correlates with binding to STAT1[48,50], although V also interacts with STAT2 via STAT1[50]. A region encompassing only amino acids 50-150 of NiV P/V/W was sufficient to block IFN-induced gene expression and to bind to STAT1[48]. This situation is different than that reported for other IFN antagonist paramyxovirus V proteins which target STAT1 or STAT2 for degradation. These proteins appear to require a carboxy-terminal, cysteine-rich domain to target STATs for degradation. Interestingly, the NiV and HeV V proteins also possess the conserved, carboxy-terminal cysteine-rich domain. Why it is that henipavirus V proteins do not target STAT1 for degradation is not clear. It is worth considering, however, that NiV and HeV are zoonotic agents to naturally infected Old World fruit bats. It will be of interest to determine whether NiV V might destabilize STAT proteins in fruit bats cells.

4.2 Impact of Nipah virus V and W proteins on induction of IFNα/β production

The impact of NiV V and W proteins on induction of an interferon response by viral and non-viral pathways has also been examined. Both V and W proteins were found to inhibit activation of IRF3-responsive promoters by Sendai virus. However, when another pathway leading to the activation of IRF-3 (the TLR3 signaling pathway) was examined, only W and not V could block. The nuclear localization of W was required to inhibit TLR3 signaling, although tagging of the V protein with a nuclear localization signal conferred the ability to also block TLR3 signaling. Signaling from TLR3 to IRF-3 requires a Toll IL-1 receptor (TIR) domain-containing adaptor protein, TRIF. The ability of W to inhibit signaling from TLR3 correlated well with its ability to block activation of gene expression by TRIF. As expected, V failed to block TRIF-mediated activation of IRF-3 responsive genes. When the downstream kinases that phosphorylates IRF-3 were examined, it was found that while both V and W could inhibit promoter activation by the IKKε, only W inhibits activation by TBK-1. These observations in which V and W have differential abilities to block virus- and TLR-3-pathways and to block IKKε or TBK-1 are consistent with recent reports that TBK-1 plays a critical role in TLR3 (and TLR4) signaling whereas either IKKε or TBK-1 play a critical role in virus induced activation of IRF-[60-62]. Although the relative importance of TLR3 signaling for initiating an interferon response remains uncertain, the data clearly indicate that W and V do not possesses exclusively overlapping functions[58].

5. BUNYAVIRUSES

The bunyaviruses are a family of enveloped viruses with segmented, negative-sense RNA genomes[63]. The family is large and diverse and includes several important emerging human viruses[64]. Many members are arboviruses, including members of the genera *Bunyavirus*, *Nairovirus* and *Phlebovirus*. One genus, *Hantavirus*, consists of viruses carried by rodents and transmitted via rodent droppings[64]. The family also includes a genus of plant viruses, the tospoviruses[64].

Rift Valley fever virus RVFV is an arthropod-borne Phlebovirus (genus in the family Bunyaviridae). RVFV is endemic in sub-Saharan Africa although outbreaks also occur in the Middle East[64]. Significant epizootics occur in association with high rainfall levels and flooding which leads to the proliferation of vector mosquitoes. Because the virus is transmitted transovarially an animal reservoir may not be needed for maintenance of RVFV[64]. RVFV causes zoonotic infection in humans, in which infections are most often transmitted from infected animals to humans in close contact with them. Other less common routes of infection include mosquito bites and contact with infectious bodily fluids. Rift Valley fever typically causes an influenza-like illness but may also cause complications such as liver necrosis, encephalitis or retinitis[64].

A role for the interferon response in the effective control of RVFV infection was provided by infections in rhesus monkeys. IV inoculation of the monkeys with virus resulted in clinical presentations ranging from lethal hemorrhagic fever to clinically unapparent. The appearance of interferon correlated with the outcome, suggesting that an early effective interferon response helped control disease outcome[65].

Reassortment studies identified determinants of virulence in the S segment. The S segment of an attenuated isolate of RVFV, clone 13, was found to attenuate the virus in mice. The clone 13 S segment was found to contain a large, in-frame deletion within the sequences encoding the NSs protein (the non-structural protein on the small (S) segment), suggesting that NSs is a virulence determinant[66]. Subsequent studies suggest that the NSs protein confers resistance to IFNα/β[67]. The attenuated clone 13 and the MP12 vaccine strain (which contains attenuating mutations on all three viral segments), became lethal in IFNα/β receptor knock-out mice, causing disease similar to wild-type virus and growing to titers in vivo similar to wild-type virus. Thus, IFN-resistance maps to the S segment, and based upon the observation that clone 13 contains a deletion in its NSs, it appears that the RVFV NSs confers IFN resistance[67].

Evidence for an interferon antagonist function of the Bunyamwera virus (BUNV) NSs has also been obtained. BUNV is the prototype for the Bunyavirus genus within the *Bunyaviridae*. A recombinant BUNV which fails to express the NSs protein was found to have defects in growth in tissue culture, to be attenuated following intracranial inoculation in mice and to induce more interferon than the NSs-expressing counterpart[68]. Expression of NSs was sufficient to inhibit virus and dsRNA-mediated activation of IRF-3 and NF-κB-dependent transcription. The attenuation of the deltaNSs virus both in terms of virus replication in cell culture and in mice was relieved when the IFNAR was absent[69]. This provided direct evidence that the NSs gene product is required for efficient replication in the presence of an

IFNα/β response. Interestingly, although the presence of NSs did not prevent activation of the cellular transcription factor IRF-3, IRF-3-dependent gene expression and IRF-3-induced apoptosis were inhibited by NSs[70].

Mechanisms by which NSs proteins inhibit the IFN response have recently been described (Figure 1). For both the phlebovirus, RVFV and the bunyavirus, bunyamwera virus appears to inhibit host cell transcription. For RVFV, viruses with wild-type NSs proteins inhibit production of both polyA+ and polyA- (primarily rRNA) in host cells. A yeast two-hybrid screen of a mouse embryo cDNA library identified the p44 subunit of the TFIIH basal transcription factor as a NSs interactor. Subsequent experiments suggest that NSs prevents the proper assembly of the TFIIH complex, thus disrupting TFIIH-dependent transcription[71]. Thus, by blocking transcription in a global way, RVFV NSs appears to effectively block IFN production.

The bunyamwera virus (BUNV) NSs protein has also been found to inhibit transcription with specific effects on RNA polymerase II. Isogenic wild-type and NSs deletion BUNVs have been constructed[68]. Efficient shut-off of host cell gene expression is observed in mammalian cells following infection with NSs-expressing virus but not with the deltaNSs virus[68], and expression of NSs is sufficient to inhibit expression from several polII-driven promoters[72]. Expression of NSs, either from virus or from a plasmid, appears to suppress the phosphorylation of the carboxy-terminal domain (CTD) of RNA polII. Specifically, the CTD contains 52 repeats of a consensus sequence YSPTSPS. Phosphorylation of serine 5 is associated with transcription initiation while phosphorylation of serine 2 is important for mRNA elongation and 3'-end processing. Expression of NSs appears to specifically affect serine 2 phosphorylation and to block the transition of polII from initiation to elongation[72].

Hantaviruses, as noted above, belong to a separate genus of the bunyavirus family. These are not arboviruses but are transmitted via rodent excreta[64]. The interaction between hantaviruses and the interferon response has also been examined by several groups. Several studies find a correlation between the interferon response to a given virus and its pathogenesis. Unlike the phleboviruses and bunyaviruses, no NSs protein has been identified in hantavirus infected cells. Thus far, no specific viral gene product has been definitively identified as an IFN antagonist.

Infection of human umbilical vein endothelial cells (HUVECs) with either the pathogenic Hantaan virus (HTNV) or the relatively nonpathogenic Tula virus (TULV) has been compared. Although both viruses induced IFN-β, infection with HTNV induced larger amounts of IFN-β than did TULV. Interestingly, MxA protein levels were higher and rose faster in TULV-infected cells than in HTNV-infected cells. HTNV replicated to higher titers in HUVECs than did TULV[73]. One possible explanation for these observations would be that HTNV encodes a more effective inhibitor of the IFN system than does TULV.

Infection of HUVECs with the HPS strain NY-1, the HFRS strain HTNV and the non-pathogenic PHV, indicated that the non-pathogenic strain induced more genes in general and more IFN-responsive genes early in infection. All three viruses induced many IFN-responsive genes later in infection. Other differences were also noted, including an upregulation of chemokine gene expression unique to HTNV infected cells[74]. These data thus implicate differential activation of the IFN response as one possible factor in hantavirus pathogenesis, although the determinants of induction versus non-induction of the IFN response is unclear.

6. FLAVIVIRUSES

The *Flaviviridae* family includes the flaviviruses, pestiviruses and hepatitis C viruses. For the purposes of illustration, this chapter will only focus on the flavivirus genus. However, hepatitis C virus is an emerging virus of significance for human health. The literature on hepatitis C virus and the interferon system is extensive and controversial. However, a number of studies do suggest a variety of mechanisms by which HCV may evade the IFN response (for example[75-77]). Readers are referred to the literature for more detail.

6.1 Dengue virus

Dengue virus is a mosquito-transmitted virus of great public health significance across the world[78]. Several lines of evidence suggest that interferons may play an important role in the outcome of dengue virus infection. For example, pre-treatment of human cells with IFNα, IFNβ or IFNγ inhibits dengue virus replication, although the antiviral effects of IFNs are diminished when treatment beings a few hours after infection[79]. The IFN-pre-treatment appears to block dengue virus replication by a PKR-independent mechanism[79,80]. Further, mice lacking IFNAR of IFNGR show increased susceptibility to dengue virus infection[81,82]. At the same time, infection of human cells induces the production of IFNs, and IFNγ appears to enhance the level of DC activation[83].

Evidence now points to the existence of dengue virus-encoded IFN antagonists. Individual dengue virus proteins were screened for their ability to act as interferon antagonists in a manner analogous to that used to identify the IFN antagonists of Nipah virus. Expression of either NS2A, NS4A or NS4B was able to rescue replication of an NDV-GFP virus from inhibition by IFN. Expression of any of these proteins also inhibited, to some extent, the ability of IFNα/β to activate expression of interferon-responsive reporter genes. Of these 3 putative interferon antagonists, NS4B appeared to have the most potent IFN antagonist activity. Further analysis suggests that NS4B specifically blocks IFNβ signaling and does so by preventing STAT1 phosphorylation (Figure 1). Consistent with this observation, infection of LLMCK2 cells with dengue virus type 1 also inhibits phosphorylation of STAT1 in response to either IFNβ or IFNγ[84].

6.2 Other flaviviruses

Data on other flaviviruses also suggests that IFNs modulate the outcome of infection and that flaviviruses other than dengue virus also counteract host IFN responses. For example, elimination of IFN signaling, by disruption of the IFNAR or IFNGR, exacerbates disease caused by the flaviviruses Murray Valley encephalitis virus[85]. Additionally, a study on Japanese encephalitis virus suggests the presence of an inhibitor of the IFNAR associated kinase Ttk2 which inhibits IFNα/β signaling although a specific viral interferon antagonist was not identified[86]. These observations and the numerous reports suggesting modulation of the IFN response by hepatitis C virus, suggest that additional IFN antagonists will be identified in other flaviviruses.

7. ALPHAVIRUSES

Members of the alphavirus genus of the family *Togaviridae* are arboviruses which are usually maintained in a cycle of mosquito to bird or small mammal and back to mosquito. Several of these viruses are significant human pathogens[87]. Several studies have described the ability of IFNα/β to inhibit alphavirus replication, and elimination of components of the IFN system alters alphavirus pathogenesis in mice[29,88-92].

Venezuelan equine encephalitis virus (VEEV) belongs to the genus *Alphavirus* in the family *Togaviridae.* VEEV antigenic subtype I (varieties AB and C) have been associated with major epidemics and equine enzootics, and horses are the major host for amplification of the virus in these scenarios. Antigenic subtypes ID to IF and II-VI are enzootic, do not typically cause disease in horses and circulate in sylvatic cycles between rodents and mosquito, although subtype IC viruses may evolve from subtype ID progenitors[93]. Experimental infections of horses demonstrates that the epizootic strains typically cause high titer viremia and encephalitis whereas enzootic strains display little to no viremia and no disease[94]. Some studies have reported an association between resistance to IFNα/β treatment and epizootic potential of some strains[92,95]. However, a more recent study calls this conclusion into question. When two recombinant, closely-related VEEVs, one an epizootic subtype IC strain and the other an enzootic subtype ID strain, were compared for sensitivity to murine IFNα/β, no significant differences were found[96]. It thus remains unclear whether VEEVs might encode an antagonist of the IFN system that may influence virulence and epidemic/epizootic potential.

A mutation in the 5' untranslated region (5'UTR) of VEEV has been described whose attenuation phenotype is dependent on an intact IFNα/β response. Mutation of G to A at position 3 of the 5'UTR resulted in a virus avirulent in mice. However, in IFNα/β receptor null mice, the mutant was a virulent as wild-type. The molecular explanation might be related to a role for the 5'UTR in regulating translation of the viral non-structural proteins or a role for the 5'UTR in regulating initiation of viral plus-strand RNA synthesis[89].

Studies on the prototype alphavirus, Sindbis virus, have demonstrated that the nsP2 protein influences levels of IFNα/β induction during infection. nsP2 is an essential component of the alphavirus replicase which possesses helicase, NTPase and protease activities. Mutations in Sindbis virus nsP2 replicons resulted in non-cytopathic replicons which were rapidly lost from some cell types, despite the presence of positive drug selection. Introduction of one such mutant nsP2-P726L into replication—competent virus resulted in an attenuated virus which induced large amounts of IFNα/β relative to wild-type virus[91].

8. CONCLUSIONS

From the examples provided above, it is clear that emerging viruses, like other viruses, frequently encode gene products that counteract the host IFN response. That these gene products are important for virulence is evidenced by the data on Ebola virus replication in the mouse, where the presence of an IFN response prevents disease but its removal (as in IFNα/β receptor knock-out mice) leads to rapidly lethal infection. It is also illustrated by the importance for RVFV virulence of an intact NSs gene. It is therefore likely that the IFN antagonist activities of other emerging viruses will play an essential role in pathogenesis. It may be hypothesized, for example, that Nipah virus, because it encodes multiple IFN antagonists, is particularly suited to counteract the host innate immune response and thus to cause disease. A final thought worth considering is the role of IFN antagonists in host range restriction. In the case of zoonotic human pathogens, the ability of

IFN antagonists to function in different species is of obvious importance. It has been demonstrated that the interferon antagonist proteins of viruses can function in a species specific manner, thus acting as host range factors[97-102]. An emerging aspect of IFN antagonists then is the role this function plays in interspecies transmission and in species-specific disease. Questions that might be asked, for example of Nipah virus, include how well do the P, V, W and C proteins function as IFN antagonists in the virus' presumed natural host, Old World fruit bats? Is disease less in this host compared with accidental hosts such as pigs and humans, and is this difference due to the potency of the IFN antagonists? Other questions that may be asked regarding emerging viruses include, does the establishment of a virus in a new species require adaptive changes in IFN antagonist function, and in the case of arboviruses, do the IFN antagonists that function in mammalian cells also serve to block innate immunity in arthropod hosts where no exact homologue of IFNs have been found? In conclusion, much has been learned, but much more remains to be explored.

REFERENCES

1. Biron, C. A. Interferons alpha and beta as Immune Regulators-A New Look. *Immunity* **14:**661-4
2. Hoebe, K. & Beutler, B. LPS, dsRNA and the interferon bridge to adaptive immune responses: Trif, Tram, and other TIR adaptor proteins. *J Endotoxin Res* **10:**130-6
3. Malmgaard, L. Induction and regulation of IFNs during viral infections. *J Interferon Cytokine Res* **24:**439-54
4. Barnes, B. J., Moore, P. A. & Pitha, P. M. Virus-specific activation of a novel interferon regulatory factor, IRF- 5, results in the induction of distinct interferon alpha genes. *J Biol Chem* **276:**23382-90
5. Barnes, B. J., Kellum, M. J., Field, A. E. & Pitha, P. M. Multiple Regulatory Domains of IRF-5 Control Activation, Cellular Localization, and Induction of Chemokines That Mediate Recruitment of T Lymphocytes. *Mol Cell Biol* **22:**5721-40
6. Barnes, B. J. et al. Global and specific targets of IRF-5 and IRF-7 during innate response to viral infection. *J Biol Chem* (2004).
7. Yoneyama, M. et al. The RNA helicase RIG-I has an essential function in double-stranded RNA-induced innate antiviral responses. *Nat Immunol* **5:**730-7
8. Basler, C. F. & Garcia-Sastre, A. Viruses and the Type I Interferon Antiviral System: Induction and Evasion. *International Reviews of Immunology* in press. (2002).
9. Garcia-Sastre, A. et al. Influenza A virus lacking the NS1 gene replicates in interferon-deficient systems. *Virology* **252:**324-30
10. Talon, J. et al. Influenza A and B viruses expressing altered NS1 proteins: A vaccine approach. *Proc Natl Acad Sci USA* **97:**4309-14
11. Markovitz, N. S., Baunoch, D. & Roizman, B. The range and distribution of murine central nervous system cells infected with the gamma(1)34.5- mutant of herpes simplex virus 1. *J Virol* **71:**5560-9

12. Chou, J., Kern, E. R., Whitley, R. J. & Roizman, B. Mapping of herpes simplex virus-1 neurovirulence to gamma 134.5, a gene nonessential for growth in culture. *Science* **250:** 1262-6
13. Brandt, T. A. & Jacobs, B. L. Both carboxy- and amino-terminal domains of the vaccinia virus interferon resistance gene, E3L, are required for pathogenesis in a mouse model. *J Virol* **75:**850-6
14. Leib, D. A. et al. Interferons regulate the phenotype of wild-type and mutant herpes simplex viruses in vivo. *J Exp Med* **189:**663-72
15. Leib, D. A., Machalek, M. A., Williams, B. R., Silverman, R. H. & Virgin, H. W. Specific phenotypic restoration of an attenuated virus by knockout of a host resistance gene [see comments]. *Proc Natl Acad Sci USA* **97:**6097-101
16. CDC. Biological and chemical terrorism: strategic plan for preparedness and response. Recommendations of the CDC Strategic Planning Workgroup. *MMWR Morb Mortal Wkly Rep* **49:**1-14
17. Lane, H. C., Montagne, J. L. & Fauci, A. S. Bioterrorism: a clear and present danger. *Nat Med* **7:**1271-3
18. CDC, S. P. B. (http://www.cdc.gov/ncidod/dvrd/spb/mnpages/dispages/filoviruses.htm, 2002).
19. Mahanty, S. & Bray, M. Pathogenesis of filoviral haemorrhagic fevers. *Lancet Infect Dis* **4:**487-98
20. WHO. Suspected viral haemorrhagic fever in Gabon. *WHO Communicable Disease and Surviellance and Response-Outbreak News* (2001).
21. WHO. Ebola haemorrhagic fever in Gabon/The Republic of the Congo. *WHO Communicable Disease and Surviellance and Response-Outbreak News* (2002).
22. Sanchez, A. et al. in *Fields Virology* (eds. Knipe, D. M., Howley, P. M. & al., e.) 1279-1304 (Lippincott Williams and Wilkins, Philadelphia, 2001).
23. Hensley, L. E., Young, H. A., Jahrling, P. B. & Geisbert, T. W. Proinflammatory response during Ebola virus infection of primate models: possible involvement of the tumor necrosis factor receptor superfamily. *Immunol Lett* **80:**169-79
24. Baize, S. et al. Inflammatory responses in Ebola virus-infected patients. *Clin Exp Immunol* **128:**163-8
25. Leroy, E. M. et al. Human asymptomatic Ebola infection and strong inflammatory response. *Lancet* **355:**2210-5
26. Villinger, F. et al. Markedly elevated levels of interferon (IFN)-gamma, IFN-alpha, interleukin (IL)-2, IL-10, and tumor necrosis factor-alpha associated with fatal Ebola virus infection. *J Infect Dis* **179 Suppl 1:**S188-91
27. Baize, S. et al. Defective humoral responses and extensive intravascular apoptosis are associated with fatal outcome in Ebola virus-infected patients. *Nat Med* 5, 423-6. (1999).
28. Feldmann, H. et al. Filovirus-induced endothelial leakage triggered by infected monocytes/macrophages. *J Virol* **70:**2208-14
29. Jahrling, P. B. et al. Evaluation of immune globulin and recombinant interferon-alpha2b for treatment of experimental Ebola virus infections. *J Infect Dis* **179 Suppl 1:**S224-34
30. Huggins, J. W., Zhang, Z. X. & Monath, T. I. Inhibition of Ebola virus replication in vitro and in vivo in a SCID mouse model by S-adenosylhomocysteine hydrolase inhibitors. *Antiviral Research Suppl.* **1:**122
31. Bray, M., Davis, K., Geisbert, T., Schmaljohn, C. & Huggins, J. A mouse model for evaluation of prophylaxis and therapy of Ebola hemorrhagic fever. *J Infect Dis* **178:**651-61
32. Bray, M. The role of the Type I interferon response in the resistance of mice to filovirus infection. *J Gen Virol* **82:**1365-73
33. Bray, M., Driscoll, J. & Huggins, J. W. Treatment of lethal Ebola virus infection in mice with a single dose of an S-adenosyl-L-homocysteine hydrolase inhibitor. *Antiviral Res* **45:**135-47

34. Bray, M., Raymond, J. L., Geisbert, T. & Baker, R. O. 3-Deazaneplanocin A induces massively increased interferon-alpha production in Ebola virus-infected mice. *Antiviral Res* **55:**151-9
35. Harcourt, B. H., Sanchez, A. & Offermann, M. K. Ebola virus inhibits induction of genes by double-stranded RNA in endothelial cells. *Virology* **252:**179-88
36. Harcourt, B. H., Sanchez, A. & Offermann, M. K. Ebola virus selectively inhibits responses to interferons, but not to interleukin-1beta, in endothelial cells. *J Virol* **73:** 3491-6
37. Gupta, M., Mahanty, S., Ahmed, R. & Rollin, P. E. Monocyte-derived human macrophages and peripheral blood mononuclear cells infected with ebola virus secrete MIP-1alpha and TNF-alpha and inhibit poly-IC-induced IFN-alpha in vitro. *Virology* **284:**20-5
38. Basler, C. F. et al. The Ebola virus VP35 protein functions as a type I IFN antagonist. *Proc Natl Acad Sci USA* **97:**12289-12294
39. Basler, C. F. et al. The Ebola virus VP35 protein inhibits activation of interferon regulatory factor 3. *J Virol* **77:**7945-56
40. Huang, Y., Xu, L., Sun, Y. & Nabel, G. The assembly of ebola virus nucleocapsid requires virion-associated proteins 35 and 24 and posttranslational modification of nucleoprotein. *Mol Cell* **10:**307
41. Bosio, C. M. et al. Ebola and Marburg Viruses Replicate in Monocyte-Derived Dendritic Cells without Inducing the Production of Cytokines and Full Maturation. *J Infect Dis* **188:**1630-8
42. Talon, J. et al. Activation of Interferon Regulatory Factor 3 (IRF-3) is Inhibited by the Influenza A Viral NS1 Protein. *J Virol* **74:**7989-7996
43. Lin, R., Heylbroeck, C., Pitha, P. M. & Hiscott, J. Virus-dependent phosphorylation of the IRF-3 transcription factor regulates nuclear translocation, transactivation potential, and proteasome-mediated degradation. *Mol Cell Biol* **18:**2986-96
44. Lamb, R. A. & Kolakofsky, D. in *Fields Virology* (eds. Knipe, D. M., Howley, P. M. & al., e.) 1305-1340 (Lippincott Williams and Wilkins, Philadelphia, 2001).
45. Field, H. et al. The natural history of Hendra and Nipah viruses. *Microbes Infect* **3:**307-14
46. Halpin, K., Bankamp, B., Harcourt, B. H., Bellini, W. J. & Rota, P. A. Nipah virus conforms to the rule of six in a minigenome replication assay. *J Gen Virol* **85:**701-7
47. Park, M. S. et al. Newcastle Disease Virus (NDV)-Based Assay Demonstrates Interferon-Antagonist Activity for the NDV V Protein and the Nipah Virus V, W, and C Proteins. *J Virol* **77:**1501-11
48. Shaw, M. L., Garcia-Sastre, A., Palese, P. & Basler, C. F. Nipah Virus V and W Proteins Have a Common STAT1-Binding Domain yet Inhibit STAT1 Activation from the Cytoplasmic and Nuclear Compartments, Respectively. *J Virol* **78:**5633-41
49. Rodriguez, J. J., Lau, J. F. & Horvath, C. M. in *American Society for Virology 21st Annual Meeting* (American Society for Virology, Lexington, Kentuky, 2002).
50. Rodriguez, J. J., Cruz, C. D. & Horvath, C. M. Identification of the nuclear export signal and STAT-binding domains of the Nipah virus V protein reveals mechanisms underlying interferon evasion. *J Virol* **78:**5358-67
51. Rodriguez, J. J., Wang, L. F. & Horvath, C. M. Hendra virus V protein inhibits interferon signaling by preventing STAT1 and STAT2 nuclear accumulation. *J Virol* **77:**11842-5
52. Didcock, L., Young, D. F., Goodbourn, S. & Randall, R. E. The V protein of simian virus 5 inhibits interferon signalling by targeting STAT1 for proteasome-mediated degradation. *J Virol* **73:**9928-33
53. Hariya, Y., Yokosawa, N., Yonekura, N., Kohama, G. & Fuji, N. Mumps virus can suppress the effective augmentation of HPC-induced apoptosis by IFN-gamma through disruption of IFN signaling in U937 cells. *Microbiol Immunol* **44:**537-41
54. Kubota, T., Yokosawa, N., Yokota, S. & Fujii, N. C terminal CYS-RICH region of mumps virus structural V protein correlates with block of interferon alpha and gamma

signal transduction pathway through decrease of STAT 1-alpha. *Biochem Biophys Res Commun* **283:**255-9
55. Nishio, M. et al. High resistance of human parainfluenza type 2 virus protein-expressing cells to the antiviral and anti-cell proliferative activities of alpha/beta interferons: cysteine-rich V-specific domain is required for high resistance to the interferons. *J Virol* **75:**9165-76
56. Young, D. F., Didcock, L., Goodbourn, S. & Randall, R. E. Paramyxoviridae use distinct virus-specific mechanisms to circumvent the interferon response. *Virology* **269:**383-90
57. Huang, Z., Krishnamurthy, S., Panda, A. & Samal, S. K. Newcastle disease virus V protein is associated with viral pathogenesis and functions as an alpha interferon antagonist. *J Virol* **77:**8676-85
58. Shaw, M. e. a. (submitted).
59. Harcourt, B. H. et al. Molecular characterization of Nipah virus, a newly emergent paramyxovirus. *Virology* **271:**334-49
60. Hemmi, H. et al. The roles of two IkappaB kinase-related kinases in lipopolysaccharide and double stranded RNA signaling and viral infection. *J Exp Med* **199:**1641-50
61. McWhirter, S. M. et al. IFN-regulatory factor 3-dependent gene expression is defective in Tbk1-deficient mouse embryonic fibroblasts. *Proc Natl Acad Sci USA* **101:**233-238
62. Perry, A. K., Chow, E. K., Goodnough, J. B., Yeh, W. C. & Cheng, G. Differential requirement for TANK-binding kinase-1 in type I interferon responses to toll-like receptor activation and viral infection. *J Exp Med* **199:**1651-8
63. Schmaljohn, C. S. & Hooper, J. W. in *Fields Virology* (eds. Knipe, D. M., Howley, P. M. & al., e.) 1581-1602 (Lippincott Williams and Wilkins, Philadelphia, 2001).
64. Nichol, S. T. in *Fields Virology* (eds. Knipe, D. M., Howley, P. M. & al., e.) 1603-1634 (Lippincott Williams and Wilkins, Philadelphia, 2001).
65. Morrill, J. C. et al. Pathogenesis of Rift Valley fever in rhesus monkeys: role of interferon response. *Arch Virol* **110:**195-212
66. Vialat, P., Billecocq, A., Kohl, A. & Bouloy, M. The S segment of rift valley fever phlebovirus (Bunyaviridae) carries determinants for attenuation and virulence in mice. *J Virol* **74:**1538-43
67. Bouloy, M. et al. Genetic evidence for an interferon antagonistic function of rift valley fever virus nonstructural protein NSs. *J Virol* **75:**1371-7
68. Bridgen, A., Weber, F., Fazakerley, J. K. & Elliott, R. M. Bunyamwera bunyavirus nonstructural protein NSs is a nonessential gene product that contributes to viral pathogenesis. *Proc Natl Acad Sci USA* **98:**664-9
69. Weber, F. et al. Bunyamwera bunyavirus nonstructural protein NSs counteracts the induction of alpha/beta interferon. *J Virol* **76:**7949-55
70. Kohl, A. et al. Bunyamwera virus nonstructural protein NSs counteracts interferon regulatory factor 3-mediated induction of early cell death. *J Virol* **77:**7999-8008
71. Le May, N. et al. TFIIH transcription factor, a target for the Rift Valley hemorrhagic fever virus. *Cell* **116:**541-50
72. Thomas, D. et al. Inhibition of RNA polymerase II phosphorylation by a viral interferon antagonist. *J Biol Chem* **279:**31471-7
73. Kraus, A. A. et al. Differential antiviral response of endothelial cells after infection with pathogenic and nonpathogenic hantaviruses. *J Virol* **78:**6143-50
74. Geimonen, E. et al. Pathogenic and nonpathogenic hantaviruses differentially regulate endothelial cell responses. *Proc Natl Acad Sci USA* **99:**13837-42
75. Gale, M., Jr. et al. Control of PKR protein kinase by hepatitis C virus nonstructural 5A protein: molecular mechanisms of kinase regulation. *Mol Cell Biol* **18:**5208-18
76. Gale, M. J., Jr., Korth, M. J. & Katze, M. G. Repression of the PKR protein kinase by the hepatitis C virus NS5A protein: a potential mechanism of interferon resistance. *Clin Diagn Virol* **10:**157-62

77. Gale, M. J., Jr. et al. Evidence that hepatitis C virus resistance to interferon is mediated through repression of the PKR protein kinase by the nonstructural 5A protein. *Virology* **230:**217-27
78. Burke, D. S. & Monath, T. P. in *Fields Virology* (eds. Knipe, D. M., Howley, P. M. & al., e.) 1043-1125 (Lippincott Williams and Wilkins, Philadelphia, 2001).
79. Diamond, M. S. et al. Modulation of Dengue virus infection in human cells by alpha, beta, and gamma interferons. *J Virol* **74:**4957-66
80. Diamond, M. S. & Harris, E. Interferon inhibits dengue virus infection by preventing translation of viral RNA through a PKR-independent mechanism. *Virology* **289:**297-311
81. Johnson, A. J. & Roehrig, J. T. New mouse model for dengue virus vaccine testing. *J Virol* **73:**783-6
82. Shresta, S. et al. Interferon-dependent immunity is essential for resistance to primary dengue virus infection in mice, whereas T- and B-cell-dependent immunity are less critical. *J Virol* **78:**2701-10
83. Libraty, D. H., Pichyangkul, S., Ajariyakhajorn, C., Endy, T. P. & Ennis, F. A. Human dendritic cells are activated by dengue virus infection: enhancement by gamma interferon and implications for disease pathogenesis. *J Virol* **75:**3501-8
84. Munoz-Jordan, J. L., Sanchez-Burgos, G. G., Laurent-Rolle, M. & Garcia-Sastre, A. Inhibition of interferon signaling by dengue virus. *Proc Natl Acad Sci USA* **100:**14333-8
85. Lobigs, M., Mullbacher, A., Wang, Y., Pavy, M. & Lee, E. Role of type I and type II interferon responses in recovery from infection with an encephalitic flavivirus. *J Gen Virol* **84:**567-72
86. Lin, R. J., Liao, C. L., Lin, E. & Lin, Y. L. Blocking of the alpha interferon-induced Jak-Stat signaling pathway by Japanese encephalitis virus infection. *J Virol* **78:**9285-94
87. Griffin, D. E. in *Fields Virology* (eds. Knipe, D. M., Howley, P. M. & al., e.) 917-962 (Lippincott Williams and Wilkins, Philadelphia, 2001).
88. Johnston, C., Jiang, W., Chu, T. & Levine, B. Identification of genes involved in the host response to neurovirulent alphavirus infection. *J Virol* **75:**10431-45
89. White, L. J., Wang, J. G., Davis, N. L. & Johnston, R. E. Role of alpha/beta interferon in Venezuelan equine encephalitis virus pathogenesis: effect of an attenuating mutation in the 5' untranslated region. *J Virol* **75:**3706-18
90. Ryman, K. D., Klimstra, W. B., Nguyen, K. B., Biron, C. A. & Johnston, R. E. Alpha/beta interferon protects adult mice from fatal Sindbis virus infection and is an important determinant of cell and tissue tropism. *J Virol* **74:**3366-78
91. Frolova, E. I. et al. Roles of nonstructural protein nsP2 and Alpha/Beta interferons in determining the outcome of Sindbis virus infection. *J Virol* **76:**11254-64
92. Jahrling, P. B., Navarro, E. & Scherer, W. F. Interferon induction and sensitivity as correlates to virulence of Venezuelan encephalitis viruses for hamsters. *Arch Virol* 51:23-35
93. Weaver, S. C. et al. Genetic determinants of Venezuelan equine encephalitis emergence. *Arch Virol Suppl*, **18:**43-64
94. Wang, E. et al. Virulence and viremia characteristics of 1992 epizootic subtype IC Venezuelan equine encephalitis viruses and closely related enzootic subtype ID strains. *Am J Trop Med Hyg* **65:**64-9
95. Spotts, D. R., Reich, R. M., Kalkhan, M. A., Kinney, R. M. & Roehrig, J. T. Resistance to alpha/beta interferons correlates with the epizootic and virulence potential of Venezuelan equine encephalitis viruses and is determined by the 5' noncoding region and glycoproteins. *J Virol* **72:**10286-91
96. Anishchenko, M. et al. Generation and characterization of closely related epizootic and enzootic infectious cDNA clones for studying interferon sensitivity and emergence mechanisms of Venezuelan equine encephalitis virus. *J Virol* **78:**1-8

97. Basler, C. et al. Sequence of the 1918 pandemic influenza virus nonstructural gene (NS) segment and characterization of recombinant viruses bearing the 1918 NS genes. *Proc Natl Acad Sci USA* **98:**2746-2751
98. Parisien, J. P., Lau, J. F. & Horvath, C. M. STAT2 Acts as a Host Range Determinant for Species-Specific Paramyxovirus Interferon Antagonism and Simian Virus 5 Replication. *J Virol* **76:**6435-41
99. Seo, S. H., Hoffmann, E. & Webster, R. G. Lethal H5N1 influenza viruses escape host antiviral cytokine responses. *Nature Medicine* **8(9):**950-4
100. Bossert, B. & Conzelmann, K. K. Respiratory syncytial virus (RSV) nonstructural (NS) proteins as host range determinants: a chimeric bovine RSV with NS genes from human RSV is attenuated in interferon-competent bovine cells. *J Virol* **76:**4287-93
101. Park, M. S., Garcia-Sastre, A., Cros, J. F., Basler, C. F. & Palese, P. Newcastle disease virus V protein is a determinant of host range restriction. *J Virol* **77:**9522-32
102. Young, D. F. et al. Single amino acid substitution in the V protein of simian virus 5 differentiates its ability to block interferon signaling in human and murine cells. *J Virol* **75:**3363-70

Chapter 10

VIRAL PATHOGENESIS AND TOLL-LIKE RECEPTORS

SUSAN R. ROSS
Department of Microbiology, Abramson Family Cancer Center, University of Pennsylvania, Philadelphia, PA, USA

1. INTRODUCTION

The role of the vertebrate adaptive immune system in viral infection and how it functions to eliminate infected cells is well-established. Similarly, as documented elsewhere in this volume, type I interferon (IFN) production by virus-infected cells is a highly characterized innate cellular response to double-stranded viral RNA and other by-products of infection. In contrast, much less is known about the interaction between viruses and the branch of the innate immune system that involves toll-like receptors (TLRs). In the past few years, this has been an area of intensive research and more recently, the identification of TLR ligands, such as double- and single-stranded nucleic acids that are known by-products of infection, have spurred greater interest in the interaction of viruses with this arm of the immune system. As described in this chapter, the immediate inflammatory response to viral gene products that signal through TLRs may play a role in pathogenesis as well as in antiviral responses. Additionally, the recent finding that innate immune responses can influence later humoral and cellular immunity may explain why certain viruses can escape immune destruction.

P. Palese (ed.), Modulation of Host Gene Expression and Innate Immunity by Viruses, 221-243.

2. TOLL-LIKE RECEPTORS AND SIGNALING

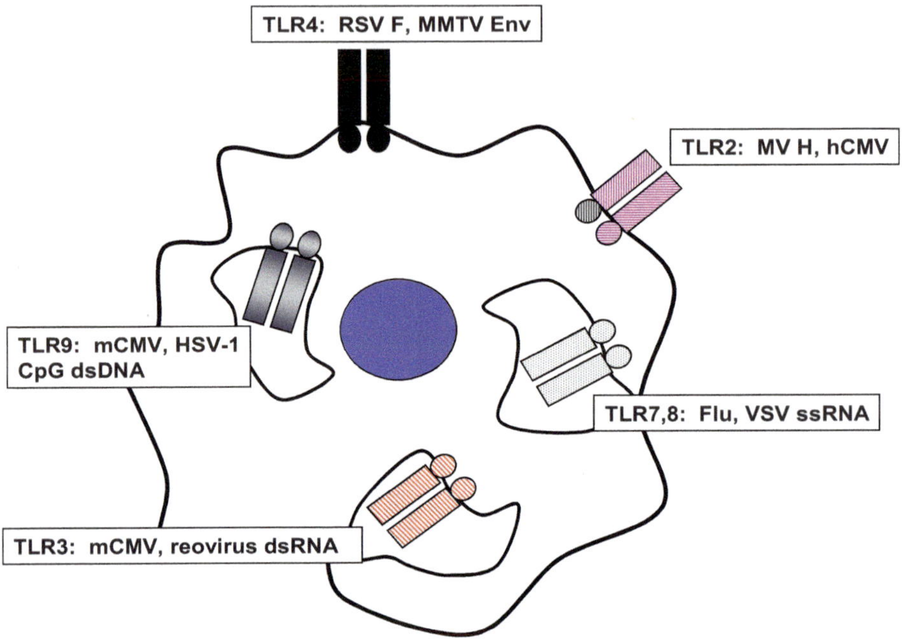

Figure 1. Subcellular localization of TLRs that interact with viral components.

Vertebrate TLRs were initially discovered through their homology to a Drosophila gene involved in embryonic development and subsequently shown to be important for antifungal immunity in this organism[1]. TLRs are present on many mammalian cell types, including the major players in innate immune responses, dendritic cells (DCs), neutrophils and macrophages. These receptors are type I transmembrane proteins found either on the cell surface (TLR1, 2, 4, 5, 6) or in the endosomal compartment (TLR3, 7, 8, 9) (Figure 1)[2-7].

TLRs contain leucine-rich repeats in their extracellular/lumenal domains and cytoplasmic segments with homology to the signaling motif of the interleukin-1 receptor (Toll-IL-1R or TIR domain). There are at least 10 known members of this family in mammals, each of which imparts innate immune responses to different microbial ligands produced by bacteria, mycoplasmas, spirochetes, fungi and viruses[8] and recent evidence suggests additional members exist[9]. The specificity for ligand recognition is thought to lie in the LRR domain and occurs through interaction with homodimers or heterodimers of different TLR molecules or co-receptors[10]. For example, TLR4, the first family member for which a ligand was identified, is the receptor for the lipopolysaccharide (LPS) component of Gram negative bacteria[11,12]. However, while TLR4 functions as a homodimer in this response, it also requires the co-receptors CD14 and MD-2. Both CD14 and MD-2 lack cytoplasmic signaling domains and are probably important for LPS-TLR4 binding[13] (Figure 2). TLR2, in contrast, is believed to heterodimerize with either TLR1 or TLR6, thereby expanding its ability to interact with different ligands[14,15]. However, direct binding has not been demonstrated for most TLRs and their ligands. Ligands that interact with the same TLR molecules do not always activate the same signal transduction pathways; such differences may be the result of the nature of the ligand/receptor interaction or the interaction of particular TLRs with additional cell surface molecules or variable downstream effector molecules.

Two major pathways are activated as a result of TLR signaling. All TLRs, as well as IL-1R, apparently use the MyD88-dependent pathway. MyD88 is a cytoplasmic adapter molecule belonging to the TLR family that lacks the transmembrane and LRR domains[16]. The MyD88 pathway signals through the IRAK-4/TRAF-6 complex leading to degradation of IκB and induction of NFκB and through the mitogen-activated protein kinases p38 and JNK (Figure 2)[17]. The outcome of this immunostimulatory signaling is NFκB-dependent transcription of cytokines such as IL-1, -6 and TNFα and increased expression of co-stimulatory molecules like CD40 and CD80. A second, MyD88-independent pathway is used by TLRs 3 and 4 that relies on a different cytoplasmic family member, TIR domain-containing adapter inducing IFN-β (TRIF)/TIR-containing adaptor molecule (TICAM-1) (Figure 2)[18,19]. In addition to using IRAK-4/TRAF-6 and inducing NF-κB-dependent genes, TIRAP can activate IRF-3 directly or result in the production of type I IFNs, which in turn leads to the expression of IFN-responsive genes (IRGs) such as IP10 (Figure 2)[20-22]. Thus, ligand binding to TLR3 and 4 has the potential to lead to both immunostimulatory and antiviral responses.

A number of microbial molecules have been identified as TLR ligands. In addition to LPS from Gram-negative bacteria, these include Gram-positive bacterial peptidoglycans, proteins and nucleic acids, including double-stranded DNA (dsDNA), and single- and double-stranded RNA (ssRNA and dsRNA) (Figure 1). Most of the pathogen-associated molecular patterns (PAMPs) found in these TLR-interacting molecules are highly conserved and critical to their function and thus, pathogens cannot mutate to avoid the innate immune response. Viral proteins and nucleic acids have also been shown to interact with various TLR family members, and like bacterial and fungal gene products, may be unable to avoid activating the innate immune response because of the conservation of these molecules.

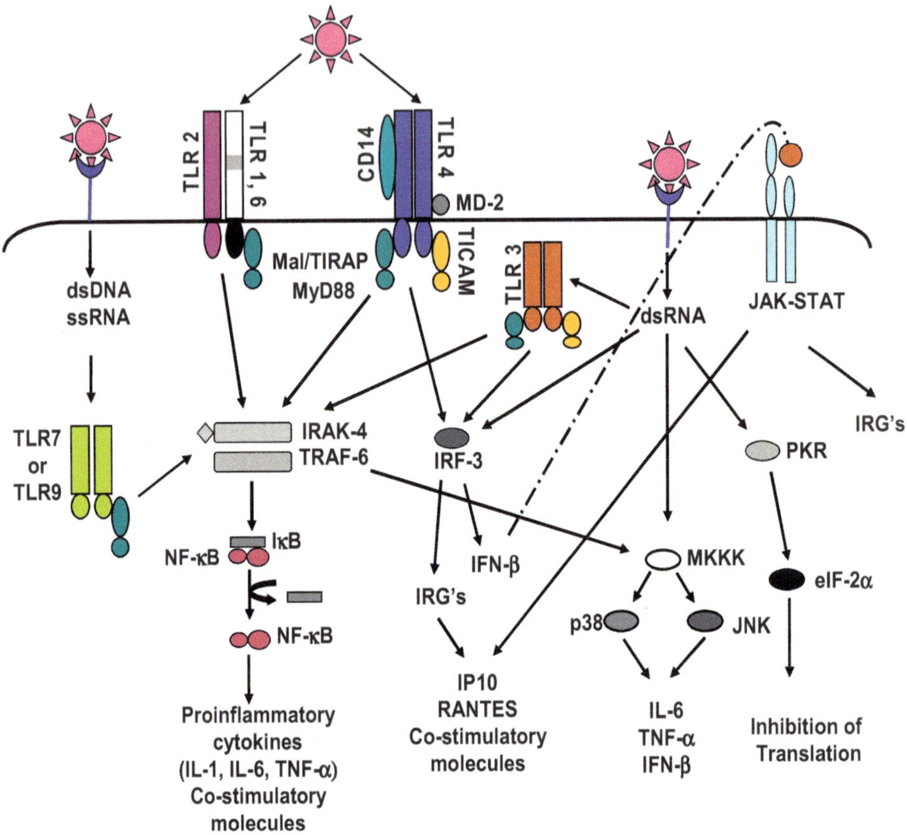

Figure 2. Signaling pathways activated by TLRs and their viral ligands.

Innate immune responses are likely involved in virus clearance. The immediate effects of TLR-mediated cell activation are the production of anti-microbial and antiviral cytokines and the secretion of chemokines that recruit macrophages, neutrophils and DCs to sites of infection. Indeed, pre-treatment of cultured bone marrow-derived dendritic cells with either TLR4 (LPS) or TLR3 (dsRNA) but not TLR9 (dsDNA) ligands inhibited mouse cytomegalovirus (mCMV) replication *in vitro* and this effect was inhibited by anti-type I IFN antibodies[23]. Recent work by several investigators has also demonstrated that activation of innate immune responses influences later adaptive immunity, that is, whether Th1 or Th2 responses predominate[24-28]. This skewing of the adaptive immune response may play a role in the ability of the host to clear virus or conversely by shifting the immune response, allow some viruses to avoid it and persist. For other viruses, the outcome of this activation may also play a role in virus-mediated pathogenesis.

3. INTERACTION OF VIRUSES WITH DIFFERENT TLRS

3.1 Viral proteins as TLR ligands

In addition to generating antiviral immune responses to infected cells, a number of viruses are able induce inflammatory responses, such as the production of cytokines or the activation of lymphocytes and other cells in the absence of actual infection. Four enveloped viruses, all of which bud from cells and thus insert viral proteins required for interaction with cellular entry receptors into the plasma membrane, have been shown to interact with TLRs via these virion surface proteins. These four viruses belong to the paromyxovirus (respiratory syncytial virus (RSV) and measles virus (MV)), retrovirus (mouse mammary tumor virus (MMTV)) and γ–herpesvirus (human cytomegalovirus (hCMV)) families (Table 1).

Table 1. Viral gene products that interact with TLRs

Virus	Family	Viral Ligand	TLR
RSV	Paramyxovirus	F protein	4
MMTV	Retrovirus	Env (SU)	4
MV	Paramyxovirus	H protein	2
hCMV	γ-Herpesvirus	gB (?)	2
HSV-1, -2	α-Herpesvirus	CpG dsDNA	9
mCMV	γ-Herpesvirus	CpG dsDNA	9
		dsRNA	3
NDV	Paramyxovirus	dsRNA (?)	3
LCMV	Arenavirus	dsRNA (?)	3
Influenza	Orthomyxovirus	ssRNA	7
VSV	Rhabdovirus	ssRNA	7
Vaccinia	Poxvirus	A46R, A52R, N1L	IRAK-4, TRAF6, I-κB

3.1.1 RSV, TLR4 and induction of antiviral immunity

RSV, a negative-sense, single-stranded RNA virus paramyxovirus, is the most prevalent cause of viral respiratory tract infections in newborns. RSV infection *in vivo* is accompanied by pro-inflammatory cytokine production by both lung epithelial and inflammatory cells and evidence has suggested that this is responsible, at least in part, for the disease pathology[29]. Re-infection with RSV is common and development of a RSV vaccination has been unsuccessful. Indeed, vaccination of children with formalin-inactivated virus enhanced lower respiratory tract disease upon later virus infection(29); this is thought to be due to induction of memory T cells that produce high levels of Th2-type cytokines[30,31]. Part of the inflammatory response appeared to be independent of active viral infection, since it also occurred with ultraviolet (*uv*)- or formalin-inactivated virus[32-34].

These latter results indicated a role for a molecule present in virions in the induction of inflammation. Like other paramyxoviruses, RSV virions contain three major proteins in addition to the genomic RNA, the attachment protein (G), the fusion protein (F) and the nucleocapsid protein (N). When these three RSV proteins were incubated with either human or mouse peripheral blood mononuclear cells (PBMCs), only purified F but not N or G proteins caused inflammatory cytokine secretion (IL-6, IL-8 and TNF-α, at levels comparable to that seen with LPS[35]. F protein-induced cytokine production was also dependent on CD14, since macrophages from mice with targeted deletion of CD14 showed no response. Similarly, macrophages from two independent strains of LPS-resistant mice with different naturally-occurring TLR4 mutations, C57BL/ScCr and C3H/HeJ, failed to produce IL-6 in response to RSV F protein. C57BL/ScCr mice have a genomic deletion of the TLR4 gene, while C3H/HeJ mice have a point mutation in the cytoplasmic domain (*lps^d^*) that renders the molecule unable to signal[11,12]. Alveolar macrophages isolated from C3H/HeJ mice intranasally infected with RSV also showed no NFκB activation when compared with wild-type BALB/c mice[34]. In contrast, influenza virus induced cytokine production in both wild-type and TLR4-mutant mice, suggesting that this virus stimulates a TLR4-independent response[36] (see section 3.2.2).

To test whether this innate response played a role in virus clearance, Kurt-Jones and colleagues inoculated C57BL/ScCr and control C57BL/ScSN mice intranasally and looked at virus titers in their lungs. Mice lacking TLR4 had impaired ability to clear virus and poorer recruitment of neutrophils and monocytes[37]. This was the first indication that signaling through TLRs might be important for antiviral immunity. However, these results have recently been called into question, because in addition to the TLR4 genomic deletion, C57BL/ScCr mice have a point mutation in the gene encoding the IL12Rβ2 chain[38]. This mutation prevents activation of the STAT4 pathway after IL-12 binding and as a result, there is no production of IFN-γ. Indeed, recent work has shown that the susceptibility of C57BL/ScCr mice to RSV is the result of impaired NK cell recruitment due to this non-functional IL-12 axis and not to the TLR4 mutation[38,39].

Although it is currently unclear what role RSV-mediated activation of TLR4 plays in experimental infection in mice, recent work has indicated that there is an association between missense mutations in the extracellular domain of TLR4 and severe RSV-induced bronchiolitis in human infants[40]. RSV bronchiolitis is associated with a Th2 response[41,42]. Thus, one possibility is that early TLR4-mediated activation by RSV in normal individuals results in Th1-type responses, thereby altering the initial inflammatory response through differential activation or recruitment of cells at the site of infection. This could alter the clinical outcome without necessarily affecting virus load in the natural host. Interestingly, it has been shown that C3H/HeJ mice, which have the non-signaling, dominant negative TLR4 but a functional IL12Rβ2, lack the early TLR4-dependent phase of NFκB activation by RSV, but retain a secondary response that is TLR4-independent[34]. Whether RSV results in Th2-type responses in C3H/HeJ mice has not yet been demonstrated.

The RSV genome is a single-stranded RNA, and during replication, a double-stranded RNA is generated. With the recent findings that the former may be the ligand for TLR3, and the latter the ligand for TLR7 (section 3.2.2), further studies are required to determine if RSV also activates these pathways. Moreover, much of the pathological inflammatory response to RSV *in vivo* is believed to result from cytokine/chemokine production by the airway epithelial cells; TLRs 1 - 6 are expressed on these cells[43]. Most of the work to date has focused on how RSV interacts with TLR4 on macrophages and it may be that important insights will be gained with regard to the disease pathology by studying these responses by epithelial cells.

3.1.2 MMTV, TLRs and subversion of the host adaptive immune response

MMTV, a milk-borne retrovirus that initially infects lymphoid targets in the Peyer's patches of the small intestine, amplifies in the systemic lymphocyte population and ultimately infects mammary epithelial cells, where it causes breast cancer by insertional activation of cellular oncogenes[44]. About 30 years ago, scientists at the Jackson Laboratory found that C3H/HeJ mice had simultaneously become LPS-resistant and exhibited significantly reduced MMTV-induced tumor formation compared to the highly related C3H/HeN or C3H/OuJ strains[45]. The genetic link between resistance to MMTV-induced tumorigenesis and a mutant TLR4 allele indicated that MMTV might utilize this receptor as part of its infection pathway, perhaps by activating its cellular targets. This was supported by more recent work showing that MMTV binding to B-lymphocytes resulted in their activation[46].

B cell activation is due, at least in part, to MMTV binding to TLR4. While subcutaneous injection of MMTV into wild-type C3H/HeN mice resulted in increased numbers of activated B cells in draining lymph nodes, this did not occur in C3H/HeJ mice or congenic BALB/cJ mice containing the defective C3H/HeJ *lps*^{d} allele[47]. Similarly, incubation of primary bone-marrow-DCs with MMTV induced their differentiation and production of inflammatory cytokines such as TNFα, IL-6, and IL-12[48]. B and dendritic cell activation occurred with *uv*- or chemically-inactivated but not heat-denatured virions, indicating that, like RSV, an MMTV protein was responsible for cell activation. Indeed, co-transfection/co-immunoprecipitation assays showed that the MMTV envelope protein binds to both TLR2 and TLR4[47].

MMTV can also use TLR2 to activate cells. In contrast to what occurred with C3H/HeJ mice, MMTV activated B and DCs from C57BL/10ScN mice, which also have genomic deletion of *Tlr4*[48] (Rassa et al., in preparation). However, MMTV failed to activate B cells or DCs from double-mutant animals generated by crossing C3H/HeJ with TLR2 knockout mice[48](Rassa et al., in preparation). The mutation in the cytoplasmic tail of the C3H/HeJ TLR4 molecule not only abrogates its ability to signal but causes it to function as a co-dominant-negative receptor[49]. Thus, it appears that dominant-negative TLR4 protein in C3H/HeJ cells suppresses MMTV-initiated signaling, either through direct heterodimerization with other TLRs, as has been reported for TLR1 and TLR6[14] or through interaction with one of the TLR adaptor proteins.

Interestingly, the TLR-dependent binding of MMTV to DCs but not B cells up-regulated expression of the MMTV entry receptor, transferrin receptor 1, a known lymphocyte activation marker[48]. Previous studies have shown that MMTV infects DCs *in vivo*[50,51]. Thus, DCs may be the initial targets of MMTV infection and the ability of this virus to activate the TLR signaling pathway may represent a mechanism to increase the expression of its own receptor, similar to MV (see next section).

There are several possible consequences of MMTV-mediated cell activation. Retroviruses require target cell activation and division for efficient entry of reverse-transcribed proviral DNA into the nucleus and integration into the chromosomes[52,53]. Lymphocyte infection at early times in C3H/HeJ mice is lower than that seen in wild-type mice, perhaps due to the reduced levels of dendritic and B cell activation in the initial rounds of infection[54] (Rassa and Ross, unpublished). Activation via TLRs also likely leads to antiviral immune responses. Jude and colleagues showed that MMTV-mediated activation of the TLR4 pathway may subvert the antiviral adaptive immune response[54]. With cultured splenocytes from wild-type but not TLR4-mutant mice, MMTV-mediated activation led to the production of the Th2-type cytokine, IL-10. These results suggested that in the absence of MMTV-mediated TLR4 signaling, an antiviral Th1 response occurs. In support of this, when milk-borne MMTV was transmitted through TLR4-

mutant mice by breast-feeding, a novel recombinant form of the virus arose after several generations that acquired its structural genes from an endogenous MMTV present in the mouse genome[55]. Similar recombinants were not generated in mice with a wild-type TLR4 gene, suggesting that a cell-mediated antiviral response in the TLR4-mutant mice resulted in the selection of structural gene-cytotoxic T lymphocyte escape mutants.

Other retroviruses may also interact with TLRs. Ardavin and colleagues previously showed that Moloney murine leukemia virus could activate B cells[46]. Using co-transfection/co-immunoprecipitation assays, a physical interaction between M-MLV envelope protein and TLR4 that is competed by the MMTV envelope has been demonstrated[47]. This suggests that the two Env molecules share a common binding domain on the TLR4 protein. Given the common requirement for cellular activation and subversion of immune responses for persistence, interaction with TLR proteins may be a shared phenotype of several retroviruses.

3.1.3 Measles virus, TLR2 and up-regulation of CD150

MV, like RSV, is a paramyxovirus. Although largely eradicated by vaccination in developed countries, MV is a major cause of childhood mortality and morbidity in underdeveloped countries. Most if not all of the pathogenic effects of MV infection are believed to be the result of virus-induced immunodepression, leading to secondary bacterial or viral infections[56]. MV infection is thought to shift immune responses from Th1 to Th2[57] and in a small percentage of individuals, leads to neurological syndromes, including post-infection encephalitis; this is believed to be the result of an inflammatory response[58].

Like RSV, MV also contains two proteins on the virion surface, the attachment or hemagglutinin protein H and the fusion protein F. Unlike RSV, for which the receptor is unknown, at least two surface proteins are known to function as cellular entry receptors for MV. The MV vaccine strain, adapted for replication on monkey kidney cells, uses CD46, a complement-regulating factor, to infect epithelial cells, while primary or lymphocyte-adapted MV isolates use CD150 or SLAM (signaling lymphocyte activation molecule) for lymphocyte entry[59,60]. A single amino acid change in the H protein at amino acid 481 (asn to tyr in the vaccine strain) is responsible for this change in receptor tropism. CD150 is a lymphocyte co-stimulatory molecule expressed on activated T, B and DCs[61].

Primary lymphocytes are clearly initial targets for MV infection. However, quiescent lymphocytes and DCs express little or no CD150. Previous work had demonstrated that MV interacted with DCs and monocytes[61] and activated NF-κB at early stages of virus infection[62].

Additionally, wild-type but not vaccine strains were shown to have a more profound effect on DCs and monocytes[63].

Recent work indicates that at least part of this activation is due to interaction of MV with TLR2 and an additional outcome is the induction of expression of the entry receptor CD150, at least on monocytes and DCs. Wild-type MV, but not the vaccine strain, activated TLR2-transfected cells and mouse macrophages from wild-type but not TLR2 knockout mice, and this activation was independent of infection, since *uv*-inactivated virus also had this effect[64]. The ability to activate monocytes mapped to amino acid 481 of the H protein and importantly, incubation of MV bearing wild-type but not vaccine strain H proteins with cultured or primary human or monocytes, resulted in increased production of a number of inflammatory cytokines, including IL-12, IL-1 and IL-6 as well as up-regulation of CD150 surface expression. This work paralleled studies showing that DCs induced to mature *ex vivo* with TLR2 or TLR4 ligands also had increased CD150 expression and supported higher levels of MV infection[65].

Thus, MV appears to subvert the antiviral innate response, because it preferentially infects those cells that become activated. Interestingly, Th1 cells express more CD150 than Th2 cells and antibody ligation of CD150 on B cells causes their apoptosis[57]. Both phenomena could play a role in the generalized immunosuppression seen after MV infection.

3.1.4 Human cytomegalovirus and TLR2-mediated cytokine production

HCMV, a γ-herpesvirus with a double-stranded DNA genome, is a widespread, opportunistic human pathogen whose cytopathic effects are largely associated with immunological competency. HCMV infects most cell types by virtue of glycoprotein B (gB) found on the virion surface; virus entry is thought to be a two-step process, involving initial interaction with heparin sulfate on the cell surface and then receptor-mediated entry by one or more host proteins. It was demonstrated a number of years ago using differential display and gene chip/microarray analysis that HCMV virions could activate a number of signaling pathways in a virus gene expression-independent manner, since this also occurred with *uv*-inactivated virions[66,67]. These studies indicated that HCMV induced rapid cytokine production and that this induction was due, at least in part, to the interaction of gB with cells[68]. It is believed that the pathogenic effects associated with HCMV infection are facilitated or directly mediated, at least in part, by this production of inflammatory cytokines. Included in the genes activated by CMV are IFN-stimulated genes (ISG) [68].

CMV virion-mediated activation of cells involves both TLR2 and CD14. In cells lacking either of these molecules, CMV was unable to activate the NF-κB pathway or trigger inflammatory cytokine production (IL-6 and -8), in contrast to cells expressing both molecules[69]. This activation did not require productive infection, since both *uv*-inactivated and defective particles induced cytokine production. It appears that at least some, if not all of this activation is mediated by the interaction of gB with a heterodimer of TLR2 and TLR1 but not TLR6 (M. Guerrero and T. Compton, personal communication). Recent work has also indicated that gB may contribute to a cellular antiviral response by the activation of IRF-3[70]. Since IRF-3 activation is thought to occur via the TICAM and ligand binding to TLR2 is not thought to activate this pathway (Figure 2), the mechanism for this effect is not known.

In addition to the pathogenic affects associated with CMV infection that may be facilitated or directly mediated by inflammatory cytokines, CMV induction of the inflammatory response may play a role in virus replication and dissemination. For example, viral gene transcription is dependent on NF-κB and the recruitment of CMV targets such as neutrophils and monocytes to the site of infection may result from cytokine production. Some of these effects may also be the result of other by-products of CMV infection, such as dsDNA (see section 3.2.1). Interestingly, recent work has also indicated that the α-herpesvirus Herpes simplex virus-1 (HSV-1) activates cytokine production by cells via an unknown interaction with TLR2 and that this had pathogenic consequences, leading to a lethal form of encephalitis, at least at high virus doses[71].

3.2 Viral nucleic acids and TLR-mediated cell activation

Viral dsRNA, generated either by replication in the case of RNA viruses or opposite-strand transcription in the case of some DNA viruses, is a well-known inducer of type I IFN production[72]. With the discovery that all three by-products of viral replication, dsRNA, ssRNA and dsDNA, are ligands for TLR3, TLR7/8 and TLR9, respectively, recent work has focused on the role of these innate immune sensors in antiviral immunity and pathogenesis. In this section, studies documenting TLR activation by mCMV, HSV and several RNA viruses are covered.

3.2.1 Viral dsDNA and TLR-mediated antiviral responses

TLR9 was identified as an innate immune receptor that is activated by unmethylated dsDNA at CpG motifs[73]. TLR9 is predominantly located in an intracellular acidic compartment and upon ligand binding, recruits MyD88 to this compartment whereupon it signals[5,7]. Low pH is required for signaling, since chemicals such as chloroquine and bafilomycin A that neutralize acidic intracellular compartments, all inhibit this signaling[74]. It is not known whether neutralization causes TLR9 to localize to an inappropriate compartment, inhibit ligand binding or prevent structural changes in this receptor required for signaling.

Three herpesviruses have now been shown to activate cells via TLR9-dependent pathways, HSV-1 and HSV-2, both of which are α-herpesviruses and mCMV, a β-herpesvirus. HSV-1, usually transmitted through oral mucosa and HSV-2, a common sexually transmitted disease in humans transmitted through urogenital mucosa, generate good immune responses in infected individuals, although both remain in a latent state in neurons. In mouse models of HSV-2 mucosal infection, $CD11b^+$ DCs present viral antigens to T cells and generate a Th1 response[75]. When plasmacytoid DCs, potent secretors of type I IFNs, were incubated with either live or *uv*-inactivated HSV-2 or HSV-1, they secreted levels of IFN α and other cytokines similar to that seen with CpG[76,77]. Additionally, this induction was dependent on both MyD88 and TLR9, since pDCs from mice with targeted mutation of either gene did not respond to CpG, HSV-2 or HSV-1.

Neutralization of the acidic compartment with either chloroquine or bafilomycin inhibited induction by HSV-2 to a degree similar to that seen with purified CpG DNA[76]. Although the HSV-1 and HSV-2 genome has a relatively normal distribution of CpG residues compared to human genomic DNA, these residues are more abundant in this virus than in other herpesviruses such as Epstein Barr Virus[78]. Moreover, the viral genome is not methylated, thus, the TLR9-mediated responses may not be specific to HSV-2 but merely the result of the presence of the unmethylated CpG ligand in an acidic compartment. One prediction from these experiments is that any virus with a dsDNA genome in which CpG residues are suppressed, such as EBV[78], would not activate DCs via TLR9; this has yet to be tested. Additionally, although these data argue a role for TLR9 in innate immune responses to HSV, mice lacking MyD88 or TLR9 were capable of controlling HSV-1 replication *in vivo* in a corneal swabbing model; indeed, the MyD88 knockout mice may have shown somewhat better control of infection[77]. Whether the herpesviruses utilize the response to enhance virus spread or subvert the adaptive immune response is also yet to be determined.

TLR9 has been implicated in controlling host response to mCMV. This virus, which shares many features with hCMV, can replicate to high titers and cause damage to multiple organs, especially in immunocompromised mice[79]. Blood monocytes and tissue macrophages serve as target cells for infection, disseminate virus and play a major role in the pathogenesis. Organ damage is also associated with high levels of circulating tumor necrosis factor-α and other cytokines, as is a generalized immunosuppression. Natural killer cells play a critical role in controlling the initial response to mCMV infection, and genetic polymorphisms in the NK receptor Ly49H are associated with the susceptibility of certain mouse strains, such as BALB/c, to this virus[80,81].

Mice with a mutant TLR9 allele encoding a structurally aberrant receptor or homozygous for a null allele of MyD88 were both shown to have poor cytokine and NK responses to mCMV infection and to have a poor survival rate when infected with high doses of virus, compared to wild-type controls[9]. Interestingly, the response to mCMV and the mortality rate of the TLR9 and MyD88 mutant mice was still better than that observed with BALB/c mice bearing the mutant NK LY49H receptor but with a wild-type TLR9-axis. Moreover, the virus dose used in these experiments may not be biologically relevant[82]. Whether activation of TLR9 via viral dsDNA is important for controlling α- or β-herpesvirus infections *in vivo* needs to be resolved before concluding that this branch of the innate immune system is important in the host response.

Although the role of TLR9 signaling in herpesvirus infection is not yet resolved, recent findings with poxviruses do indicate that dsDNA or other TLR ligands activate innate immune responses. Poxviruses have dsDNA genomes that encode a large number of genes involved in avoiding immune responses[83] and at least vaccinia virus (VV), one family member, is relatively CpG-rich[78]. Interestingly, included in these anti-immune response genes are A46R and A52R, which inhibit signaling through MyD88-activated pathways[84]. Both proteins contain TIR domains but lack a highly conserved region commonly found in the cytoplasmic tails of TIR-containing proteins that is critical for signaling[85]. A52R binds to both IRAK2 and tumor necrosis factor-associated factor 6 (TRAF6), two downstream components involved in TIR signaling (Figure 2)[86]. Moreover, VV mutants lacking A52R are moderately attenuated in a murine intranasal model of infection, indicating that this gene does play a role in blocking an immune response[87]. More recently, a known VV virulence factor, N1L was shown to have homology to A52R and to inhibit signaling of multiple pathways including those initiated through TLRs by interacting with the I-κB kinase complex; this protein also blocked IFN regulatory factor signaling (Figure 1)[88]. It is not yet known how A46R works to block IL-1R signaling, whether it specifically targets TLR family members or what its role is in *in vivo* infection.

The presence of viral proteins that block IL-1R, IL-18R or TLR signaling suggests that VV elicits an antiviral innate immune response. Previous work has shown that type I IFNs are required to control VV infection[89]. Thus, blocking TLR-mediated activation of IRF3 in response to virus infection may prevent IFN production. The idea that these proteins block antiviral innate immune responses is also supported by the observation that VV infection does not induce differentiation of DCs, but instead blocks their poly (I:C)-induced maturation[90]. It is interesting to note that the A52R and N1L proteins block overlapping cell-signaling pathways by targeting different proteins. Thus, unlike many viral proteins which function as decoy receptors for specific cytokines or chemokines, these poxvirus proteins have the ability to block activation by many different receptors whose signaling converges on the same pathway. It has not been demonstrated yet whether VV dsDNA activation of TLR9 during virus infection is one of the responses inhibited by A46R, A52R or N1L. It has been shown, however, that cells from mice with mutations in the TICAM-1 locus do not respond to VV [91].

How does viral genomic DNA reach the compartments where TLR9 is located? Although several enveloped viruses, particularly RNA viruses, are known to enter cells via an acidic compartment, this is not clearly the case for herpesviruses. Although entry into most cell types is believed to occur at the plasma membrane, there have been reports that entry into certain cell types is dependent on intracellular acidic compartments[92,93]. Even in cells where entry occurs in an acidic compartment, the membrane fusion events would release the virus contents to the cytoplasmic side of the compartment membrane, away from the ligand binding domain of TLR9. One possibility is that some virus particles that enter cells in a non-productive manner and are thereby degraded in an acidic compartment, are responsible for this activation. Alternatively, virus DNA-mediated activation of TLR9 could occur when infected cells are phagocytosed by monocytes, DCs and macrophages. In this case, although the response to virus can be called innate at the cellular level, it requires actively-infected cells and thus would not be an innate response in a temporal sense.

3.2.2 Viral RNA and TLR-mediated responses

Not surprisingly, given the large number of viruses with either dsRNA or ssRNA genomes or in which virus genome replication or transcription generate dsRNA, TLR3 and TLR7/8 have also been implicated in innate antiviral immune responses. Initial experiments showed that purified reovirus genomic dsRNA resulted in the production of cytokines by macrophages from wild-type by not TLR3-null mice[94]. Moreover, LPS- or poly I:C-mediated activation of TLR4 or TLR3, respectively, resulted in the production of type I IFN (Figure 1) in macrophages or NIH3T3 cells and this in turn had a negative effect on replication of mouse γ-herpesvirus 68 *in vitro*[23]. Mice with a mutation in TICAM, an essential adaptor molecule in TLR3 or TLR4 signaling, showed higher levels of infection and greater mortality when infected with mCMV than wild-type mice, although this defect was still greater in BALB/c mice lacking the NK receptor LY49H but having a functional TLR response[91]. Additionally, macrophages from TICAM-null mice showed higher levels of VV; since both mCMV and VV produce dsRNA during transcription of their genome, this inability to control infection may be the result of impaired signaling through TLR3. Indeed, TLR3 has been directly implicated in the control of mCMV infection, since mice null for this gene showed somewhat attenuated control of infection when inoculated with high virus doses, although not to the same extent as MyD88-null, TLR9-mutant or BALB/c mice[9].

Interestingly, there is no direct evidence as of yet for TLR3 playing a role in the response to viruses with dsRNA genomes. In tissue culture, Newcastle disease virus, a paramyxovirus with a dsRNA genome, still induced type 1 IFNs, as well as other inflammatory cytokines in DCs from TLR3-/- mice[95]. Indeed whether TLR3-mediated activation induces any antiviral response has recently been called into question[82]. In four different infectious virus models, one with a double-stranded RNA genome (reovirus); two with ssRNA genomes that go through dsRNA intermediates during replication (lymphocytic choriomeningitis virus, vesicular stomatitis virus (VSV)); and one which generates dsRNA during transcription (mCMV), TLR-null mice generated effective antiviral responses and showed no impairment of the adaptive immune response. Unlike the previous studies that indicated a role for TLR3 in the restriction of mCMV replication[91], these experiments used virus doses similar to what would occur during natural infection.

More recently, ssRNA has been identified as a ligand for TLR7 in both mice and humans; humans, but not mice, apparently also have a second receptor TLR8[6,36,96]. Previously, it had been demonstrated that small antiviral compounds such as imiquod, R-848 and loxoribine used TLR7 in an MyD88-dependent pathway to induce production of IFN-α and other cytokines, and moreover, that this induction was dependent on recognition of ligand by TLR7 in an acidic compartment[97,98]. Two viruses with ssRNA genomes were tested in mice deficient either for TLR7 or MyD88, influenza and VSV. In both *ex vivo* cultures of DC and *in vivo* infections, the mutant mice demonstrated reduced IFN-α production in response to virus[36,96]. In addition, unlike the case with viruses whose membrane surface proteins interacted with TLRs, only infectious influenza virus was capable of mediating this response[96]. Both viruses also required an acidic compartment for this response, since no cytokine production occurred in DC treated with chloroquine.

Both influenza and VSV show pH-dependent entry into cells, because triggering of fusion peptide exposure requires acid pH in the endosome where membrane fusion occurs[99,100]. To control for effects on viral entry,

Sendai virus, which fuses at the cell membrane and a hybrid VSV bearing the RSV F protein on the surface, were also tested in this assay; both showed pH-dependent activation of DC[96].

However, as with viruses that enter through the plasma membrane, entry through fusion with the endosomal membrane releases virus nucleocapsid and ssRNA into the cytoplasm. Like TLR9, the ligand binding domain of TLR7 is likely located in the endosomal compartment and thus, would not be accessible to cytoplasmic nucleic acid[4]. As in the case of viral dsDNA that activates cells via TLR9, under normal infection conditions, ligand interaction most likely occurs when virus-infected cells are engulfed by inflammatory cells and degraded in lysosomes. Alternatively, engulfment of antibody-coated virus may be targeted to endosomal or lysosomal compartments by Fc receptors, leading to activation of TLRs 3, 7/8 or 9, which are all found in this compartment[36].

Although these data point to a role for TLR7's involvement in the innate immune response to viruses with ssRNA genomes, whether this response is important for controlling infection *in vivo* has not yet been determined. Whether other viruses with ssRNA genomes, such as retroviruses, also activate this pathway has also not been tested. Interestingly, however, there is evidence that the signals transduced by activation of TLR pathways can result in increased transcription of human immunodeficiency virus[101], leading to a possible subversion of the innate immune response by this virus.

4. CONCLUSIONS

A number of different viral gene products are now known to interact with different TLR family members. In some cases, induction of the innate response may limit infection by different viruses, especially when the infectious dose is high. Because these receptors recognize elements of the virus that are highly conserved, instead of evolving to avoid this response, some viruses may have adapted to utilizing the response, as a means of facilitating entry or replication, producing gene products such as transcription factors that increase virus gene expression or by down-modulating or subverting subsequent antiviral immune responses. Future studies are likely to uncover more effects of virus-induced TLR activation on pathogenesis.

ACKNOWLEDGEMENTS

I thank my colleagues for discussion of the issues in this chapter, especially Drs. John Rassa, Tatyana Golovkina and Lorraine Albritton. I apologize to the many authors whose work was not cited due to space considerations.

REFERENCES

1. Belvin, M. P. and Anderson, K. V., 1996, A conserved signaling pathway: the Drosophila Toll-Dorsal pathway. *Annu. Rev. Cell. Dev. Biol.* **12**:393-416
2. Akira, S., Takeda, K., and Kaisho, T., 2001, Toll-like receptors: critical proteins linking innate and acquired immunity. *Nat. Immunol.* **2**:675-680
3. Medzhitov, R., 2001, Toll-like receptors and innate immunity. *Nat. Rev. Immunol.* **1**:135-145
4. Nishiya, T. and DeFranco, A. L., 2004, Ligand-regulated chimeric receptor approach reveals distinctive subcellular localization and signaling properties of the Toll-like receptors. *J. Biol. Chem.* **279**:19008-19017
5. Ahmad-Nejad, P., Hacker, H., Rutz, M., Bauer, S., Vabulas, R. M., and Wagner, H., 2002, Bacterial CpG-DNA and lipopolysaccharides activate Toll-like receptors at distinct cellular compartments. *Eur. J. Immunol.* **32**:1958-1968
6. Heil, F., Hemmi, H., Hochrein, H., Ampenberger, F., Kirschning, C., Akira, S., Lipford, G., Wagner, H., and Bauer, S., 2004, Species-specific recognition of single-stranded RNA via toll-like receptor 7 and 8. *Science* **303**:1526-1529
7. Latz, E., Schoenemeyer, A., Visintin, A., Fitzgerald, K. A., Monks, B. G., Knetter, C. F., Lien, E., Nilsen, N. J., Espevik, T., and Golenbock, D. T., 2004, TLR9 signals after translocating from the ER to CpG DNA in the lysosome. *Nat. Immunol.* **5**:190-198
8. Akira, S. and Hemmi, H., 2003, Recognition of pathogen-associated molecular patterns by TLR family. *Immunol. Lett.* **85**:85-95
9. Tabeta, K., Georgel, P., Janssen, E., Du, X., Hoebe, K., Crozat, K., Mudd, S., Shamel, L., Sovath, S., Goode, J., Alexopoulou, L., Flavell, R. A., Beutler, B., 2004, Toll-like receptors 9 and 3 as essential components of innate immune defense against mouse cytomegalovirus infection. *Proc. Nat. Acad. Sci. USA* **101**:3516-3521
10. Sandor, F., Latz, E., Re, F., Mandell, L., Repik, G., Golenbock, D., Espevik, T., Kurt-Jones, E. A., and Finberg, R. W., 2003, Importance of extra- and intracellular domains of TLR1 and TLR2 in NFκB signaling. *J. Cell Biol.* **162**:1099-1110
11. Poltorak, A., He, X., Smirnova, I., Liu, M.-Y., van Huffel, C., Ku, X., Birdwell, D., Alejos, E., Silva, M., Galanos, C., Freudenberg, M., Ricciarid-Castagnoli, P., Layton, B., and Beutler, B., 1998, Defective LPS signaling in C3H/HeJ and C57BL/10ScCr mice: mutations in *Tlr4* gene. *Science* **282**:2085-2088
12. Qureshi, S. T., Lariviere, L., Leveque, G., Clermont, S., Moore, K. J., Gros, P., and Malo, D., 1999, Endotoxin-tolerant mice have mutations in Toll-like receptor 4 (*Tlr4*). *J. Exp. Med.* **189**:615-625
13. Shimazu, R., Akashi, S., Ogata, H., Nagai, Y., Fukudome, K., Miyake, K., and Kimoto, M., 1999, MD-2, a molecule that confers lipopolysaccharide responsiveness on toll-like receptor 4. *J. Exp. Med.* **189**:1777-1782
14. Ozinsky, A., Underhill, D. M., FOntenot, J. D., Hajjai, A. M., Smith, K. D., Wilson, C. B., Schroeder, L., and Aderem, A., 2000, The repertoire for pattern recognition of

pathogens by the innate immune system is defined by cooperation between Toll-like receptors. *Proc. Natl. Acad. Sci. USA* **97**:13766-13771

15. Hajjar, A. M., O'Mahony, D. S., Ozinsky, A., Underhill, D. M., Aderem, A., Klebanoff, S. J., and Wilson, C. B., 2001, Functional interactions between Toll-like receptor (TLR) 2 and TLR1 or TLR6 in reponse to phenol-soluble modulin. *J. Immunol.* **166**:15-19
16. Medzhitov, R., Preston-Hurlburt, P., Kopp, E., Stadlen, A., Chen, C., Ghosh, S., and Janeway, C. A. Jr., 1998, MyD88 is an adaptor protein in the hToll/IL-1 receptor family signaling pathways. *Mol. Cell.* **2**: 253-258
17. McGettrick, A. F. and O'Neill, L. A. J., 2004, The expanding family of MyD88-like adaptors in Toll-like receptor signal transduction. *Mol. Immunol.* **41**:577-582
18. Yamamoto, M., Sato, S., Mori, K., Hoshino, K., Takeuchi, O., Takeda, K., and Akira, S., 2002, A novel Toll/IL-1 receptor domain-containing adapter that preferentially activates the IFN-β promoter in the Toll-like receptor signaling. *J. Immunol.* **169**:6668-6672
19. Oshiumi, H., Matsumoto, M., Funami, K., Akazawa, T., and Seya, T., 2003, TICAM-1, an adaptor molecule that participates in Toll-like receptor 3-mediated interferon-β induction. *Nature Immunol.* **4**:161-167
20. Kawai, T., Takeuchi, O., Fujita, T., Inoue, J.-I., Muhlradt, P. F., Sato, S., Hoshino, K., and Akira, S., 2001, Lipopolysaccharide stimulates the Myd88-independent pathway and results in activation of IFN-regulatory factor 3 and the expression of a subset of lipopolysaccharide-inducible genes. *J. Immunol.* **167**:5887-5894
21. Hoshino, K., Kaisho, T., Iwabe, T., Takeuchi, O., and Shizuo, A., 2002, Differential involvement of IFN-ß in Toll-like receptor-stimulated dendritic cell activation. *Int. Immunol.* **14**:1225-1231
22. Toshchakov, V., Jones, B. W., Perera, P. Y., Thomas, K., Cody, M. J., Zhang, S., Williams, B. R., Major, J., Hamilton, T. A., Fenton, M. J., and Vogel, S. N., 2002, TLR4, but not TLR2, mediates IFN-β-induced STAT1α/β-dependent gene expression in macrophages. *Nat. Immunol.* **3**:392-398
23. Doyle, S. E., Vaidya, S. A., O'Connell, R., Dadgostar, H., Dempsey, P. W., Wu, T.-T., Rao, G., Sun, R., Haberland, M. E., Modlin, R. L., and Cheng, G., 2002 , IRF3 mediates a TLR3/TLR4-specific antiviral gene program. *Immunity* **17**:251-263
24. Sieling, P. A., Chung, W., Duong, B. T., Godowski, P. J., and Modlin, R. L., 2003, Toll-like receptor 2 ligands as adjuvants for human Th1 responses. *J. Immunol.* **170**:194-200
25. Dabbagh, K., Dahl, M. E., Stepick-Biek, P., and Lewis, D. B., 2002, Toll-like receptor 4 is required for optimal development of Th2 immune responses: role of dendritic cells. *J. Immunol.* **168**:4524-4530
26. Kaisho, T., Hoshino, K., Iwabe, T., Takeuchi, O., Yasui, Y., and Akira, A., 2002, Endotoxin can induce MyD88-deficient dendritic cells to support Th2 cell differentiation. *Int. Immunol.* **14**:695-700
27. Barton, G. M. and Medzhitov, R., 2002, Control of adaptive immune responses by Toll-like receptors. *Curr. Opin. Immunol.* **14**:380-383
28. Schnare, M., Barton, G. M., Holt, A. C., Takeda, K., Akira, S., and Medzhitov, R., 2001, Toll-like receptors control activation of adaptive immune responses. *Nat. Immunol.* **2**:947-950
29. Graham, B. S., 1996, Immunological determinants of disease caused by respiratory syncytial virus. *Trends Microbiol.* **4**:290-293
30. Connors, M., Giese, N. A., Kulkarni, A. B., Firestone, C.-Y., Holmes, K. L., Morse III, H. C., and Murphy, B. R., 1994, Enhanced pulmonary histopathology induced by respiratory syncytial virus (RSV) challenge of formalin-inactivated RSV-immunized BALB/c mice is abrogated by depletion of interleukin-4 (IL-4) and IL-10. *J. Virol.* **68**:55321-5325
31. Waris, M. E., Tsou, C., Erdman, D. D., Zaki, S. R., and Anderson, L. J., 1996, Respiratory syncytial virus infection in BALB/c mice previously immunized with

formalin-inactivated virus induces enhanced pulmonary inflammatory response with a predominant Th2-like cytokine pattern. *J. Virol.* **70**:2852-2860

32. Becker, S., Quay, J., and Soukup, J., 1991, Cytokine (tumor necrosis factor, IL-6, and IL-8) production by respiratory syncytial virus-infected human alveolar macrophages. *J. Immunol.* **147**:4307-4312
33. Tripp, R. A. and Anderson, L. J., 1998, Cytoxic T-lymphocyte precursor frequencies in BALB/c mice after acute respiratory syncytial virus (RSV) infection or immunization with a formalin-inactivated RSV vaccine. *J. Virol.* **72**:8971-8975
34. Haeberle, H. A. , Takizawa, R., Casola, A., Brasier, A. R., Dieterich, H. J., Van Rooijen, N., Gatalica, Z., and Garofalo, R. P., 2002, Respiratory syncytial virus-induced activation of nuclear factor-kappaB in the lung involves alveolar macrophages and toll-like receptor 4-dependent pathways. *J. Infect. Dis.* **186**:1199-1206
35. Kurt-Jones, E. A., Popova, L., Kwinn, L., Haynes, L. M., Jones, L. P., Tripp, R. A., Walsh, E. E., Freeman, M. W., Golenbock, D. T., Anderson, L. J., and Finberg, R. W., 2000, Pattern recognition receptors TLR4 and CD14 mediate response to respiratory syncytial virus. *Nat. Immunol.* **1**:398-401
36. Diebold, S. S., Kaisho, T., Hemmi, H., Akira, S., and Reis e Sousa, C., 2004 , Innate antiviral responses by mean of TLR7-mediated recognition of single-stranded RNA. *Science* **303**:1529-1531
37. Haynes, L. M., Moore, D. D., Kurt-Jones, E. A., Finberg, R. W., Anderson, L. J., and Tripp, R. A., 2001, Involvement of toll-like receptor 4 in innate immunity to respiratory syncytial virus. *J. Virol.* **75**:10730-10737
38. Poltorak, A., Merlin, T., Nielsen, P. J., Sandra, O., Smirnova, I., Schupp, I., Boehm, T., Galanos, C., and Freudenberg, M. A., 2001, A point mutation in the IL-12R 2 gene underlies the IL-12 unresponsiveness of Lps-defective C57BL/10ScCr mice. *J. Immunol.* **167**:2106-2111
39. Ehl, S., Bischoff, R., Ostler, T., Vallbracht, S., Schulte-Monting, J., Poltorak, A., and Freudenberg, M., 2004, The role of Toll-like receptor 4 versus interleukin-12 in immunity to respiratory syncytial virus. *Eur. J. Immunol.* **34**:1146-1153
40. Tal, G., Mandelberg, A., Dalal, I., Cesar, K., Somekh, E., Tal, A., Oron, A., Itskovich, S., Ballin, A., Houri, S., Beigelman, A., Lider, O., Rechavi, G., and Amariglio, N., 2004, Association between common Toll-like receptor 4 mutations and severe respiratory syncytial virus disease. *J. Infect. Dis.* **189**:2057-2063
41. Aberle, J. H., Aberle, S. W., Dworzak, M. N., Mandl, C. W., Rebhandl, W., Vollnhofer, G., Kundi, M., and Popow-Kraupp, T., 1999, Reduced interferon-gamma expression in peripheral blood mononuclear cells of infants with severe respiratory syncytial virus disease. *Am. J. Respir. Crit. Care Med.* **160**:1263-1268
42. Roman, M., Calhoun, W. J., Hinton, K. L., Avendano, L. F., Simon, V., Escobar, A. M., Gaggero, A., and Diaz, P. V., 1997, Respiratory syncytial virus infection in infants is associated with predominant Th-2-like response. *Am. J. Respir. Crit. Care Med.* **156**:190-195
43. Becker, M. N., Diamond, G., Verghese, M. W., and Randell, S. H., 2000, CD14-dependent lipopolysaccharide-induced beta-defensin-2 expression in human tracheobronchial epithelium. *J. Biol. Chem.* **275**:29731-29736
44. Ross, S. R., 2000, Using genetics to probe host-virus interactions: the mouse mammary tumor virus model. *Microbes and Inf.* **2**:1215-1223
45. Outzen, H. C., Morrow, D., and Shultz, L. D., 1985, Attenuation of exogenous murine mammary tumor virus virulence in the C3H/HeJ mouse substrain bearing the Lps mutation. *J. Natl. Canc. Inst.* **75**:917-923
46. Ardavin, C., Luthi, F., Andersson, M., Scarpellino, L., Martin, P., Diggelmann, H., and Acha-Orbea, H., 1997, Retrovirus-induced target cell activation in the early phases of infection: the mouse mammary tumor virus model. *J. Virol.* **71**:7295-7299

47. Rassa, J. C., Meyers, J. L., Zhang, Y., Kudaravalli, R., and Ross, S. R., 2002, Murine retroviruses activate B cells via interaction with Toll-like receptor 4. *Proc. Natl. Acad. Sci. USA* **99**:2281-2286
48. Burzyn, D., Rassa, J. C., Kim, D., Nepomnaschy, I., Ross, S. R., and Piazzon, I., 2004, Toll-like receptor 4-dependent activation of dendritic cells by a retrovirus. *J. Virol.* **78**:576–584
49. Vogel, S. N., Johnson, D., Perera, P., Medvedev, A., Lariviere, L., Qureshi, S. T., and Malo, D., 1999, Functional characterization of the effect of the C3H/HeJ defect in mice that lack an *Lps*^{n} gene: In vivo evidence for a dominant negative mutation. *J. Immunol.* **162**:5666-5670
50. Martin, P., Ruiz, S. R., Martinez del Hoyo, G., Anjuere, F., Vargas, H. H., Lopez-Bravo, M., and Ardavin, C., 2002, Dramatic increase in lymph node dendritic cell numbers during infection by the mouse mammary tumor virus occurs by a CD62L-dependent blood-borne DC recruitment. *Blood* **99**:1282-1288
51. Vacheron, S., Luther, S. J., and Acha-Orbea, H., 2002, Preferential infection of immature dendritic cells and B cells by mouse mammary tumor virus. *J. Immunol.* **168**:3470-3476
52. Harel, J., Rassart, E., and Jolicoeur, P., 1981, Cell cycle dependence of synthesis of unintegrated viral DNA in mouse cells newly infected with murine leukemia virus. *Virol.* **110**:202-207
53. Roe, T., Reynolds, T. C., Yu, G., and Brown, P. O., 1983, Integration of murine leukemia virus DNA depends on mitosis. *EMBO J.* **12**:2099-2108
54. Jude, B. A., Pobezinskaya, Y., Bishop, J., Parke, S., Medzhitov, R. M., Chervonsky, A. V., and Golovkina, T. V., 2003, Subversion of the innate immune system by a retrovirus. *Nat. Immunol.* **4**:573-578
55. Hook, L. M., Agafonova, Y., Ross, S. R., Turner, S. J., and Golovkina, T. V., 2000, Genetics of mouse mammary tumor virus-induced mammary tumors: linkage of tumor induction to the gag gene. *J. Virol.* **74**:8876-8883
56. Schneider-Schaulies, S. and ter Meulen, V., 2002, Triggering of and interference with immune activation: interactions of measles virus with monocytes and dendritic cells. *Viral Immunol.* **15**:417-428
57. Schneider-Schaulies, S., Klagge, I. M., and ter Meulen, V., 2003, Dendritic cells and measles virus infection. *Curr. Top. Microbiol. Immunol.* **276**:77-101
58. Schneider-Schaulies, J., Meulen, V., and Schneider-Schaulies, S., 2003, Measles infection of the central nervous system. *J. Neurovirol.* **9**:247-252
59. Dorig, R. E., Marcil, A., Chopra, A., and Richardson, C. D., 1993, The human CD46 molecule is a receptor for measles virus (Edmonston strain). *Cell* **75**:295-305
60. Tatsuo, H., Ono, N., Tanaka, K., and Yanagi, Y., 2000, SLAM (CDw150) is a cellular receptor for measles virus. *Nature* **406**:893-897
61. Klagge, I. M. and Schneider-Schaulies, S., 1999, Virus interactions with dendritic cells. *J. Gen. Virol.* **80**:823-833
62. Helin, E., Vainionpaa, R., Hyypia, T., Julkunen, I., and Matikainen, S., 2001, Measles virus activates NF-kappa B and STAT transcription factors and production of IFN-alpha/beta and IL-6 in the human lung epithelial cell line A549. *Virology* **290**:1-10
63. Schnorr, J. J., Xanthakos, S., Keikavoussi, P., Kaempgen, E., ter Meulen, V., and Schneider-Schaulies, S., 1997, Induction of maturation of human blood dendritic cell precursors by measles virus is associated with immunosuppression. *Proc. Natl. Acad. Sci. USA* **94**:5326-5331
64. Bieback, K., Lien, E., Klagge, I. M., Avota, E., Schneider-Schaulies, J., Duprex, W. P., Wagner, H., Kirschning, C. J., ter Meulen, V., and Schneider-Schaulies, S., 2002, Hemagglutinin protein of wild-type measles virus activates toll-like receptor 2 signaling. *J. Virol.* **76**:8729-8736
65. Murabayashi, N., Kurita-Taniguchi, M., Ayata, M., Matsumoto, M., Ogura, H., and Seya, T., 2002, Susceptibility of human dendritic cells (DCs) to measles virus (MV) depends

on their activation stages in conjunction with the level of CDw150: role of Toll stimulators in DC maturation and MV amplification. *Microbes and Inf.* **4**:785-794
66. Zhu, H., Cong, J. P., and Shenk, T., 1997, Use of differential display analysis to assess the effect of human cytomegalovirus infection on the accumulation of cellular RNAs: induction of interferon-responsive RNAs. *Proc. Natl. Acad. Sci. USA* **94**:13985-13990
67. Zhu, H., Cong, J. P., Mamtora, G., Gingeras, T., and Shenk, T., 1998, Cellular gene expression altered by human cytomegalovirus: global monitoring with oligonucleotide arrays. *Proc. Natl. Acad. Sci. USA* **95**:14470-14475
68. Simmen, K. A., Singh, J., Luukkonen, B. G., Lopper, M., Bittner, A., Miller, N. E., Jackson, M. R., Compton, T., and Fruh, K., 2001, Global modulation of cellular transcription by human cytomegalovirus is initiated by viral glycoprotein B. *Proc. Natl. Acad. Sci. USA* **98**:7140-7145
69. Compton, T., Kurt-Jones, E. A., Boehme, K. W., Belko, J., Latz, E., Golenbock, D. T., and Finberg, R. W., 2003, Human cytomegalovirus activates inflammatory cytokine responses via CD14 and Toll-like receptor 2. *J. Virol.* **77**:4588-4596
70. Boehme, K. W., SIngh, J., Perry, S. T., and Compton, T., 2004, Human cytomegalovirus elicits a coordinated cellular antiviral response via envelope glycoprotein B. *J. Virol.* **78**:1202-1211
71. Kurt-Jones, E. A., Chan, M., Zhou, S., Wang, J., Reed, G., Bronson, R., Arnold, M. M., Knipe, D. M., and Finberg, R. W., 2004, Herpes simplex virus 1 interaction with Toll-like receptor 2 contributes to lethal encephalitis. *Proc. Natl. Acad. Sci USA* **101**:1315-1320
72. Sen, G. C., 2001, Viruses and interferons. *Annu. Rev. Microbiol.* **55**:255-281
73. Hemmi, H., Takeuchi, O., Kawai, T., Kaisho, T., Sato, S., Sanjo, H., Matsumoto, M., Hoshino, K., Wagner, H., Takeda, K., and Akira, S., 2000, A Toll-like receptor recognizes bacterial DNA. *Nature* **408** :740-745
74. Leadbetter, E. A., Rifkin, I. R., Hohlbaum, A. M., Beaudette, B. C., Shlomchik, M. J., and Marshak-Rothstein, A. , 2002, Chromatin-IgG complexes activate B cells by dual engagement of IgM and Toll-like receptors. *Nature* **416**:603-607
75. Zhao, X. Y., Deak, E., Soderberg, K., Linehan, M., Spezzano, D., Zhu, J., Knipe, D. M., and Iwasaki, A., 2003, Vaginal submucosal dendritic cells, but not Langerhans' cells, induce protective Th1 responses to herpes simplex virus-2. *J. Exp. Med.* **197**:153-162
76. Lund, J., Sato, A., Akira, S., Medzhitov, R., and Iwasaki, A., 2003, Toll-like receptor 9-mediated recognition of Herpes simplex virus-2 by plasmacytoid dendritic cells. *J. Exp. Med.* **198**:513-520
77. Krug, A., Luker, G. D., Barchet, W., Leib, D. A., Akira, S., and Colonna, M., 2004, Herpes simplex virus type 1 activates murine natural interferon-producing cells through Toll-like receptor 9. *Blood* **103**:1433-1437
78. Karlin, S., Doerfler, W., and Cardon, L. R., 1994, Why is CpG suppressed in the genome of virtually all small eukaryotic viruses but not in those of large eukaryotic viruses? *J. Virol.* **68**:2889-2897
79. Krmpotic, A., Bubic, I., Polic, B., Lucin, P., and Jonjic, S., 2003, Pathogenesis of murine cytomegalovirus infection. *Microbes Infect.* **5**:1263-1277
80. Brown, M. G., Dokun, A. O., Heusel, J. W., Smith, H. R., Beckman, D. L., Blattenberger, E. A., Dubbelde, C. E., Stone, L. R., Scalzo, A. A., and Yokoyama, W. M., 2001, Vital involvement of a natural killer cell activation receptor in resistance to viral infection. *Science* **292**:934-937
81. Lee, S. H., Girard, S., Macina, D., Busa, M., Zafer, A., Belouchi, A., Gros, P., and Vidal, S. M., 2001, Susceptibility to mouse cytomegalovirus is associated with deletion of an activating natural killer cell receptor of the C-type lectin superfamily. *Nat. Genet.* **28**:42-45

82. Edelmann, K. H., Richardson-Burns, S., Alexopoulou, L., Tyler, K. L., Flavell, R. A., and Oldstone, M. B. A., 2004, Does Toll-like receptor 3 play a biological role in virus infections? *Virol.* **322**:231-238
83. Seet, B. T., Johnston, J. B., Brunetti, C. R., Barrett, J. W., Everett, H., Cameron, C., Sypula, J., Nazarian, S. H., Lucas, A., and Grant McFadden, G., 2003, Poxviruses and immune evasion. *Annu. Rev. Immunol.* **21**:377-423
84. Bowie, A., Kiss-Toth, E., Symons, J. A., Smith, G. L., Dower, S. K., and O'Neill, L. A., 2000, A46R and A52R from vaccinia virus are antagonists of host IL-1 and Toll-like receptor signaling. *Proc. Natl. Acad. Sci. USA* **97**:10162-10167
85. Xu, Y., Tao, X., Shen, B., Horng, T., Medzhitov, R., Manley, J. L., and Tong, L., 2000, Structural basis for signal transduction by the Toll/interleukin-1 receptor domains. *Nature* **408**: 111-115
86. Harte, M. T., Haga, I. R., Maloney, G., Gray, P., Reading, P. C., Bartlett, N. W., Smith, G. L., Bowie, A., and O'Neill, L. A., 2003, The poxvirus protein A52R targets Toll-like receptor signaling complexes to suppress host defense. *J. Exp. Med.* **197**:343-351
87. Merchant, M., Caldwell, R. G., and Longnecker, R., 2000, The LMP2A ITAM is essential for providing B cells with development and survival signals in vivo. *J. Virol.* **74**:9115-9124
88. DiPerna, G., Stack, J., Bowie, A. G., Boyd, A., Kotwal, G., Zhang, Z., Arvikar, S., Latz, E., Fitzgerald, K. A., and Marshall, W. L., 2004, Poxvirus protein N1L targets the I-κB kinase complex, inhibits signaling to NF-κB by the tumor necrosis factor superfamily of receptors, and inhibits NF-κB and IRF3 signaling by Toll-like receptors. *J. Biol. Chem.* **in press**.
89. van den Broek, M. F., Muller, U., Huang, S., Zinkernagel, R. M., and Aguet, M., 1995, Immune defense in mice lacking type I and/or type II interferon receptors. *Immunol. Rev.* **148**:5-18
90. Engelmayer, J., Larsson, M., Subklewe, M., Chahroudi, A., Cox, W. I., Steinman, R. M., and Bhardwaj, N., 1999, Vaccinia virus inhibits the maturation of human dendritic cells: a novel mechanism of immune evasion. *J. Immunol.* **163**:6762–6768
91. Hoebe, K., Du, X., Georgel, P., Janssen, E., Tabeta, K., Kim, S. O. , Goode, J., Lin, P., Mann, N., Mudd, S., Crozat, K., Sovath, S., Han, J., and Beutler, B., 2003, Identification of Lps2 as a key transducer of MyD88-independent TIR signalling. *Nature* **424**:743-748
92. Akula, S. M., Naranatt, P. P., Walia, N. S., Wang, F. Z., Fegley, B., and Chandran, B., 2003, Kaposi's sarcoma-associated herpesvirus (human herpesvirus 8) infection of human fibroblast cells occurs through endocytosis. *J. Virol.* **77**:7978-7990
93. Nicola, A. V., McEvoy, A. M., and Straus, S. E., 2003, Roles for endocytosis and low pH in herpes simplex virus entry into HeLa and Chinese hamster ovary cells. *J. Virol.* **77**:5324-5332
94. Alexopoulou, L., Holt, A. C., Medzhitov, R., and Flavell, R. A., 2001, Recognition of double-stranded RNA and activation of NF-kappaB by Toll-like receptor 3. *Nature.* **413**:732-738
95. Honda, K., Sakaguchi, S., Nakajima, C., Watanabe, A., Yanai, H., Matsumoto, M., Ohteki, T., Kaisho, T., Takaoka, A., Akira, S., Seya, T., and Taniguchi, T., 2003, Selective contribution of IFN-alpha/beta signaling to the maturation of dendritic cells induced by double-stranded RNA or viral infection. *Proc. Natl. Acad. Sci. USA* **100**:10872-10877
96. Lund, J. M., Alexopoulou, L., Sato, A., Karow, M., Adams, N. C., Gale, N. W., Iwasaki, A., and Flavell, R. A., 2004, Recognition of single-stranded RNA viruses by Toll-like receptor 7. *Proc. Natl. Acad. Sci. USA* **101**:5598-5603
97. Hemmi, H., Kaisho, T., Takeuchi, O., Sato, S., Sanjo, H., Hoshino, K., Horiuchi, T., Tomizawa, H., Takeda, K., and Akira, S., 2002, Small antiviral compounds activate immune cells via the TLR7 MyD88-dependent signaling pathway. *Nat. Immunol..* **3**:196-200

98. Heil, F., Ahmad-Nejad, P., Hemmi, H., Hochrein, H., Ampenberger, F., Gellert, T., Dietrich, H., Lipford, G., Takeda, K., Akira, S., Wagner, H., and Bauer, S., 2003, The Toll-like receptor 7 (TLR7)-specific stimulus loxoribine uncovers a strong relationship within the TLR7, 8 and 9 subfamily. *Eur. J. Immunol.* **33**:2987-2997
99. Hernandez, L. D., Hoffman, L. R., Wolfsberg, T. G., and White, J. M., 1996, Virus-cell and cell-cell fusion. *Annu. Rev. Cell. Dev. Biol.* **12**:627-661
100. Skehel, J. J. and Wiley, D. C., 2000, Receptor binding and membrane fusion in virus entry: the influenza hemagglutinin. *Annu. Rev. Biochem.* **69**:531-569
101. Equils, O., Faure, E., Thomas, L., Bulut, Y., Trushin, S., and Arditi, M., 2001, Bacterial lipopolysaccharid activates HIV long terminal repeat through toll-like receptor 4. *J. Immunol.* **166**:2342-2347

Chapter 11

DIGESTING ONESELF AND DIGESTING MICROBES

Autophagy as a Host Response to Viral Infection

MONTRELL SEAY*, SAVITHRAMMA DINESH-KUMAR*, and BETH LEVINE#

**Department of Molecular, Cellular, and Developmental Biology, Yale University, New Haven, CT, USA; #Department of Medicine, University of Texas Southwestern Medical Center, Dallas, TX, USA*

1. INTRODUCTION

The cellular pathway of autophagy is as ancient as the origins of eukaryotic life. Derived from the Greek and meaning to eat ("phagy") oneself ("auto"), the term autophagy refers to a lysosomal pathway of self-digestion, involving dynamic membrane rearrangement to sequester cargo for delivery to the lysosome, where the sequestered material is degraded and recycled. For decades, it has been known that autophagy is the primary intracellular catabolic mechanism for the degradation and recycling of long-lived cellular proteins and organelles. For decades, it has also been known that the recycling function of autophagy is an important adaptive response to nutrient deprivation and other forms of environmental stress. However, only recently have we discovered that autophagy may also be an important mechanism for the degradation of intracellular pathogens and that autophagy may also be important in cellular protection against the stress of microbial infection. Not surprisingly, we have also recently learned that some successful intracellular pathogens have devised strategies either to block host autophagy or to subvert the host autophagic process to foster their own replication. In this chapter, we will review recent progress in understanding

P. Palese (ed.), Modulation of Host Gene Expression and Innate Immunity by Viruses, 245-279.

the interrelationships between viruses, autophagy, and innate immunity (Figures 1 & 2).

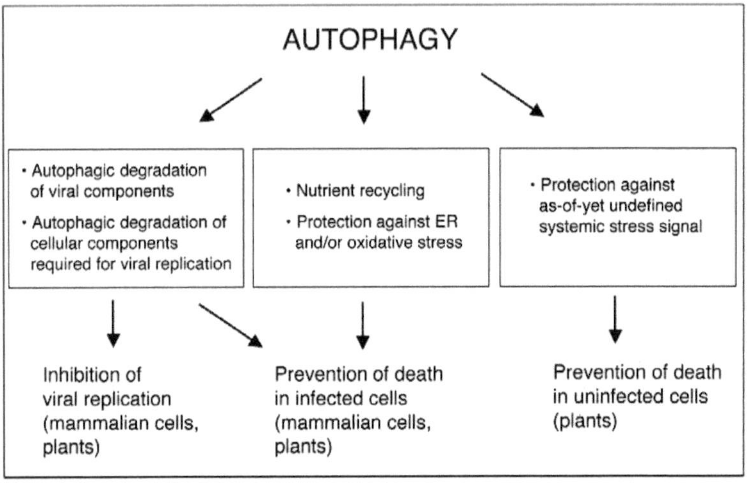

Figure 1. Conceptual overview of protective roles of autophagy in mammalian and plant viral infections. Areas in boxed regions represent potential mechanisms by which autophagy exerts each type of protective effect. See text for details.

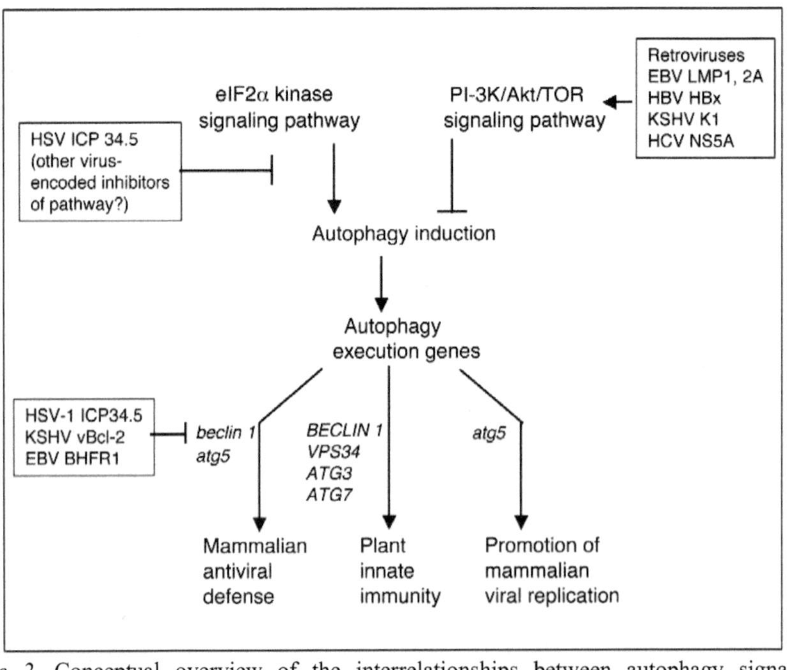

Figure 2. Conceptual overview of the interrelationships between autophagy signaling pathways, autophagy genes, autophagy functions in viral infections, and viral inhibitors of autophagy. See text for details.

2. INTRODUCTION TO THE MOLECULAR AND CELL BIOLOGY OF AUTOPHAGY

Before discussing the interrelationships between viruses, autophagy, and innate immunity, we will provide a brief overview of the molecular and cell biology of autophagy. While this subject has been covered extensively in a recent book and numerous recent review articles[2-5] we will highlight the aspects of this subject that may have particular relevance for viral infections. The process of autophagy was first described more than forty years ago, however, for many decades our understanding of autophagy was based largely on morphological observations from electron microscopy (reviewed in [6]). The field has expanded considerably within the last 15 years after the cloning and molecular characterization of the yeast *AuTophagGy (ATG)* - related genes (reviewed in [7]). The analysis of sequenced genomes of higher eukaryotes has identified *ATG* homologues in mammals, *C. elegans*, *Drosophila*, *Dictyostelium*, and plants, and many of these genes have been shown to be essential for autophagy function in higher eukaryotes (reviewed in [2]) (Table 1). In addition to the identification of the autophagy genes, significant progress has been made in the past decade in understanding some of the signaling events that regulate autophagy (reviewed in [8,9]). Interestingly, some of the signaling molecules that regulate autophagy, as well as some of the autophagy genes, play a role in the host antiviral innate immune response (Table 1, Figure 2).

Table 1. Orthologs of yeast autophagy-related genes in higher eukaryotes, including genes that play a role in the host response to viral infections[a].

Gene designation							Ref	Protein characteristics
ATG	*At/Nb*[a]	*Ce*	*Dd*	*Dm*	*Hs*	*Mm*		
Regulation of induction								
1	---	*unc-51*	*DdATG1*	---	---	---	20, 59, 125	Protein kinase
Vesicle nucleation								
6	***BECLIN 1***	*bec-1*	*DdATG6*	---	*beclin 1*	***beclin 1***	20, 23, 24, **37-40,** 59, 126	Component of PtdIns 3-kinase complex
Vesicle expansion and completion								
3	***ATG3***	---	---	*DrAUT1*	*hatg3*	---	**40**, 42, 127, 128	E2-like enzyme conjugates PE to Atg8
4	---	---	---	*ATG4/AUT2*	---	*atg4B*	48, 129, 130	Cysteine protease cleaves at C terminus of Atg8
5	---	---	*DdATG5*	---	*hatg5*	***atg5***	**38**, **51**, 58, 131-133	Conjugated to Atg12 through internal lysine
7	***ATG7***	*M7.5*	*DdATG7*	---	*HsGSA7* *hatg7*	*atg7*	20, **40**, 58, 61, 134, 135	E1-like enzyme activates ubiquitin-like protein Atg8 and Atg12
8	---	*lgg-1*	*DdATG8*	---	*MAP1LC3*[b]	*atg8* *APG8L*	11, 20, 46, 59, 128	Ubiquitin-like protein conjugated to PE
10	---	---	---	---	---	*atg10*	136, 137	E2-like enzyme conjugates Atg12 to Atg5

2.1 Formation and structure of autophagosomes

The initial step of autophagy is the formation and elongation of the isolation membrane. The isolation membrane invaginates and sequesters cytoplasmic constituents including mitochondria, endoplasmic reticulum (ER) and ribosomes, and the edges of the membrane fuse with each other to form a double-membrane structure called an autophagosome. The outer membrane of the autophagosome fuses to the lysosome/vacuole with subsequent delivery of the inner vesicle or autophagic body into the lumen of the degradative compartment. The source of the autophagosomal membrane is still unclear, but presently, it is thought that the pre-autophagosomal structure (PAS) acts as the site of vesicle formation during autophagy[10-12]. The PAS is thought to form *de novo*, but the source of the vesicle membrane is not known. It seems likely that the "typical" autophagosomes observed during viral infection that contain a mix of virions and self-cytoplasmic constituents, originate from the PAS. However, it is not yet known whether the PAS also serves as the site of vesicle formation for the formation of "atypical" autophagic-like double-membrane vesicles that function as replication sites for certain RNA viruses (e.g. poliovirus, mouse hepatitis virus, equine arterivirus) (reviewed in [13]). More likely, these double-membrane vesicles arise directly from the endoplasmic reticulum (ER)[13].

Autophagosomes are lipid-rich, protein-poor vesicles that vary in size and membrane thickness depending on the organism and cell type. The composition and abundance of proteins sequestered within autophagosomes reflects the relative composition and abundance of proteins in the surrounding cytoplasm[14]. This observation has led to the concept that autophagosomes indiscriminately sequester cytoplasmic content. However, in yeast, there are well-established pathways of specific autophagy, including the biosynthetic cytoplasm-to-vacuole targeting pathway and pexophagy (reviewed in [15]), and in mammalian cells, mitochondria-specific autophagy has been reported[16]. Although molecular determinants of cargo recognition have been identified in yeast pathways of specific autophagy, virtually nothing is known about the specificity of cargo recognition in higher eukaryotes. In circumstances where there is degradation of viruses observed inside "typical" autophagosomes that also contain cellular constituents, the sequestration step may lack specificity. However, in circumstances where viruses utilize components of the autophagic machinery for the formation of "autophagic-like" double-membrane structures that exclusively contain viral constituents, the sequestration step is likely to have exquisite specificity. The identification of the viral and cellular determinants of this specificity will be an important advance in understanding the cell biology of these types of RNA virus infections and may eventually lead to the identification of novel antiviral therapeutic targets.

2.2 Regulation of autophagy

Autophagy is tightly regulated by nutritional, hormonal, and other environmental cues. It occurs as a cellular response to extracellular stimuli (e.g. nutrient starvation, hypoxia, overcrowding, high temperature, hormonal or chemotherapeutic treatment), and intracellular stimuli (e.g. accumulation of damaged, superfluous or unwanted organelles, accumulation of misfolded proteins, invasion of microorganisms). Although it is not yet known whether different stimuli act through parallel, convergent, or divergent pathways to trigger autophagy, significant progress has been made within the past decade in identifying different signaling molecules that function in the positive (e.g. eIF2α kinases, Class III PI-3 kinases, PTEN, death-associated protein kinases) or negative (e.g. Tor, insulin-like growth factor signals, Class I PI-3 kinase, Rho/Ras family of GTPases) regulation of autophagy (reviewed in [8,9]).

The identification of a role for these signaling molecules in autophagy regulation has implications for understanding antiviral immunity, and more speculatively, generates hypotheses about novel principles of virus-host interactions. The recently defined evolutionarily conserved role of the eIF2α kinase signaling pathway in autophagy induction suggests that autophagy regulation may contribute to the antiviral function of the interferon-inducible eIF2α kinase, PKR. PKR and other eIF2α kinases induce a general translational arrest by phosphorylating the serine 51 residue of eIF2α (reviewed in [17]). Genetic studies in yeast and mammalian cells have also shown that the eIF2α kinase signaling pathway is required for starvation and herpes simplex virus-induced autophagy[18]. While further analyses are required to dissect the relative contributions of autophagy induction vs. translational arrest in mediating the antiviral effects of PKR, these findings link a new cellular function (i.e. autophagy) with interferon signaling.

Although the eIF2α kinase signaling pathway is the only as-of-yet defined autophagy regulatory signaling pathway that has known antiviral functions, it is interesting to note that most autophagy regulatory signals play a role in other important cellular processes, including cell growth control, cell death, and aging. Some of these effects may be the consequence of divergent downstream targets of these regulatory signals and some of these effects may be directly mediated through autophagy. As will be discussed below, given the evidence for a role of autophagy in innate immunity, it is likely that viruses have evolved different strategies to antagonize host autophagy, which, at least in the case of herpes simplex virus (see Sections 4.1 and 5) include the targeting of upstream autophagy regulatory signals[18]. Therefore, it is tempting to speculate that some of the effects of viruses on cell growth control and cell death may be either direct or indirect consequences of the evolutionary pressure that viruses face to modulate host autophagy.

As one example, the insulin-like/Class I PI-3K/Akt signaling pathway inhibits autophagy[19,20], promotes oncogenesis (reviewed in [21]), and decreases lifespan (most likely through autophagy-inhibitory effects)[20]. Certain retroviruses have recruited the catalytic subunit of PI3-K and its downstream target Akt and these viral gene products function as oncoproteins (reviewed in [22]). The emerging link between these signaling molecules and autophagy inhibition raises the interesting hypothesis that the initial acquisition of these molecules by viruses was perhaps related to the selective advantage of autophagy inhibition in viral growth. In view of recent evidence supporting a role of autophagy in tumor suppression[23-26], the presence of these genes in retroviral genomes could contribute to oncogenesis at least, in part, through inhibition of autophagy signaling, as well as through modulation of other downstream pathways.

2.3 Autophagy genes

The *ATG* genes encode proteins important for responding to upstream signaling pathways as well as proteins needed for the generation, maturation, and recycling of autophagosomes (reviewed in [4,5,15,27]). The Atg proteins can be grouped into four functional groups, including a protein kinase cascade important for responding to upstream signals, a lipid kinase signaling complex important for vesicle nucleation, ubiquitin-like conjugation pathways important for vesicle expansion, and a recycling pathway important for the disassembly of Atg protein complexes from matured autophagosomes. The role of some of the Atg proteins, including ones that act in the lipid kinase signaling complex and in the ubiquitin-like conjugation pathways, has been studied in plant and mammalian viral infections (see Table 1).

2.3.1 Protein kinase signaling

Autophagy is a dynamic process that is tightly regulated by protein kinases and phosphatases. One of the first *ATG* genes identified in yeast, *ATG1*, encodes a serine/threonine kinase[28]. The Atg1 kinase maintains a weak interaction with a hyperphosphorylated Atg protein, Atg13, in nutrient-rich conditions. Upon starvation conditions or stress, Atg13 is dephosphorylated resulting in a tighter association with Atg1[29]. Atg13 binding is essential for autophagy since *atg13* mutants unable to bind to Atg1 are completely defective in autophagy[29]. Atg17, which is also thought to play a role in Atg1 activation, also interacts with Atg1[29]. Downstream targets of Atg1 have not been identified, although Atg1 interacts with other proteins independently of its kinase activity[30]. The upstream kinase, Tor (target of rapamycin), indirectly or directly results in Atg13 hyperphosphorylation, which is one presumptive mechanism by which Tor kinase inhibits autophagy. Of note, the Atg1 component of the yeast autophagy induction complex plays a conserved role in autophagy in higher eukaryotes. However, as-of-yet, the role of Atg1 in antiviral immunity has not been evaluated.

2.3.2 Lipid kinase signaling

VPS34, which encodes a phosphatidylinositol-3 kinase (PI3-K), phosphorylates the 3' hydroxyl group inosotiol ring of phosphoinositides[31]. Although there is only one PI3-K in yeast, there are three classes of PI3-K in higher eukaryotes; class III PI3-K has been shown to be analogous to yeast VPS34. The importance of Class III PI3-K signaling in autophagy has been demonstrated pharmacologically and genetically. The nucleotide derivative, 3-methyladenine, inhibits Class III PI3-K activity and is widely used to inhibit autophagosome formation in mammalian cells[32,33]. A null mutation in *VPS34* causes defects in autophagosome formation in yeast[34,35] and microinjection of an inhibitory antiVps34 antibody blocks autophagy in cultured mammalian cells[36].

Vps34 functions through the association with other Atg proteins in a large complex that includes Vps15, Atg6/Vps30 and Atg14[35]. This complex is thought to be important in vesicle nucleation by mediating the localization of other Atg proteins at the PAS[10,35]. Vps34 and Atg6/Vps30 are conserved in higher eukaryotes. Importantly, the mammalian (*beclin 1*) and plant (*BECLIN 1*) homologues of yeast *ATG6/VPS30* have been the most extensively studied *ATG* genes in viral infections. As will be discussed in more detail below, mammalian *beclin 1* restricts viral replication, protects against virus-induced cell death, and is a target of inhibition by different virally-encoded gene products[37-39]. Furthermore, both plant *BECLIN 1* and its binding partner, Class III PI3-K/VPS34 prevent the spread of programmed cell death during the plant antiviral hypersensitive response[40]. Thus, the lipid kinase complex plays an evolutionarily conserved role in antiviral innate immunity.

2.3.3 Ubiquitin-like conjugation reactions

Autophagic vesicle expansion and completion involves conjugation machinery analogous to the ubiquitin conjugation needed for proteasome-mediated protein degradation. Autophagy utilizes an E1-like enzyme (Atg7), two E2-like enzymes (Atg10 and Atg3) that facilitate the conjugation, and activation and localization of different ubiquitin-like modifiers (Atg5 and Atg8). The conjugation modification of Atg proteins is necessary for the formation of an autophagosome of appropriate size and shape[41]. However, the precise molecular functions of the conjugation reactions are not known and remain a critical unanswered question in autophagy research.

The first conjugation system involves the lipidation of Atg8, a ubiquitin-like protein whose close mammalian homologues have three-dimensional structures very similar to ubiqutin[42-45]. Both Atg8 and the mammalian homomlogue LC3 are cleaved post-translationally by the cysteine endopeptidase Atg4[46,47]. The cleavage of Atg8/Lc3 is essential for conjugation and further maturation of the autophagosomes[48]. In yeast and mammalian systems, the cleaved Atg8/LC3 is immediately activated by Atg7, an E1-like enzyme; transferred to Atg3, an E2-like enzyme; and finally conjugated to the lipid molecule phosphatidylethanolamine (PE) [42,47,49,50].

The second ubiquitin-like reaction is the conjugation of Atg12 to Atg5. Atg12 is an ubiquitin-like protein that is activated by Atg7 (E1-like enzyme), transferred to Atg10 (E2-like enzyme), and subsequently conjugated to Atg5 through an isopeptide bond[41,49]. The conjugation of Atg5 to Atg12 is necessary for autophagosome formation but not necessary for localization to the PAS[10].

Almost all of the components of the autophagy machinery that participate in the protein conjugation systems have orthologs in at least some higher eukaryotes. However, only mammalian Atg5 has been studied in the context of its role in viral infections. The contrasting phenotypes of *atg5* null cells infected with two different RNA viruses illustrate two distinct mechanisms by which viruses interact with the autophagic machinery. The murine coronavirus, mouse hepatitis virus, which replicates in association with double-membrane vesicles, has severely impaired growth in *atg5* null embryonic stem (ES) cells[51], suggesting that *atg5* is required for the formation of coronavirus replication complexes. In contrast, the prototype alphavirus, Sindbis virus, replicates to higher titers in *atg5* null murine embryonic fibroblasts (MEFs) than in wildtype controls[38], suggesting that the autophagic machinery functions to restrict Sindbis virus replication.

2.3.4 Atg protein recycling

In yeast, Atg proteins that act at the stage of vesicle formation are not associated with the completed autophagosome, with the exception of Atg8. This suggests that Atg proteins are retrieved at some point prior to, or upon, vesicle completion, and then reutilized in the generation of new autophagosomes. The process of recycling requires the action of Atg2 and Atg18, which allow the recycling of Atg9, the only transmembrane protein that is part of the autophagic machinery[52,53]. Atg9 and Atg18 have orthologues in higher eukaryotes, but their function in antiviral responses has not been studied.

The unique association of Atg8/LC3 with the mature autophagosome has led to an important technical advance in autophagy research. Atg8/LC3 is presently the most widely used and reliable marker for labeling autophagosomes[10, 11,20,50,54] and with the recent availability of transgenic mice that express GFP-tagged LC3[54], it is now possible to study autophagy induction *in vivo* during viral infections.

3. INTRODUCTION TO THE BIOLOGICAL FUNCTIONS OF AUTOPHAGY

In addition to its emerging role in innate immunity, autophagy plays a role in diverse other biological processes, including survival during starvation, differentiation and development, tissue homoeostasis, aging, cell growth control, and certain forms of programmed cell death. These biological functions of autophagy have been reviewed in detail elsewhere[2,55,56]. In this section, we will however, briefly discuss selected biological functions of autophagy that have relevance either to understanding the mechanisms by which autophagy protects cells against virus infections (Figure 1) or to understanding the potential consequences for the host of viral evasion of autophagy (Figure 3).

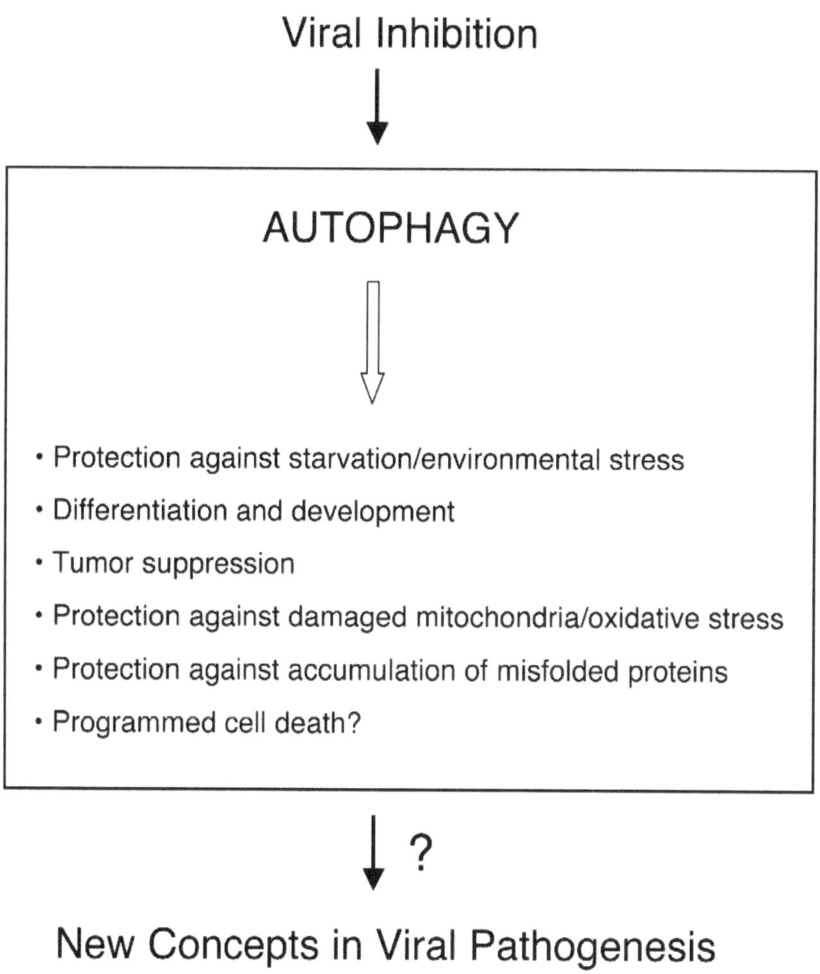

Figure 3. Conceptual overview of the biological functions of autophagy (other than antiviral defense). Viral inhibition of autophagy may block these functions, representing novel potential mechanisms of viral pathogenesis. See text for details.

3.1 Role of autophagy in protection against nutrient starvation

Perhaps the primordial function of autophagy is its ability to recycle nutrients and help sustain life during periods of starvation. Several decades ago, starvation was noted to be a potent inducer of autophagy in rodent liver (reviewed in [57]), leading to the hypothesis that autophagy is an adaptive response to starvation. Following the identification of the conserved autophagy genes, genetic studies in different species have confirmed that autophagy genes are required for the maintenance of eukaryotic life in the face of limited environmental nutrient supply. This principle was first demonstrated in yeast, i.e. all *ATG* gene mutant yeasts grow normally in nutrient rich conditions, but unlike wild-type yeasts, die rapidly during carbon or nitrogen starvation[28]. Similarly, *Dictyostelium discoideum* that lack *ATG* genes also grow normally in the presence of their food, nonpathogenic bacteria, but die rapidly when subjected to starvation[58,59]. During nitrogen or starvation, *atg7* and *atg9* mutant plants display two phenotypes that are thought to result from a defective ability to mobilize nutrients through autophagic delivery, including enhanced chlorosis (yellowing of leaves due to a loss of chlorophyll) and accelerated leaf senescence[60,61]. In addition, mammalian cells deleted of the autophagy genes, *atg5* or *beclin 1*, also undergo accelerated cell death in response to starvation as compared to their wild-type counterparts[62].

The pro-survival function of autophagy during starvation is thought to be related directly related to its ability to recycle nutrients to generate a sufficient pool of amino acids required for the synthesis of essential proteins. While the eIF2α kinase signaling pathway shuts off general translation during starvation, at least in yeast, Gcn2 signaling simultaneously stimulates the transcription of essential starvation response genes, including autophagy genes[63,64]. Thus, this signaling pathway provides a coordinated method to effectively generate new amino acids by autophagy and redirect the host cell synthetic machinery to use its limited amino acid supply specifically for the synthesis of essential starvation response proteins.

Although a downstream transcription factor like yeast Gcn4 (which is downstream of yeast eIF2α and transcriptionally transactivates autophagy genes), has not yet been identified for mammalian PKR signaling initiated during virus infection, it seems likely that there are functionally homologous molecules that direct virus-infected cells to mount an adaptive and selective transcriptional and translational response during virus infection. Even in the absence of this postulated arm of PKR signaling, the mere recycling of nutrients in virus-infected cells would be predicted to have a beneficial function for the host. Although few studies have compared cellular amino acid pools during nutrient starvation and virus infection, acute viral replication involves the parasitism of not only the host cell's translational machinery, but also the host cell's translational building blocks. Therefore, it seems likely that acute viral replication induces what can be thought of as a state of "pseuodostarvation". According to this model, the prediction is that the nutrient recycling function of autophagy plays a similar protective function during viral infection as it plays during nutrient deprivation.

3.2 Role of autophagy in differentiation and development

Differentiation and development both require cells to undergo significant phenotypic changes and must entail a mechanism for the breakdown and recycling of obsolete cellular components. Genetic studies have revealed an essential role for components of the autophagic machinery in differentiation and developmental processes in several different organisms, including sporulation in yeast, multicellular development in *Dictyostelium*, dauer development in *C. elegans*, and embryonic development in mice (reviewed in [2]). In addition, the mammalian autophagy gene, *beclin 1*, appears to play a role in epithelial cell differentiation, since the mammary glands in *beclin 1* heterozygous-deficient mice display striking morphological abnormalities[23]. Since viral gene products can inhibit the autophagy function of Beclin 1 (see below) and potentially other autophagy proteins[65], it is possible that autophagy blockade represents a mechanism by which viruses can affect cellular differentiation. For example, the Bcl-2-like BHRF1 protein encoded by EBV binds to Beclin 1[39] blocks its autophagy function[39], and also perturbs epithelial cell differentiation[66]. Further studies are needed to determine the role of Beclin 1 binding in the perturbation of epithelial cell differentiation by BHRF1, as well as to investigate the effects of other viral inhibitors of autophagy on cellular differentiation and multicellular development.

3.3 Role of autophagy in cell growth control

Many different viruses, including retroviruses, gammaherpesviruses, papillomaviruses, and hepatitis viruses are oncogenic. Studies of the mechanisms of viral oncogenesis have largely focused on the ability of viruses to alter mitogenic signaling, cell cycle regulation, and/or apoptosis. However, new evidence is emerging that autophagy plays a role in tumor suppression and that autophagy is antagonized by gene products encoded by certain oncogenic viruses. Accordingly, it will be important to evaluate the role of viral inhibition of autophagy in viral oncogenesis.

Normal cell growth requires a well-coordinated balance between the cell's biosynthetic machinery (e.g. protein synthesis and organelle biogenesis) and its degradative processes (e.g. protein degradation and organelle turnover). In the 1970's, it was first proposed that protein catabolism through autophagy is a major determinant of cell growth[67,68]. According to this model, both cell mass and the rate of cell growth is a balance between the amount of protein synthesized and the amount of autophagic protein degradation. Although this model has received little attention in recent years, interest in the role of autophagy in cell growth control has reemerged in light of new biochemical and genetic links between autophagy and the negative regulation of tumorigenesis.

As stated above in Section 2.2, several different oncogenic signaling molecules, including members of the insulin signaling pathway (e.g. Class I PI-3K, Akt) and members of the Rho and Ras family of GTPases negatively regulate autophagy in mammalian cells and the PTEN tumor suppressor positively regulates autophagy (reviewed in [69]). Furthermore, the autophagy inhibitor, Tor, is an important positive regulator of cell growth in diverse organisms, and the Tor inhibitor, rapamycin, has promising anti-tumor effects in human clinical trials (reviewed in [70]). Oncogenic viruses have developed multiple different strategies to activate autophagy-inhibitory signaling pathways. These strategies include encoding viral oncoproteins that represent activated forms of the corresponding cellular proto-oncogene (reviewed in [22]) or upregulating Rho/Ras or Class I PI-3K/Akt/TOR signaling through alternative mechanisms[71-77].

Components of the autophagic machinery may also play a direct role in tumor suppression. The *beclin 1* gene is monallelically deleted in a high percentage of cases of human breast, ovarian, and prostate cancer (reviewed in [78]) and has tumor suppressor function in cultured mammary carcinoma cells[79,80]. Heterozygous disruption of *beclin 1* in mice increases the frequency of spontaneous tumorigenesis (including papillary lung carcinomas, B cell lymphomas, and hepatocellular carcinomas) and accelerates the development of hepatitis B virus-induced pre-malignant lesions[23,24]. In addition, *atg5* null ES cells are more tumorigenic in mice than their wild-type counterparts and result in teratomas that are less well-differentiated[81]. Together, these findings lead to the concept that autophagy genes may represent a novel class of tumor suppressor genes and that genetic disruption of autophagy may represent a novel mechanism of tumorigenesis. As will be discussed in more detail below, two different classes of viral gene products have been identified thus far that bind to Beclin 1 and inhibit its autophagy function, including the alphaherpesvirus-encoded neurovirulence protein, HSV-1 ICP34.5, and the gammaherpesvirus-encoded Bcl-2-like proteins, KSHV vBcl-2 and EBV BHRF1[38,39]. The gammaherpesviruses are oncogenic viruses that are etiologically linked to a variety of different malignancies, including lymphoma, nasopharyngeal carcinoma, and Kaposi's sarcoma. At present, the precise role of gammaherpesvirus Bcl-2-like proteins in viral oncogenesis is uncertain. Nonetheless, given the well-defined role of cellular Bcl-2 in oncogenesis and the emerging evidence that Beclin 1 is a tumor suppressor protein, it will be important to evaluate whether viral Bcl-2 antagonism of Beclin 1 function plays a role in gammaherpesvirus oncogenesis. Of note, preliminary data indicates that KSHV may also encode other gene products that interact with other components of the autophagic machinery[65]. Thus, oncogenic gammaherpesviruses may have multiple mechanisms to disarm host autophagy. It will be of interest to determine whether other oncogenic DNA viruses, especially human papillomavirus, also directly inhibit the host autophagic machinery.

3.4 Role of autophagy in lifespan extension

In many tissues in the adult organism (especially post-mitotic cells), protein and organelle turnover by autophagy plays an essential cellular homeostatic or housekeeping function, removing damaged or unwanted organelles and proteins. For many decades, it has been presumed that this homeostatic function of autophagy represents an anti-aging mechanism, perhaps by reducing reactive oxidative species and other toxic intracellular substances that contribute to genotoxic stress (reviewed in [82]). The conserved effects of protein caloric restriction (a dietary inducer of autophagy) on lifespan extension has provided further fuel for this concept (reviewed in [83]). Recent genetic studies, especially those performed in *C. elegans*, provide more direct evidence for a role of both autophagy regulatory signals and components of the autophagic machinery in anti-aging pathways. Loss-of-function mutations in autophagy-inhibitory insulin-like signaling pathway extend lifespan (reviewed in [84]), and inactivation of the *C. elegans* ortholog of yeast autophagy gene, *ATG6/VPS30*, blocks this lifespan extension[20].

While the precise mechanisms by which autophagy extends lifespan are unknown, one theory is that autophagy selectively removes damaged mitochondria, resulting in decreased levels of intracellular reactive oxygen species and cellular protection against oxidative damage. Viral infections, as well as the inflammatory response to viral infections, can damage mitochondria and/or increase the intracellular generation of reactive oxygen species, and these effects may contribute to viral pathogenesis. For example, the mitochondrial damage that occurs in HIV infection (even in the absence of antiretroviral treatment) is thought to be a major contributory factor to the metabolic abnormalities and cardiomyopathy that occur in patients with

AIDS (reviewed in [85]). As another example, in a transgenic mouse model of hepatitic C virus (HCV) infection, oxidative stress in the absence of inflammation has been implicated in HCV-associated hepatocarcinogenesis[86]. Similarly, studies in transgenic mice and cultured cells indicate that pre-S1/S2 mutant hepatitis B virus surface antigens, which accumulate in late stages of HBV infection, cause oxidative stress and DNA damage[87]. Therefore, it is possible that the mechanisms by which autophagy functions as an anti-aging pathway may be relevant to potential roles that autophagy may play in protecting cells against adverse sequelae of oxidative stress during virus infection.

3.5 Role of autophagy in preventing diseases associated with protein aggregates

Diseases associated with an accumulation of misfolded and aggregated proteins, including neurodegenerative disorders and α_1-anti-trypsin liver disease, are associated with an increase in the accumulation of autophagic vacuoles (reviewed in [56]). In these diseases, it has both been argued that autophagy plays a protective role (i.e. by removing protein aggregates and damaged mitochondria) and a pathologic role (i.e. by promoting liver dysfunction in α_1-anti-trypsin deficiency through excessive mitochondrial autophagy[88] or by promoting autophagic cell death). Although both of these roles may be operative in different diseases or even in different facets of a single disease, recent studies provide compelling evidence that autophagy plays a protective role against the toxic effects associated with protein aggregation. For example, mutant α-synuclein (associated with early onset Parkinson's disease), and aggregate-prone proteins with polyglutamine expansions (associated with Huntington's disease) are targeted for autophagic degradation[88,89]. Rapamycin, which stimulates autophagy, not only enhances the clearance of aggregate-prone proteins but also reduces the appearance of the aggregates and the cell death associated with expression of mutant Huntington's proteins[89]. Furthermore, induction of autophagy with rapamycin protects against neurodegeneration in both a fly and mouse model of Huntington's disease[90].

Recent advances have also been made in understanding the mechanisms by which autophagy is induced in response to misfolded protein aggregates. In cell models, transgenic mice, and samples from human brains of patients with Huntington's disease, mTOR is sequestered into polyglutamine aggregates. This sequestration impairs its kinase activity, leading to induction of autophagy[90]. Although it has not yet been evaluated, it is likely that the accumulation of misfolded protein aggregates also induces autophagy through activation of the ER stress response, which is mediated by the eIF2α kinase, PKR-like ER resident kinase (PERK) [91,92], since other stress stimuli (e.g. starvation and virus infection) that activate other eIF2α kinases (e.g. Gcn2 and PKR) induce autophagy through this same signaling pathway[18].

These observations are potentially relevant to understanding the role of autophagy in protection against virus-induced diseases in which protein misfolding and ER stress are thought to play pathogenetic roles. Similar to genetic neurodegenerative disorders, there is increasing evidence that murine retrovirus-associated spongiform-like neuronal degeneration is also associated with protein misfolding and ER stress. For example, viral envelope proteins from avirulent strains are processed normally and fail to induce ER stress, whereas envelope proteins from neurovirulent strains are misfolded and activate ER stress response pathways[93-95]. In addition, it has been proposed that the mechanism by which pre-S mutant HBV surface antigens promote oxidative stress and DNA damage is through the accumulation of misfolded mutant proteins and activation of ER stress[87,96]. Thus, based on recent studies with non-viral associated neurodegenerative disorders, the prediction is that autophagy induction might be beneficial in attenuating diseases associated with misfolded viral proteins, such as retrovirus-associated spongiform encephalopathy and hepatitis B virus-induced liver damage. As a corollary, the possibility that these viruses might possess mechanisms to evade host autophagy could be an exacerbating factor in the pathogenesis of these infections.

An interesting question is whether viral protein aggregates trigger autophagy by mechanisms that are similar to those involved in autophagy induction initiated by cellular protein aggregates. Different viral glycoproteins are known to activate the ER stress-related eIF2α kinase, PERK[97,98], although a role for PERK in autophagy induction has not yet been formally demonstrated. It is completely unknown, however, whether viral protein aggregates, like polyglutamine aggregates in Huntington's disease, sequester and thereby inactivate the autophagy-inhibitory kinase, mTOR. If so, this would represent a highly novel mechanism by which viruses trigger intracellular innate immune responses.

4. AUTOPHAGY AND INNATE IMMUNITY TO VIRUSES

Autophagy is emerging as a newly described mechanism of antiviral innate immunity that is targeted by viral virulence gene products. Although there are not yet many published articles in this area, there are several observations that support this concept. First, during herpes simplex virus infection, the interferon-inducible antiviral PKR signaling pathway regulates the autophagic degradation of cellular and viral components[18,99]. Second, mammalian autophagy execution genes, including *beclin 1* and *atg5*, regulate Sindbis virus replication and Sindbis virus-induced cell death[37,38]. Third, plant autophagy execution genes, including *BECLIN 1*, Class III *PI3-K/VPS34*, *ATG3*, and *ATG7*, restrict tobacco mosaic virus replication and limit the spread of cell death during the innate immune response[40]. In this section, each of these observations will be described in more detail.

4.1 PKR-dependent autophagy degrades cellular and viral components

The interferon-inducible dsRNA-dependent protein kinase R (PKR) plays an important role in innate immunity against viral infections. PKR activation leads to phosphorylation of the α subunit of eukaryotic initiation factors 2 (eIF2 α) and a subsequent shutdown of host and viral protein synthesis and viral replication (reviewed in [100]). To avoid this translational shutdown, many viruses have evolved different strategies to antagonize PKR function. These include interference with the dsRNA-mediated activation of PKR or PKR dimerization; blockade of the kinase catalytic site or PKR-substrate interactions; alterations in the levels of PKR; direct regulation of eIF2α phosphorylation; and effects on components downstream of eIF2α (reviewed in [100,101]). The importance of viral antagonism of PKR function in viral pathogenesis has been most clearly demonstrated using a herpes simplex virus type 1 (HSV-1) model system[102,103]. The HSV-1 neurovirulence protein, ICP34.5, binds to protein phosphatase 1α and causes it to dephosphorylate eIF2α, thereby negating the activity of PKR[104,105]. A neuroattenuated HSV-1 mutant lacking ICP34.5 exhibits wild-type replication and virulence in mice genetically lacking *pkr*[103], proving that the ICP34.5 gene product mediates neurovirulence by antagonizing PKR-dependent functions.

In addition to regulating host translation during viral infection, the PKR signaling pathway also regulates the autophagic degradation of host proteins[18]. As mentioned above, molecules in the yeast eIF2α kinase signaling pathway (e.g. the eIF2α kinase, Gcn2, the eIF2α Ser51 residue, and the transcriptional transactivator, Gcn4) are required for nitrogen starvation-induced autophagy. Interestingly, the autophagy defect of *gcn2* null yeast can be rescued by mammalian *pkr* transformation[18]. Direct evidence that viruses can induce PKR-dependent autophagy has been provided by studies done with herpes simplex virus type 1 (HSV-1) infection in genetically engineered MEFs[18]. A herpes simplex virus (HSV-1) mutant virus lacking the ICP34.5 inhibitor of PKR signaling (termed HSV-1Δ 34.5), but not wild-type HSV-1, is able to induce the autophagic breakdown of long-lived cellular proteins in wild-type MEFs. However, HSV-1Δ34.5 infection is not able to induce the autophagic breakdown of long-lived cellular proteins in MEFs lacking *pkr* or with a nonphosphorylatable mutation in Ser-51 of eIF2α. These findings indicate the PKR-dependent signaling events regulate the autophagic breakdown of host proteins during viral infection and that this function of PKR is antagonized by the HSV-1 ICP34.5 neurovirulence gene product. As discussed in section 3.1, the breakdown of cellular proteins may help protect host cells against the effects of "pseudostarvation" induced by viral infection.

More recent studies indicate that PKR-dependent signaling also regulates the breakdown of viral components during HSV-1 infection[99]. Ultrastructural analyses of wild-type and *pkr*-deficient MEFs and sympathetic neurons infected with wild-type and HSV-1Δ34.5 demonstrate that HSV-1 is degraded in autophagosomes by a *pkr*-dependent process. In wild-type cells infected with wild-type HSV-1, the majority of intracytoplasmic virions are either randomly dispersed in the cytoplasm or are contained within "viral vacuoles," a structure thought to represent an important intermediate in the egress of HSV-1 from the nucleus out of the cell[106]. In contrast, in wild-type cells infected with HSV-1Δ34.5, most cytoplasmic virions are localized within autophagosomes that contain a mix of different cytoplasmic constituents (Figure 4A). *Pkr*-deficient MEFs or *pkr*-deficient neurons infected with HSV-1Δ34.5 have very few autophagosomes, and appear similar to wild-type HSV-1-infected wild-type cells, with randomly dispersed intracytoplasmic virions and numerous viral vacuoles. Together, these observations demonstrate that HSV-1 is degraded by autophagy, that HSV-1 ICP34.5 antagonizes the cellular autophagic degradation of HSV-1, and that this process requires PKR.

Recent biochemical analyses confirm that PKR signaling and HSV-1 ICP34.5 regulate viral protein degradation[99]. HSV-1 protein degradation is significantly accelerated in wild-type MEFs infected with HSV-1Δ34.5 as compared to wild-type MEFs infected with wild-type HSV-1, indicating that ICP34.5 delays viral protein degradation. However, in autophagy-deficient $pkr^{-/-}$MEFs or eIF2α S51A mutant MEFs, the rate of HSV-1 protein degradation is similar in HSV-1- and HSV-1Δ34.5-infected cells, indicating that HSV-1 protein degradation is positively regulated by the PKR signaling pathway.

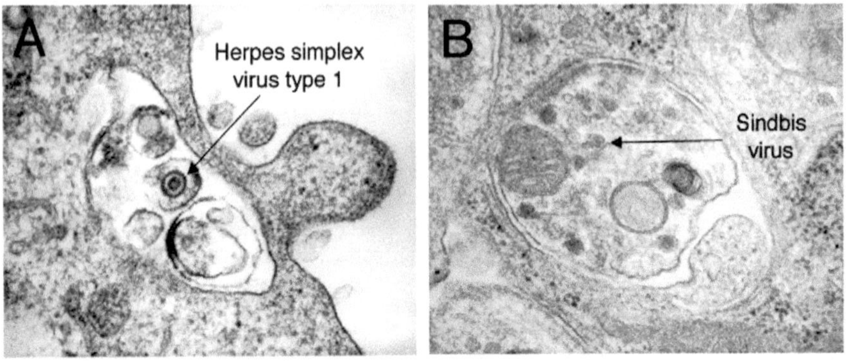

Figure 4. Electron micrographs demonstrating autophagic degradation of herpes simplex virus type 1 (A) and Sindbis virus (B) in neurons. The HSV-1 strain is a mutant lacking the virus-encoded PKR inhibitor, ICP34.5. Similar structures are not observed in neurons infected with wild-type HSV-1 (containing ICP34.5), or in HSV-1-infected or Sindbis virus-infected *pkr* neurons. See text for details.

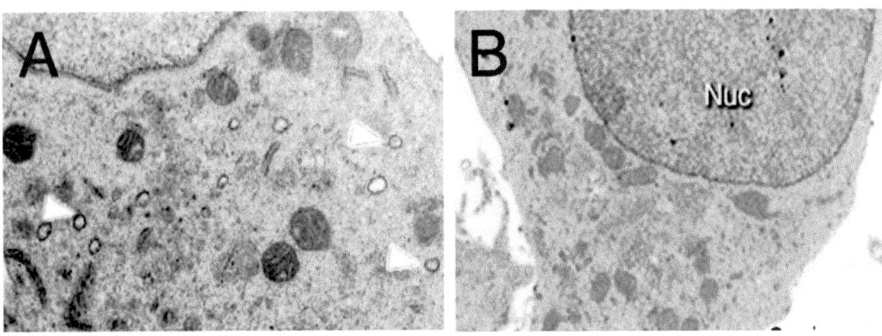

Figure 5. Electron micrographs demonstrating murine hepatitis virus replication complexes associated with double-membrane vesicles (white arrowheads) in wild-type ES cells (A) but not in *atg* ES cells (B). Adapted with permission from[51]. See text for details.

Thus, the eIF2α kinase-dependent autophagy signaling pathway not only regulates the degradation of long-lived cellular proteins but also regulates the degradation of viral proteins. Accordingly, it seems logical to speculate that PKR-dependent autophagic degradation of viruses inhibits viral replication and is an antiviral defense mechanism. However, the relative contributions of the effects of PKR on viral protein synthesis and the effects of PKR on viral protein degradation in the regulation of the HSV-1 replication have not yet been assessed. For this purpose, it will be necessary to selectively inhibit the autophagic protein degradation machinery and/or have HSV-1 mutant viruses that selectively block specific downstream functions regulated by PKR. It will also be important to determine whether PKR-dependent autophagy degrades and inhibits the replication of viruses other than HSV-1. Preliminary observations indicate that PKR-dependent autophagy does lead to the degradation of another neurotropic virus, the enveloped, positive-strand RNA virus in the alphavirus genus, Sindbis virus[107] (Figure 4B).

4.2 Mammalian autophagy genes play a role in host antiviral defense

The role of PKR in antiviral innate immunity is well established, but PKR regulates many different cellular processes, and it is not yet known exactly what role autophagy induction plays in the antiviral effects of PKR. However, the concept that autophagy is important in innate immunity is more directly supported by studies involving components of the mammalian and plant autophagic machinery.

The first identified mammalian autophagy gene product, Beclin 1, was isolated in a yeast two-hybrid screen, in the context of studies of the mechanism by which the antiapoptotic protein, Bcl-2, protects mice against lethal Sindbis virus encephalitis[37]. Similar to the neuroprotective effects of Bcl-2[108], enforced neuronal expression of wild-type Beclin 1 in a recombinant chimeric Sindbis virus vector reduces Sindbis virus replication, reduces Sindbis virus-induced apoptosis, and protects mice against lethal Sindbis virus encephalitis[37]. Mutations in the Bcl-2-binding domain of Beclin 1 and mutations in other regions of Beclin 1 that block its autophagy function also block its protective effects during Sindbis virus infection[37,62]. Thus, it appears that both the interaction with Bcl-2 and the autophagy function may be required for the antiviral effects of Beclin 1. However, further studies are required to define the precise mechanism of how Beclin 1 inhibits viral replication and virus-induced apoptosis and to identify the precise role of Bcl-2-Beclin 1 interactions in these processes.

Preliminary studies with *beclin 1* null ES cells and *atg5* null MEFs indicate a role for these two endogenous autophagy genes in innate immunity against Sindbis virus infection. Sindbis virus replicates to higher titers and results in accelerated death in *beclin 1* null ES cells as compared to wild-type control ES cells and in *atg5* null MEFs as compared to wild-type control MEFs[38]. In the case of Sindbis virus infection, it is not known whether the acceleration of virus-induced death in *beclin 1* null or *atg5* null cells is a result of increased viral replication or of independent effects of *atg* gene deficiency on cell death. However, studies comparing HSV-1 infection in wild-type and *beclin 1*$^{-/-}$ES cells suggest that *beclin 1* can protect against virus-induced cell death in the absence of inhibitory effects on viral replication[38].

Together, the studies of Sindbis virus infection in neurons overexpressing *beclin 1* or in cultured cells lacking *beclin 1* or *atg5* demonstrate a role for mammalian autophagy genes in both restricting viral replication and in protection against virus-induced cell death. It will be important to examine whether other autophagy genes have a similar antiviral function and to examine whether autophagy genes also protect against other types of virus infections. The mechanisms by which autophagy genes exert protective effects in Sindbis virus infection are not yet known. Presumably, the autophagic breakdown of viral components leads to decrease viral yields. However, it is also possible that autophagy leads to the breakdown of cellular components required for viral replication. As noted above, the protective effects against cell death may be secondary to inhibitory effects on viral replication. Alternatively, the protective effects may relate to the nutrient recycling functions or "damage control" functions of autophagy, or in the case of *beclin 1*, to interactions between autophagy proteins and anti-apoptotic pathways. It is also possible that autophagy may protect against cell death by degrading specific viral proteins (e.g. the Sindbis virus E1 and E2 envelope glycoprotein's[109]) that are involved in triggering the apoptotic pathway.

The protective effects of *beclin 1* and *atg5* on virus-induced cell death are consistent with the "pro-survival" effects of autophagy during nutrient starvation and other forms of environmental stress. It is not yet clear how to reconcile these pro-survival effects with the view that autophagy represents an alternative form of non-apoptotic programmed cell death (reviewed in [2,25,110]). While the primary basis for this view has been morphologic correlations between the presence of autophagic vacuoles and dying cells (reviewed in [2,111]), recent genetic experiments establish a more direct role for autophagy genes in certain types of programmed cell death. Mammalian *Atg7* and *beclin 1* RNAi blocks cell death in fibroblast and macrophage cell lines treated with the caspase inhibitor, zVAD[112], and *atg 6* , *7* , and *12* RNAi blocks salivary gland destruction during *Drosophila* development[113]. Thus, the relationship between autophagy, cell survival, and cell death is quite complex and likely varies according to the cell type and the specific physiological or pathophysiological setting. It remains to be determined whether autophagy genes primarily play a protective role in preventing cell death during virus infection, or whether they also participate directly in cell death that is induced by certain viruses.

4.3 Plant autophagy genes play a role in the plant innate immune response

The plant homologue of *beclin 1* also functions in antiviral host defense in plants. Similar to mammalian *beclin 1*, plant *BECLIN 1* restricts viral replication; tobacco mosaic virus replication is increased in *BECLIN 1*-silenced tobacco plants as compared to vector-treated control plants[40]. However, in contrast to mammalian *beclin 1, which* prevents the death of Sindbis virus- and HSV-1-infected cells, plant *BECLIN 1* plays an interesting role in preventing the death of uninfected cells[40].

In plants, the innate immune response during virus infection is characterized by a hypersensitive response which is a programmed cell death response that occurs around the infected areas (reviewed in [114-116]). This hypersensitive response limits virus spread and confers pathogen resistance. It is triggered by a pathogen-encoded avirulence protein, which is recognized by a specific plant cognate resistance protein, termed an R protein. In plants lacking R proteins, there is uncontrolled virus spread and pathogen sensitivity.

A tobacco mosaic virus infection model has recently been used to study the role of autophagy genes in plant innate immunity. During tobacco mosaic virus infection, the hypersensitive response is triggered by tobacco mosaic virus protein, TMV p50, which is the helicase domain of the viral replicase[117]. TMV p50 is recognized by an R protein (called the N protein) of *N. benthamiana*, which is composed of a Toll/interleukin 1 receptor domain, a nucleotide binding domain, and a leucine-rich repeat domain[118]. Therefore, in tobacco plants containing the N protein ($N^{+/+}$), there is local cell death in cells that are either infected with TMV or that express the TMV p50 protein but no systemic illness is observed.

BECLIN 1 silencing in $N^{+/+}$ tobacco plants reveals a striking role for this autophagy gene in limiting the spread of cell death during the hypersensitive response[40]. During TMV infection of *BECLIN 1*-silenced $N^{+/+}$ plants, cell death begins as discreet and defined foci but continues to spread beyond the site of TMV infection until there is death of the entire inoculated leaf and other uninoculated leaves. A similar spreading cell death phenotype is seen with local expression of the TMV p50 protein, suggesting that the cell death occurs in response to a specific signal triggered by the pathogen encoded avirulence protein, and is not due to increased TMV replication or altered virus movement. In addition, in plants that lack the *N* gene, *BECLIN 1* silencing does not lead to cell death after TMV infection. Moreover, *BECLIN 1* silencing also results in spreading cell death during the hypersensitive response triggered by bacterially-encoded pathogen avirulence proteins. Together, these observations demonstrate that the spreading cell death phenotype in *BECLIN 1*-silenced plants is mediated by *R* gene-mediated innate immune responses and that *BECLIN 1* is an important negative regulator of cell death during the plant innate immune response.

A similar role for other autophagy genes in limiting the spread of programmed cell death during TMV infection has also been observed. As discussed in Section 2.3.3, PI3-K/Vps34 is a protein that physically interacts with Atg6/Beclin 1 in yeast and mammals and is essential for proper autophagosome formation. Interestingly, silencing of the plant class III *PI-3K/VPS34* in $N^{+/+}$ plants results in a spreading cell death phenotype during TMV infection that is similar to that observed with *BECLIN 1* silencing. As discussed in Section 2.3.3, yeast and mammalian *ATG3* and *ATG7* are essential for conjugation reactions needed for autophagosome formation, and silencing of the plant homologues of these genes also results in a spreading cell death phenotype after TMV infection. Thus, multiple different autophagy genes, including those that act in the vesicle nucleation stage (e.g. *BECLIN 1*, Class III *PI3-K/VPS34*) and those that act in the vesicle expansion stage (e.g. *ATG3* and *ATG7*) are necessary to prevent the spread of cell death during the plant innate immune response.

While plant autophagy genes protect uninfected cells against death whereas mammalian autophagy genes protect infected cells against death during virus infection, the plant data nonetheless further support a "pro-survival", rather than "pro-death" function of autophagy genes during viral infections. At present, it is not yet clear how autophagy genes protect uninfected cells against death during the plant hypersensitive response. One possibility is that the absence of autophagy genes in uninfected cells somehow modifies the *R* gene-mediated signal transduction pathway in a way that instructs uninfected cells to die. An alternative, perhaps more likely, possibility is that the absence of autophagy genes in uninfected cells renders them more susceptible to pro-death signals emitted from infected cells. Regardless of the mechanism, this newly defined role for autophagy genes in preventing the spread of cell death during plant innate immunity has significant implications for understanding the role of autophagy in systemic protection against viral infections. An important question is whether autophagy genes play a similar role during animal virus infections.

5. EVASION OF AUTOPHAGY BY VIRUSES

The evolutionarily conserved function of both mammalian and plant autophagy genes in restricting viral replication and/or protection against cell death suggests an essential role for autophagy in innate immunity. This concept is further supported by recent observations indicating that the herpes simplex virus neurovirulence protein, ICP34.5, possesses multiple mechanisms to disarm host autophagy. It can both antagonize the PKR signaling pathway required for autophagy induction and inhibit the function of one component of the autophagic machinery, Beclin 1.

As discussed in Section 4.1, ICP34.5 blocks PKR-dependent, eIF2α Ser-51-dependent autophagic degradation of cellular and viral components in HSV-1-infected MEFs and neurons[99]. One predicted mechanism by which ICP34.5 blocks PKR-dependent autophagy is through its known ability to promote the dephosphorylation of eIF2α via interactions with PP1α[105]. However, new evidence also suggests a second potential mechanism. Roizman et al. isolated the mammalian autophagy protein, Beclin 1, in a yeast two-hybrid screen using ICP34.5 as a bait[119]. Subsequent studies have shown that ICP34.5 directly interacts with Beclin 1 in mammalian cells and inhibits the ability of Beclin 1 to rescue autophagy in autophagy-defective *atg6* null yeast and in autophagy-defective human MCF7 breast carcinoma cells[38]. Since ICP34.5 binds to Beclin 1 via a domain that is distinct from its PP1α-binding domain, it should be possible to construct HSV-1 viruses containing mutations in ICP34.5 that help differentiate between the role of PP1α-binding (and eIF2α phosphorylation) and the role of Beclin 1-binding in HSV-1 ICP34.5-mediated neurovirulence.

Besides HSV-1 ICP34.5, there are numerous other viral proteins or RNAs that suppress PKR signaling through a variety of different mechanism (reviewed in [100,101]). For example, vaccinia virus E3, influenza virus NS1, HSV-1 Us11, reovirus σ3, and rotavirus NSP3 encode double-stranded (ds) RNA-binding proteins that prevent PKR activation. Adenovirus VAI RNAs and HIV Tar RNAs bind to dsRNA substrates and inhibit PKR. Hepatitis C virus NS5A protein inhibits the dimerization of PKR, and influenza virus recruits a cellular protein, $P58^{IPK}$, that directly interacts with PKR and inhibits its dimerization. The vaccinia virus K3L, hepatitis C virus E2, and HIV Tat proteins act as pseudosubstrates of PKR. As-of-yet, the role of these other viral RNAs and proteins in autophagy inhibition has not been investigated. However, given the evolutionarily conserved requirement for an intact eIF2α kinase signaling pathway in autophagy induction, the prediction is that these other viral inhibitors of PKR, like HSV1 ICP34.5, also function as antagonists of host autophagy. Further studies are needed to test this prediction and to study the role of this predicted antagonism of host autophagy in the pathogenesis of diseases caused by these other important viral pathogens that encode putative autophagy inhibitors.

Not only may other viruses antagonize the autophagy function of PKR, but other viral gene products may also antagonize the autophagy function of specific mammalian *atg* genes. Beclin 1 was originally isolated in a yeast two-hybrid screen with the cellular anti-anti-apoptotic protein, Bcl-2[37]. Subsequently, Beclin 1 has also been shown to interact with viral Bcl-2-like proteins encoded by different gammaherpesviruses, including EBV-encoded BHRF1, KSHV-encoded v-Bcl-2, and murine γHV68-encoded M11[39]. Like ICP34.5, these viral proteins can also inhibit the autophagy function of Beclin 1 in yeast and mammalian assays. In addition, preliminary evidence indicates that other KSHV-encoded proteins may interact with other specific Atg proteins[65].

An as-of-yet explored area is whether viruses also inhibit autophagy by activating autophagy inhibitory signaling pathways. As noted in Section 2.2 and 3.2, the Class I PI-3K/Akt signaling pathway negatively regulates autophagy in both mammalian cells and *C. elegans*[19,20,120] and several different viruses activate this pathway. Certain oncogenic retroviruses encode the catalytic subunit of PI3-K and Akt (reviewed in [22]). In addition, the EBV latent membrane proteins, LMP1 and 2A, the hepatitis B virus protein, HBx, the Kaposi's sarcoma virus protein, K1, and the hepatitis C virus protein, NS5A all activate the PI3-K/Akt signaling pathway[72-77]. Presumably, such activation plays a role in autophagy inhibition, although this has not yet been formally tested.

While further studies are required to more precisely define the interactions between viral gene products and autophagy regulatory signals and autophagy proteins, there is, however, accumulating evidence that viruses do target multiple different steps of the host autophagy pathway. This observation strongly suggests an evolutionary advantage for viruses to inhibit host autophagy, and by extrapolation, a beneficial role for host autophagy in defense against viral infections.

6. SUBVERSION OF AUTOPHAGY BY VIRUSES

Some viruses appear to have even further outsmarted host autophagy. Rather than merely devising strategies to block host autophagy, certain positive-strand RNA viruses have figured out ways to "co-opt" elements of the autophagy pathway to promote their own replication. This subject has been recently reviewed in detail elsewhere[13] and will therefore only be briefly summarized in this section.

As early as 1965, electron microscopic studies of poliovirus-infected cells demonstrated the presence of large numbers of membranous vesicles that were postulated to develop by an autophagic-like mechanism[121]. More recently, work by Kirkegaard et al, has extended these findings to further show that poliovirus replication complexes are associated with double-membrane vesicles that resemble autophagosomes, in that they (1) have similar double membrane-bound morphology; (2) have low buoyant density; and (3) label with the autophagosome marker, GFP-LC3, and the lysosome marker, LAMP1[13,122]. Unlike classical autophagosomes, these autophagic-like vesicles do not appear to have a destructive role or mature into degradative compartments. In support of this, treatment with autophagy inducers, rapamycin or tamoxifen, both increase, rather than decrease poliovirus growth[13]. Furthermore, these double-membrane vesicles are also different from classical autophagosomes in that they contain Sec13 and Sec31, components of the anterograde transport system that bud from the ER[123].

Therefore, it is possible that poliovirus-induced vesicles arise from an alternate source rather than the PAS, but still share some of the same characteristics of classical autophagosomes (e.g. GFP-LC3 and labeling, augmentation with rapamycin treatment). Similar to the replication vacuoles that are associated with certain intracellular bacterial pathogens (e.g. *Legionella pneumophila*)[124], the poliovirus-induced double-membrane vesicles likely originate from the ER[123]. Furthermore, these poliovirus-induced vesicles seem to have an alternate function than autophagosomes (i.e. they are pro-replicative, rather than degradative compartments). These observations suggest that poliovirus may promote its own replication by inducing dynamic membrane rearrangements that share in common certain features of the autophagy pathway (e.g. formation of sequestering double-membrane vesicles, presence of overlapping markers) but avoid, other unwanted features of the autophagy pathway (e.g. maturation into degradation compartments). Of note, specific poliovirus proteins, including 2BC and 3A, have been identified that are sufficient for the induction of these "autophagic-like" double-membrane bound vesicles[122]. However, the mechanisms by which these proteins induce the formation of such vesicles are not yet known.

A recent study with the coronavirus, murine hepatitis virus, has provided more direct evidence that components of the autophagic machinery can be utilized for RNA virus replication[51]. MHV replication complexes localize to double-membrane vesicles (that are also thought to arise from the ER) (Figure 5A) and they co-localize with certain autophagy proteins, including LC3 and Atg12. In MHV-infected *atg5*$^{-/-}$ES cells, double-membrane vesicles are not detected (Figure 5B), and viral replication is dramatically reduced. These observations provide the first genetic demonstration that proteins necessary for autophagic vacuole formation are also required for maximal levels of viral replication. Thus, MHV, and potentially other viruses that replicate in association with double-membrane vesicles (e.g. poliovirus, equine arterivirus), utilize components of cellular autophagy to foster their own growth. Presumably, the Atg protein conjugation system (involving

Atg5) that plays a role in autophagic vesicle expansion and completion also plays a role in the formation of double-membrane vesicles involved in viral replication. It is not yet known whether the entire autophagic machinery or only selective components of the autophagic machinery are used for the formation of double-membrane vesicles that are associated with viral replication complexes.

These observations with poliovirus and MHV represent two examples of how viruses can "subvert" elements of the host autophagy pathway to promote their own intracellular growth. In these infections, RNA replication complexes are observed in association with "autophagic-like" double-membrane vesicles but not in association with degradative autophagosomes. It is not clear whether this represents fundamental differences in the host pathways leading to the formation of "autophagic-like" double-membrane vesicles and classical degradative autophagosomes, the diversion of the autophagic machinery towards the formation of "autophagic-like" double-membrane vesicles from the formation of classical degradative autophagosomes, or specific viral mechanisms to antagonize the maturation of "autophagic-like" double-membrane vesicles into mature degradative autophagosomes. However, interestingly, MHV infection does lead to the induction of *atg-5*-dependent long-lived cellular protein degradation, ruling out the hypothesis that the autophagic machinery is entirely diverted to form membranes required for viral replication complexes. Perhaps MHV possesses as-of-yet defined mechanisms to shield its replication complexes from autophagic degradation.

7. CONCLUSIONS

Although research in this area is still in a stage of infancy, it seems likely that the lysosomal degradation pathway of autophagy plays an evolutionarily conserved role in antiviral immunity. The interferon-inducible, antiviral PKR signaling pathway positively regulates autophagy, and both mammalian and plant autophagy genes restrict viral replication and protect against virus-induced cell death. Given this role of autophagy in innate immunity, it is not surprising that viruses have evolved numerous strategies to inhibit host autophagy. Different viral gene products can either modulate autophagy regulatory signals or directly interact with components of the autophagy execution machinery. Moreover, certain RNA viruses have managed to "co-apt" the autophagy pathway, selectively utilizing certain components of the dynamic membrane rearrangement system to promote their own replication inside the host cytoplasm.

In addition to this newly emerging role of autophagy in innate immunity, autophagy plays an important role in many other fundamental biological processes, including tissue homeostasis, differentiation and development, cell growth control, and the prevention of aging. Accordingly, the inhibition of host autophagy by viral gene products has important implications not only for understanding mechanisms of immune evasion, but also for understanding novel mechanisms of viral pathogenesis. It will be interesting to dissect the role of viral inhibition of autophagy in acute, persistent, and latent viral replication, as well as in the pathogenesis of cancer and other medical diseases.

ACKNOWLEDGEMENTS

The work done in the authors' laboratories was supported by NIH RO1 grants AI51367 and AI44157 to B.L, and an NSF plant genome grant DBI-0116076 and NIH RO1 grant GM62625 to S.P.D-K.

REFERENCES

1. (Ed), D. K. *Autophagy* (Landes Biosciences, Georgetown, TX, 2003).
2. Levine, B. & Klionsky, D. J. Development by self-digestion: molecular mechanisms and biological functions of autophagy. *Developmental Cell* **6:**463-477
3. Levine, B. & Klionsky, D. J. Development by self-digestion: molecular mechanisms and biological functions of autophagy. *Developmental Cell* **6:**463-477
4. Yoshimori, T. Autophagy: a regulated bulk degradation process inside cells. *Biochem Biophys Res Comm* **313:**453-458
5. Wang, C. W. & Klionsky, D. J. The molecular mechanism of autophagy. *Mol Med* **9:**65-76
6. Duve, C. d. & Wattiaux, R. Functions of lysosomes. *Annu Rev Physiol* **28:**435-492
7. Klionsky, D. J. et al. A unified nomenclature for yeast autophagy-related genes. *Dev. Cell* **5:**539-545
8. Codogno, P. & Meijer, A. J. in *Autophagy* (ed. Klionsky, D.) 26-47 (Landes Biosciences, Georgetown, TX, 2004).
9. Petiot, A., Pattingre, S., Arico, S., Melez, D. & Codogno, P. Diversity of Signaling Controls of Macroautophagy in Mammalian Cells. *Cell Struct Function* **27:**431-441
10. Suzuki, K. et al. The pre-autophagosomal structure organized by concerted functions of APG genes is essential for autophagosome formation. *EMBO J* **20:**5971-5981
11. Kim, J., Huang, W. P., Stromhaug, P. E. & Klionsky, D. J. Convergence of multiple autophagy and cytoplasm to vacuole targeting components to a perivacuolar membrane compartment prior to de novo vesicle formation. *J Biol Chem* **277:**763-773
12. Noda, T., Suzuki, K. & Ohsumi, Y. Yeast autophagosomes: de novo formation of a membrane structure. *Trends Cell Biol* **12:**231-235
13. Kirkegaard, K., Taylor, M. P. & Jackson, W. T. Cellular autophagy: surrender, avoidance and subversion by microrganisms. *Nature Rev in Microbiol* **2:**301-314

14. Kopitz, J., Kisen, G. O., Gordon, P. B., Bohley, P. & Seglen, P. O. Nonselective autophagy of cytosolic enzymes by isolated rat hepatocytes. *J Cell Biol* **111:**941-953
15. Klionsky, D. J. & Emr, S. D. Autophagy as a regulated pathway of cellular degradation. *Science* **290:**1717-1721
16. Elmore, S. P., Qian, T., Grissom, S. F. & Lemasters, J. J. The mitochondrial permeability transition initiates autophagy in rat hepatocytes. *FASEB J* **15:**2286-2287
17. Williams, B. PKR: a sentinel kinase for cellular stress. *Oncogene* **18:**6112-6120
18. Talloczy, Z. et al. Regulation of starvation- and virus-induced autophagy by the eIF2*a* kinase signaling pathway. *Proc Natl Acad Sci USA* **99:**190-195
19. Petiot, A., Ogier-Denis, E., Blommaart, E. F., Meijer, A. J. & Codogno, P. Distinct classes of phosphatidylinositol 3'-kinases are involved in signaling pathways that control macroautophagy in HT-29 cells. *J Biol Chem* **275:**992-998
20. Melendez, A. et al. Autophagy genes are essential for dauer development and lifespan extension in *C. elegans*. *Science* **301:**1387-1391
21. Blume-Jensen, P. & Hunter, T. Oncogenic kinase signalling. *Nature* 411, 355-365 (2001).
22. Aoki, M. & Vogt, P. K. Retroviral oncogenes and TOR. *Curr Top Microbiol Immunol* **279:**321-338
23. Qu, X. et al. Promotion of tumorigenesis by heterozygous disruption of the *beclin 1* gene. *J Clin Invest* **112:**1809-1820
24. Yue, Z., Jin, S., Yang, C., Levine, A. J. & Heintz, N. Beclin 1, an autophagy gene essential for early embryonic development, is a haploinsufficient tumor suppressor. *Proc Natl Acad Sci USA* **100:**15077-15082
25. Gozuacik, D. & Kimchi, A. Autophagy as a cell death and tumor suppressor mechanism. *Oncogene* **23:**2891-2906
26. Edinger, A. L. & Thompson, C. B. Defective autophagy leads to cancer. *Cancer Cell* **4:**422-424
27. Ohsumi, Y. Molecular dissection of autophagy: two ubiquitin-like systems. *Nat. Rev. Mol. Cell. Biol.* **2:**211-216
28. Tsukada, M. & Ohsumi, Y. Isolation and characterization of autophagy-defective mutants of *Saccharomyces cerevisiae*. *FEBS Lett* **333:**169-174
29. Kamada, Y. et al. Tor-mediated induction of autophagy via an Apgl protein kinase complex. *J Cell Biol* **150:**1507-1513
30. Abeliovich, H., Zhang, C., Dunn, W. A. J., Shokat, K. M. & Klionsky, D. J. Chemical genetic analysis of Apg1 reveals a non-kinase role in the induction of autophagy. *Mol Cell Biol* **14:**477-49
31. Schu, P. V. et al. Phosphatidylinositol 3-kinase encoded by yeast VPS34 gene essential for prtoein sorting. *Science* **260:**88-91
32. Seglen, P. O. & Gordon, P. B. 3-methyladenine: specific inhibitor of autophagic/lysosomal protein degradation in isolated rat hepatocytes. *Proc. Natl. Acad. Sci. USA* **79:**1889-1892
33. Blommaart, E. F., Krause, U., Schellens, J. P., Vreeling-Sindelarova, H. & Meijer, A. J. The phosphatidylinositol 3-kinase inhibitors wortmannin and LY294002 inhibit autophagy in isolated rat hepatocytes. *Eur J Biochem* **243:**240-246
34. Kiel, J. A. K. W. et al. The *Hansenula polymorpha PDD1* gene product, essential for the selective degradation of peroxisomes, is a homologue of *Saccharomyces cerevisiae* Vps34p. *Yeast* **15:**741-754
35. Kihara, A., Noda, T., Ishihara, N. & Ohsumi, Y. Two distinct Vps34 phosphatidylinositol 3-kinase complexes function in autophagy and carboxypeptidase Y sorting in *Saccharomyces cerevisiae*. *J Cell Biol* **152:**519-530
36. Eskelinen, E. L. et al. Inhibition of autophagy in mitotic animal cells. *Traffic* **3:**878-893
37. Liang, X. H. et al. Protection against fatal Sindbis virus encephalitis by Beclin, a novel Bcl-2-interacting protein. *J Virol* **72:**8586-8596
38. Talloczy, Z. & Levine, B. Unpublished data. (2004).

39. Pattingre, S., Liang, X. H. & Levine, B. Unpublished data. (2004).
40. Liu, Y., Schiff, M., Talloczy, Z., Levine, B. & Dinesh-Kumar, S. P. Autophagy genes are essential for limiting the spread of programmed cell death associated with plant innate immunity. *submitted* (2004).
41. Mizushima, N. et al. A protein conjugation system essential for autophagy. *Nature* **395:** 395-398
42. Ichimura, Y. et al. A ubiquitin-like system mediates protein lipidation. *Nature* **408:**488-492
43. Knight, D. et al. The X-ray crystal structure and putative ligand-derived peptide binding properties of gamma-aminobutyric acid receptor type A receptor-associated protein. *J Biol Chem* **277:**5556-5561
44. Stangler, T., Mayr, L. M. & Willbold, D. Solution structure of human GAGA(A) receptor-associated protein GABARAP: implications for biological function and its regulation. *J Biol Chem* **277:**13363-13366
45. Sugawara, K. et al. The cyrstal structure of microtubule-associated protein light chain 3, a mammalian homolgue of *Saccharomyces cerevisiae* Atg8. *Genes Cells* **9:**611-618
46. Kirisako, T. et al. Formation process of autophagosome is traced with Apg8/Aut7p in yeast. *J Cell Biol* **147:**435-446
47. Tanida, I. et al. HsAtg4B/HsApg4B/autophagin-1 cleaves the carboxyl termini of three humans Atg8 homologues and delipidates LC3- and GABARAP-phospholipid conjugates. *J Biol Chem* **279:**36268-36276
48. Kirisako, T. et al. The reversible modifcation regulates the membrane-binding state of Apg8/Aut7 essential for autophagy and the cytoplasm to vacuole targeting pathway. *J Cell Biol* **151:**263-276
49. Tanida, I. et al. Apg7p/Cvt2p: A novel protein-activating enzyme essential for autophagy. *Mol Cell Biol* **10:**1367-1379
50. Kabeya, Y. et al. LC3, a mammalian homologue of yeast Apg8p, is localized in autophagosome membranes after processing. *EMBO J* **19:**5720-5728
51. Prentice, E., Jerome, W. G., Yoshimori, T., Mizushima, N. & Denison, M. R. Coronavirus replication complex formation utilizes components of cellular autophagy. *J Biol Chem* **279:**10136-10141
52. Noda, T. et al. Apg9/Cvt7p is an integral membrane protein required for transport vesicle formation in the Cvt and autophagy pathways. *J Cell Biol* **148:**465-480
53. Reggiori, F., Tucker, K. A., Stromhaug, P. E. & Klionsky, D. J. The Atg1-Atg13 complex regulates Atg9 and Atg23 retrieval transport from the preautophagosomal structure. *Dev Cell* **6:**79-90
54. Mizushima, N., Yamamoto, A., Matsui, M., Yoshimori, T. & Ohsumi, Y. In vivo analysis of autophagy in response to nutrient starvation using transgenic mice expressing a fluorescent autophagy marker. *submitted* (2003).
55. Klionsky, D. J. *Autophagy* (Landes Bioscience, Georgetown, Texas, 2004).
56. Shintani, T. & Klionsky, D. J. Autophagy in health and disease: a double-edged sword. *Science* In press. (2004).
57. deDuve, C. & Wattiaux, R. Functions of lysosomes. *Annu Rev Physiol* **28:**435-492
58. Otto, G. P., Wu, M. Y., Kazgan, N., Anderson, O. R. & Kessin, R. H. Macroautophagy is required for multicellular development of the social amoeba *Dictyostelium discoideum*. *J Biol Chem* **278:**17636-17645
59. Otto, G. P., Wu, M. Y., Kazgan, N., Anderson, O. R. & Kessin, R. H. Dictyostelium macroautophagy mutants vary in the severity of their developmental defects. *J Biol Chem* **279:**15621-15629
60. Hanaoka, H. et al. Leaf senescence and starvation-induced chlorosis are accelerated by the disruption of an Arabidopsis autophagy gene. *Plant Physiol* **129:**1181-1193

61. Doelling, J. H., Walker, J. M., Friedman, E. M., Thompson, A. R. & Vierstra, R. D. The APG8/12-activating enzyme APG7 is required for proper nutrientrecycling and sensescence in *Arabidopsis thaliana. J Biol Chem* **277:**33105-33114
62. Levine, B. Unpublished data. (2004).
63. Dever, T. Translation initiation: adept at adapting. *Trends Biochem Sci* **10:**398-340
64. Natarajan, K. et al. Trascriptional profiling shows that Gcn4p is a master regulator of gene expression during amino acid starvation in yeast. *Mol Cell Biol* **21:**4347-4368
65. Jung, J. (2004).
66. Dawson, C. W., Eliopoulos, A. G., Dawson, J. & Young, L. S. BHRF1, a viral homologue of the Bcl-2 oncogene, disturbs epithelial cell differentiation. *Oncogene* **10:**69-77
67. Gunn, J. M., Clark, M. G., Knowles, S. E., Hopgood, M. F. & Ballard, F. J. Reduced rates of proteolysis in transformed cells. *Nature* **266:**58-60
68. Amenta, J. S., Sargus, M. J., Venkatesan, S. & Shinozuka, H. Role of the vacuolar apparatus in augmented protein degradation in cultured fibroblasts. *J Cell Physiol* **94:**77-86
69. Furuya, N., Liang, X. H. & Levine, B. in *Autophagy* (ed. Klionsky, D. J.) 244-253 (Landes Bioscience, Georgetown, Texas, 2004).
70. Huang, S. & Houghton, P. J. Inhibitors of mammalian target of rapamycin as novel antitumor agents: from bench to clinic. *Curr Opin Investig Drugs* **3:**295-304
71. Payne, E., Bowles, M. R., Don, A., Hancock, J. F. & McMillan, N. A. Human papillomavirus type 6b virus-like particles are able to activate the Ras-MAP kinase pathway and induce cell proliferation. *J Virol* **75:**4150-4157
72. Scholle, F., Bendt, K. M. & Raab-Traub, N. Epstein-Barr virus LMP2A transforms epithelial cells, inhibits cell differentiation, and activates Akt. *J Virol* **74:**10681-10689
73. Fukuda, M. & Longnecker, R. Latent membrane protein 2A inhibits transforming growth factor-beta 1-induced apoptosis through the phosphatidylinositol 3-kinase/Akt pathway. *J Virol* **78:**1697-1705
74. Morrison, J. A., Gulley, M. L., Pathmanathan, R. & Raab-Traub, N. Differential signaling pathways are activated in the Epstein-Barr virus-associated malignancies nasopharyngela carcinoma and Hodgin lymphoma. *Cancer Res* **64:**5251-5260
75. Chung, T. W., Lee, Y. C. & Kim, C. H. Hepatitis B viral HBx induces matrix metalloproteinase-9 gene expression trhough activation of ERK and PI-3K/Akt pathways: involvement of invasive potential. *FASEB J* **18:**1123-1125
76. Tomlinson, C. C. & Damania, B. The K1 protein of Kaposi's sarcoma-associated herpesvirus activates the Akt signaling pathway. *J Virol* **78:**1918-1927
77. Street, A., Macdonald, A., Crowder, K. & Harris, M. The hepatitis C virus NS5A protein activates a phosphoinosite 3-kinase-dependent survival signaling cascade. *J Biol Chem* **279:**12232-12241
78. Aita, V. M. et al. Cloning and genomic organization of *beclin 1*, a candidate tumor suppressor gene on chromosome 17q21. *Genomics* **59:**59-65
79. Liang, X. H. et al. Induction of autophagy and inhibition of tumorigenesis by beclin 1. *Nature* **402:**672-676
80. Liang, X. H., Yu, J., Brown, K. & Levine, B. Beclin 1 contains a leucine-rich nuclear export signal that is reguired for its autophagy and tumor suppressor function. *Canc Res* **61:**3443-3449
81. Qu, X. & Levine, B. *Unpublished data* (2004).
82. Cuervo, A. M. Autophagy and aging--when "all you can eat" is yourself. *Sci Aging Knowledge Environ* **36:**pe25
83. Bergamini, E., Cavallini, G., Donati, A. & Gori, Z. The anti-ageing effects of caloric restriction may involve stimulation of macroautophagy and lysosomal degradation, and can be intensifid pharmacologically. *Biomed Pharmacother* **57:**203-208

84. Guarente, L. & Kenyon, C. Genetic pathways that regulate ageing in model organisms. *Nature* **408:**255-262
85. Cossarizza, A., Troiano, L. & Mussini, C. Mitochondria and HIV infection: the first decade. *J Biol Regul Homeost Agents* **16:**18-24
86. Moriya, K. et al. Oxidative stress in the absence of inflammation in a mouse model for hepatitis C virus-associated carcinogenesis. *Cancer Res* **61:**4365-4370
87. Hsieh, Y. H. et al. Pre-S mutant surface antigens in chronic hepatitis B virus infection induce oxidative stress and DNA damage. *Carcinogenesis* June 3 (epub ahead of print) (2004).
88. Teckman, J. H., An, J. K., Blomenkamp, K., Schmidt, B. & Perlmutter, D. Mitochondrial autophagy and injury in the liver in α_1-antitrypsin deficiency. *Am J Physiol Gastrointestin Liver Physiol* **286:**G851-G8562
89. Ravikumar, B., Duden, R. & Rubinsztein, D. C. Aggregate-prone proteins with polyglutamine and polyalanine expansions are degraded by autophagy. *Hum Mol Genet* **11:**1107-1117
90. Ravikumar, B. et al. Inhibition of mTOR induces autophagy and reduces toxicity of polyglutamine expansions in fly and mouse models off Huntington disease. *Nat Genet* **36:**585-595
91. Shi, Y. et al. Identification and characterization of pancreatic eukaryotic initiation factor 2 alpha-subunit kinase, PEK, involved in translational control. *Mol Cell Biol* **18:**7499-7509
92. Harding, H. P., Zhang, Y. & Ron, D. Protein translation and folding are coupled by an endoplasmic-reticulum-resident kinase. *Nature* **397:**271-274
93. Dimcheff, D. E., Askovic, S., Baker, A. H., Johnson-Fowler, C. & Portis, J. L. Endoplasmic reticulum stress is a determinant of retrovirus-induced spongiform neurodegeneration. *J Virol* **77:**12617-12629
94. Kim, H. T. et al. Activation of endoplasmic reticulum stress signaling pathway is associated with neuronal degeneratio in MoMuLV-ts1-induced spongiform encephalopathy. *Lab Invest* **84:**816-827
95. Dimcheff, D. E., Faasse, M. A., McAtee, F. J. & Portis, J. L. Endoplasmic reticulum (ER) stress induced by a neurovirulent mouse retrovirus is associated with prolonged BiP binding and retention of a viral protein in the ER. *J Biol Chem* **279:**33782-33790
96. Wang, H. C. et al. Different types of ground glass hepatocytes in chronic hepatitis B virus infection contain specific pre-S mutants that may induce endoplasmic reticulum stress. *Am J Pathol* **163:**2441-2449
97. Jordan, R., Wang, L., Graczyk, T. M., Block, T. M. & Romano, P. R. Replication of a cytopathic strain of bovine viral diarrhea virus activates PERK and induces endoplasmic reticulum stress-mediated apoptosis of MDBK cells. *J Virol* **76:**9588-9599
98. Liu, N. et al. Possible involvment of both endoplasmic reticulum- and mitochondria-dependent pathways in MoMuLV-ts1-induced apoptosis in astrocytes. *J Neurovirol* **10:**189-198
99. Talloczy, Z., Virgin, H. & Levine, B. PKR-dependent autophagic degradation of herpes simplex virus type 1. *Submitted* (2004).
100. GaleJr, M. & Katze, M. G. Molecular mechanisms of interferon resistance mediated by viral-directed inhibition of PKR, the interferon-induced protein kinase. *Pharmacol Ther* **78:**29-46
101. Tan, S. L. & Katze, M. G. HSV.com: maneuvering the internetworks of viral neuropathogenesis and evasion of the host defense. *Proc Natl Acad Sci USA* **97:**5684-5686
102. Chou, J., Kern, E. R., Whitley, R. J. & Roizman, B. Mapping of herpes simplex virus-1 neurovirulence to g134.5, a gene nonessential for growth in culture. *Science* **250:**1262-1266

103. Leib, D. A., Machalek, M. A., Williams, B. R. G., Silverman, R. H. & Virgin, H. W. Specific phenotypic restoration of an attenuated virus by knockout of a hostresistance gene. *Proc. Natl. Acad. Sci. USA* **97:**6097-6101
104. Chou, J., Chen, J. J., Gross, M. & Roizman, B. Association of a M(r) 90,000 phosphoprotein with protein kinase PKR in cells exhibiting enhanced phosphorylation of translation initiation factor eIF-2 alpha and premature shutoff of protein synthesis after infection with gamma 134.5- mutants of herpes simplex virus 1. *Proc Natl Acad Sci USA* **23:**10516-10520
105. He, B., Gross, M. & Roizman, B. The γ (1)34.5 protein of herpes simplex virus 1 complexes with protein phosphatase 1α to dephosphorylate the α subunit of the eukaryotic translation initiation factor 2 and preculde the shutoff of protein synthesis by double-stranded RNA-activated protein kinase. *Proc Natl Acad Sci USA* **94:**843-848
106. Morgan, C., Rose, H. M., HOlden, M. & Jones, E. P. Electron microscopic observations on the development of herpes simplex virus. *J Exp Med* **110:**643-656
107. Virgin, H. & Levine, B. *Unpublished data* (2004).
108. Levine, B., Goldman, J. E., Jiang, H. H., Griffin, D. E. & Hardwick, J. M. Bcl-2 protects mice against fatal alphavirus encephalitis. *Proc Natl Acad Sci USA* **93:**4810-4815
109. Joe, A., Foo, H., Kleeman, L. & Levine, B. The transmembrane domains of Sindbis virus envelope glycoproteins induce cell death. *J Virol* **72:**3935-3943
110. Bursch, W., Ellinger, A., Gerner, C. & Schultze-Hermann, R. in *Autophagy* (ed. Klionsky, D. J.) 290-302 (Landes Bioscience, Georgetown, Texas, 2004).
111. Bursch, W. The autophagosomal-lysosomal compartment in programmed cell death. *Cell Death Differ* **8:**569-581
112. Yu, L. et al. Regulation of an *ATG7-beclin 1* program of autophagic cell death by caspase 8. *Science* **304:**1500-1502
113. Berry, D. L., Schuldiner, O., York, K. & Baehrecke, E. H. in *American Society for Cell Biology 44th Annual Meeting* (Washinton, D.C., 2004).
114. Dangl, J. L. & Jones, J. D. Plant pathogens and integrated defence responses to infection. *Nature* **411:**826-833
115. Jones, D. A. & Takemoto, D. Plant innate immunity - direct and indirect recognition of general and specific pathogen-associated molecules. *Curr Opin Immunol* **16:**48-62
116. Lam, E. Controlled cell death, plant survival and development. *Nat Rev Mol Cell Biol* **5:**305-315
117. Erickson, F. L. et al. The helicase domain of the TMV replicase proteins induces the N-mediated defense response in tobacco. *Plant J* **18:**67-75
118. Whitham, S. et al. The product of the tobacco mosaic virus resistance gene N: similarity to toll and the interleukin-1 receptor. *Cell* **78:**1101-1115
119. Roizman, B. (2004).
120. Arico, S. et al. The tumor suppressor PTEN positively regulates macroautophagy by inhibiting the phosphatidylinositol 3-kinase/protein kinase B pathway. *J Biol Chem* **276:**35243-35246
121. Dales, S., Eggers, H. J., Tamm, I. & Palade, G. E. Electron microscopic study of the formation of poliovirus. *Virology* **26:**379-389
122. Suhy, D. A., Giddings, T. H. & Kirkegaard, K. Remodeling the endoplasmic reticulum by poliovirus infection and by individual viral proteins: an autophagy-like origin for virus-induced vesicles. *J Virol* **74:**8953-8955
123. Rust, R. C. et al. Cellular COOPII proteins are involved in production of the vesicles that form the poliovirus replication complex. *J Virol* 75:9808-9818
124. Swanson, M. (2004).
125. Matsuura, A., Tsukada, M., Wada, Y. & Ohsumi, Y. Apg1p, a novel protein kinase required for the autophagic process in *Saccharomyces cerevisiae*. *Gene* **192:**245-250

126. Kametaka, S., Okano, T., Ohsumi, M. & Ohsumi, Y. Apg14p and Apg6/Vps30p form a protein complex essential for autophagy in the yeast, Saccharomyces cerevisiae. *J Biol Chem* **273:**22284-22291
127. Juhasz, G., Csikos, G., Sinka, R., Erdelyi, M. & Sass, M. The Drosophila homolog of Aut1 is essential for autophagy and development. *FEBS Lett* **543:**154-158
128. Tanida, I., Tanida-Miyake, E., Komatsu, M., Ueno, T. & Kominami, E. Human Apg3p/Aut1p homologue is an authentic E2 enzyme for multiple substrates, GATE-16, GABARAP, and MAP-LC3, and facilitates the conjugation of hApg12p to hAPG5p. *J Biol Chem* Feb. 1, 2000; Manscript M200385200 (2002).
129. Thumm, M. & Kadowaki, T. The loss of Drosophila *APG4/AUT2* function modifies the phenotypes of *cut* and Notch signaling pathway pathway mutants. *Mol Genet Genomics* **266:**657-663
130. Hemelaar, J., Lelyveld, V. S., Kessler, B. M. & Ploegh, H. L. A single protease, Apg4B, is specific for the autophagy-related ubiquitin-like proteins GATE-16, MAP1-LC3, GABARAP, and Apg8L. *J Biol Chem* **278:**51841-51850
131. Mizushima, N. et al. Dissection of autophagosome formation using Apg5deficient mouse embryonic stem cells. *J Cell Biol* **152:**657-667
132. Mizushima, N., Sugita, H., Yoshimori, T. & Ohsumi, Y. A new protein conjugation system in human. The counterpart of the yeast Apg12p conjugation system essential for autophagy. *J. Biol. Chem.* **273:**33889-33892
133. Kametaka, S., Matsuura, A., Wada, Y. & Ohsumi, Y. Structural and functional analyses of APG5, a gene involved in autophagy in yeast. *Gene* **178:**139-143
134. Tanida, I., Tanida-Miyake, E., Ueno, T. & Kominami, E. The human homolog of Saccharomyces cerevisiae Apg7p is a protein-activating enzyme for multiple substrates including human Apg12p, GATE-16, GABARAP, and MAP-LC3. *J Biol Chem* **276:**1701-1706
135. Yuan, W., Stromhaug, P. E. & Dunn, W. A., Jr. Glucose-induced autophagy of peroxisomes in *Pichia pastoris* requires a unique E1-like protein. *Mol. Biol. Cell* **10:**1353-1366
136. Mizushima, N., Yoshimori, T. & Ohsumi, Y. Mouse Apg10 as an Apg12conjugating enzyme: analysis by the conugation-mediated yeast two-hybrid method. *FEBS Lett* **532:**450-454
137. Shintani, T. et al. Apg10p, a novel protein-conjugating enzyme essential for autophagy in yeast. *EMBO J* **18:**5234-5241
138. Tanida, I. et al. Murine Apg12p has a substrate preference for murine Apg7p over three Apg8p homologs. *Biochem Biophys Res Comm* **292:**256-262
139. Mizushima, N., Noda, T. & Ohsumi, Y. Apg16p is required for the function of the Apg12p-Apg5p conjugate in the yeast autophagy pathway. *EMBO J* **18:**3888-3896
140. Guan, J. et al. Cvt18/Gas12 is required for cytoplasm-to-vacuole transport, pexophagy, and autophagy in *Saccharomyces cerevisiae* and *Pichia Pastoris*. *Mol Biol Cell* **12:**3821-3838

Chapter 12

GENETIC VARIATION IN HOST DEFENSES AND VIRAL INFECTIONS

EUNHWA CHOI[*] and STEPHEN J. CHANOCK[#]
[*]*Department of Pediatrics, Seoul National University College of Medicine, Seoul, Korea;* [#]*Section on Genomic Variation, Pediatric Oncology Branch, National Cancer Institute, Bethesda, MD, USA*

1. INTRODUCTION

The dynamic between the host and viral pathogens represents a complex interaction, usually characterized by exposure, infection and possibly disease, all modulated by environmental circumstances. Host response to viral infection involves multiple immunological pathways, from initial pattern recognition molecules of innate immunity to both cellular and antibody-mediated immunity (Figure 1). The age of genomics has begun to afford an opportunity to examine global responses using microarray expression systems to catalog the change in gene expression and more recently, the profile of proteomics. Yet, the coordinated response of expressed genes and proteins is driven by the sequence of specific genes in the human genome, in which germ-line variation can alter the expression or function of the gene(s).

The argument over the importance of host genetic and infectious diseases has raged for generations. Rare examples of known genetic mutations, such as the X-linked lymphoproliferative syndrome (OMIM 308240) have shed light on specific immunologic pathways. The example of the highly penetrant, rare Mendelian disorder represents one end of the spectrum of genetic contributions to viral disease[1,2]. On the other side lies the complex disease paradigm; common diseases arise from a combination of common genetic variants, all of low penetrance[2-5]. To investigate both,

P. Palese (ed.), Modulation of Host Gene Expression and Innate Immunity by Viruses, 281-294.

we now have the blueprint to begin to unravel the contribution of genetic variation, which in turn can lead to an accurate assessment of the importance of host genetics in response to viral infection[5]. Furthermore, for viral infections, it is possible to investigate susceptibility, disease status or vaccine response.

With the completion of a draft sequence of the human genome, we now know the sequence of bases and have begun to identify the common genetic variants. An unexpected finding has been the extent of common genetic variation in the human genome[5-7]. Though it represent less than 0.2% overall, there are still roughly 10 to 15 million single nucleotide polymorphisms, SNPs, which are single base mutations with a minor allele frequency greater than 1% in at least one tested population. The molecular evolution of human populations has resulted in common genetic variants, mainly SNPs with a minor allele frequency of greater than 10% (perhaps comprising as many 5 million throughout the genome), which have been selected and maintained in populations[8]. There can be major differences in the allele frequencies of SNPs, which reflect major differences in selective pressures, such as infectious diseases like malaria or tuberculosis. Still, many of the SNPs are silent (that is they have no functional consequence) and have been carried as part of an inherited block of DNA across generations[7,9-11]; the term haplotype is used to designate sets of SNPs linked to each other and passes as a "unit" from generation to generation.

This review will summarize recent studies that have established the foundation for further investigation. Until recently, studies have concentrated on common genetic variants in leukocyte antigens, cytokines, innate immune molecules and receptors. Extensive effort has been focused on identifying and validating the importance of genetic variants that influence susceptibility and outcomes in HIV-1 infection, also known as AIDS-restricted genes[12]. Additional studies have identified host genetic variants important for hepatitis B (HBV) and C (HCV) viruses, and respiratory syncytial virus (RSV).

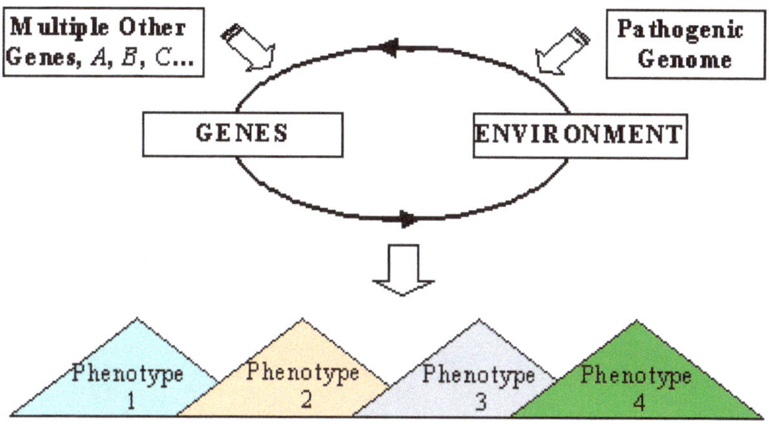

Figure 1. Schematic representation of complex interaction between host and pathogen in the development of infectious diseases.

2. HOST GENETIC VARIATION AND HIV-1 INFECTION

Genetic association studies conducted in large cohorts of HIV-1infected subjects have provided evidence that common genetic variants in more than a dozen genes can influence susceptibility and outcome (Table 1). So far, the most significant findings have concentrated on common genetic variants that regulate critical steps in HIV-1 cell entry and the host response; accordingly, the genes of interest have been drawn from cytokine defenses, chemokines and innate immunity. Certainly more variants in additional genes are expected to be confirmed. In turn, these additional markers will contribute to the complex interaction of genes that restrict HIV-1 related outcomes, including acquisition of infection and eventually form the foundation for determining a more comprehensive profile of known variants. This is particularly important because, so far, published studies have reported the effect of genetic variants of one gene at a time, but in the future, it will be necessary to begin to address the complexity of analyzing sets of genes, in an effort to dissect the differential effect of specific variants[5]. A daunting task will be to develop a systematic approach that accounts for gene-gene interaction and as well as the differential effect of variants within a single gene.

Table 1. Association of HLA alleles and candidate genes with three major viral infectious diseases; human immunodeficiency virus, hepatitis B virus, and respiratory syncytial virus.

		Impact on Viral Infections	
Virus	Gene/Genotype	Susceptibility	Outcome
Human Immunodeficiency Virus (HIV)	Δ32 *CCR5*	Resistance	Delayed progression to AIDS
	CCR2 V64I		Delayed progression to AIDS
	SDF1-3'A		Influence on progression in late-stage
	RANTES-28G		Delayed progression of HIV-1 disease
Hepatitis B virus (HBV)	*HLA*-DRB1*1302	Resistance	Associated with HBV persistence
	HLA-DQA1*0501		
	HLA-DQB1*0301		
	HLA-DRB1*1102		
	HLA-B8,SCO1,DR3	Susceptibility	Nonresponsiveness to HBV vaccine
Respiratory syncytial virus (RSV) bronchiolitis	*SP-A2* 1A3 allele	Susceptibility	Associated with severe bronchiolitis
	IL8 promoter		
	IL4 promoter haplotypes		

An illustrative example of the contribution of host genetics to infectious disease is the relationship between the CC chemokine receptor 5 gene (*CCR5*) polymorphisms and the HIV-1 infection[13-16]. Differences in susceptibility to HIV-1 infection as well as time to progression to AIDS have been associated with common genetic variants in the chemokine, *CCR5,* now known to be a critical co-receptor for HIV-1. *CCR5* Δ32 allele is a 32 base-pair (bp) deletion variant that causes truncation and loss of CCR5 receptors on lymphoid cell surfaces of homozygous individuals. In vitro analysis has confirmed the functional importance of this co-receptor, which when altered by the polymorphism inhibits viral replication.

Individuals homozygous for *CCR5* Δ32 allele have been shown to be resistant to HIV-1 infection, as indicated by decreased incidence of acquisition of infection. Individuals who are heterozygous for *CCR5* Δ32 still show a delayed progression to AIDS[13-16]. Interestingly, it has been noted that decreased circulating level of HIV-1 RNA in plasma during early chronic infection are associated with individuals who are heterozygous for *CCR5* Δ32 compared to individuals with wild-type *CCR5*. The distribution and frequency of the *CCR5* Δ32 allele varies by populations across the world; for instance, populations of African and Asian ancestry, this variant is rare. Thus, its protective effect against HIV transmission is mainly limited to populations of Caucasian ancestry. Like all other genes, there are many additional genetic variants in *CCR5*, which reflect the long and complex history of selection and adaptation in this gene[17]. For instance, a haplotype of SNPs in the promoter polymorphism of *CCR5* gene, known as P1, has been associated with the rapid progression to AIDS[18]. Interestingly, these informative variants of *CCR5* lie in the 5' end and a survey of genetic variants across the gene has provided strong evidence for natural selection, in particular in the 5' region, on the basis of higher than expected excess of intermediate frequency variant alleles[17]. Variation in this region probably represents adaptive change in response to an earlier pathogen and not HIV-1 itself.

Since the entry of HIV into cells is mediated by interactions between the viral envelope glycoproteins, the CD4 receptor, and HIV-1 co-receptors, variants in additional genes utilized by HIV-1 have been identified. Subsequent studiers have identified common genetic variants in known ligands of *CCR5,* namely, MIP-1a, MIP-1b, and RANTES (encoded by *CCL3, CCL4* and *CCL5*, respectively)[19-22]. Comparable to *CCR5*, common genetic variants in each of these genes can alter HIV-1 cell entry and have been associated with disease progression and HIV-1 transmission. RANTES inhibits CCR5-mediated entry of R5 HIV-1, thereby inhibiting HIV-1 replication. Individuals homozygous for the AC haplotype of *CCL5* have an increased risk for acquiring HIV-1 as well as acceleration of disease progression, particularly evident in individuals of European but not in African ancestry[22]. However, in Japan, individuals with the AG-*CCL5* haplotypes were associated with a delay in disease progression, thus underscoring the significance of population-specific genetic factors for HIV-1[22]. The In1.1C allele of *CCL5* has also been reported to increase susceptibility to HIV-1 infection and accelerate progression to AIDS in both African-American and European-American cohorts. The In1.1C is located in an intronic regulatory sequence element that exhibits differential allele binding to nuclear proteins and a down-regulation of gene transcription. Down-regulation of RANTES could increase the number of binding targets for HIV-1, thus promoting HIV-1 replication.

SDF-1, encoded by *CXCL12*, is the major ligand for the 7 trans-membrane receptor, *CXCR4*. The influence of *SDF1* genotypes on disease progression was shown among a group of high-risk exposed individuals who were never infected with HIV-1 in a cohort characterized by high-risk sexual practices[23]. Progression to AIDS is delayed for individuals with the *SDF1*-3'A/3'A genotype and the protective effect was prominent in later stages of HIV-1 infection. The current view is that this variant could interfere with the appearance of T cell-tropic HIV-1 strains.

A common polymorphism in *CCR2*, 64I, which lies within the first trans-membrane region, has been associated with a delay in AIDS defining illnesses[24]. The *CCR2*-64I single nucleotide polymorphism encodes a conservative substitution in a gene that is located on the same chromosome (3p21) as *CCR5*. Although *CCR2*-64I exerts no influence on the acquisition of HIV-1 infection, HIV-1infected individuals carrying the *CCR2*-64I allele progressed to AIDS 2 to 4 years later than individuals homozygous for wild type[11]. It has been postulated to reduce *CXCR4* activity.

The importance of genetic variants can alter the specific strain of HIV-1 virus because distinct co-receptors can interact with different strains for HIV-1 entry[25]. Nonsyncytium inducing (NSI) HIV-1 viruses are present throughout the disease process while syncytium inducing (SI) viruses are frequently found in progressive or late-stage HIV disease. *CCR5* has been recognized as the major coreceptor for macrophage-tropic HIV-1 (NSI strains or 'R5 HIV-1') cell entry. On the other hand, SI viruses primarily utilize the chemokine receptor *CXCR4* and are termed 'X4 HIV-1'. However, most primary SI isolates use *CXCR4* in conjunction with *CCR5* (R5X4 HIV-1).

Genes in the host response have also been studied extensively and a few validated examples demonstrate the complexity of the host defenses. Both IL-10 and interferon-gamma are cytokines that inhibit HIV-1 replication. Variants in the promoter of the *IL10* gene, which alter the expression of the gene, limit infection[26]. The *IL10* promoter SNPs have been shown to decrease expression of the cytokine. In a similar manner, common variants in the interferon-gamma gene, *IFNG*, have been reported to accelerate the course of HIV-1 infection, but by a mechanism not sufficiently appreciated[27]. Investigation of the mannose binding lectin (*MBL2*), a C-type collectin, has yielded conflicting results for analyses of transmission and progression of HIV-1. A common genetic variant in the low affinity Fc gamma receptor, FCGR3A, which functionally increases activity, has been associated with Kaposi Sarcoma, and perhaps, infection with the Kaposi sarcom herpes virus, KSHV[28].

The importance of the human major histo-compatibility complex, HLA, has been demonstrated in HIV-1 infection[29-31]. Two major observations are evident. First, specific HLA types, such as *HLA-B*35* have been associated with rapid progression to AID; the gene dose effect has been inferred from the fact that homozygotes for *HLA-B*35* progress even faster than heterozygotes[30]. Secondly, diversity in HLA alleles is protective against AIDS progression. Heterozygosity provides a wider spectrum of recognition of viral epitopes compared to those limited by homozygosity. Long-term studies have shown that individuals who are homozygous for class I gene (such as A, B or C) progress to AIDS at a faster rate[29,31].

3. ASSOCIATION OF GENETIC POLYMORPHISMS AND HEPATITIS C INFECTION

The clinical manifestations of hepatitis C virus are protean and include acute and chronic infection. Moreover, this infection, common throughout the world, can lead to cirrhosis and hepatocellular carcinoma. To date, investigators have examined genetic variation in HLA and select chemokines and cytokine defense genes. The chemokine literature is conflicting, with no clear evidence that CCR5 or CCR2 common variants (i.e., the same ones evaluated in HIV-1 reported above) are associated with response to interferon-gamma therapy[32-34]. Similarly, analyses of IL10, a potent antiviral cytokine, have yet to confirm a clear effect of the promoter polymorphisms, especially in response to interferon-alpha therapy[35,36].

Recently, it has been shown that the genes encoding the inhibitory NK receptor *KIR2DL3* and the *HLA-C1* directly influence resolution of HCV infection[37]. Interestingly, it was observed in both individuals of Caucasian and African ancestry with expected low doses of HCV but not those with high doses. This suggests that the coordination of NK and human leukocyte antigens (*KIR2DL3* and *HLA-C1*) are important in antiviral immunity but can be overwhelmed by high-dose exposure.

4. ASSOCIATION BETWEEN *HLA* ALLELES AND HBV INFECTION

Like most other viruses, the clinical manifestations of HBV can vary greatly, between populations, and even within families; these vary from a clinically asymptomatic condition to acute hepatitis or chronic liver disease. The majority of adults with primary HBV infection (90-95%) can successfully clear the virus and only 5-10% of adults become chronic HBV carriers. Among the chronically infected individuals, 20-30% leads to liver cirrhosis and about 5% develop hepatocellular carcinoma through a long-term disease progression. On the other hand, more than 90% develop chronic infection following perinatal transmission. The reasons for this variation in the natural history of HBV infection are not fully understood, but viral and immunological factors (viral load, viral mutations, and host immunity) could play important roles in modulating both the antiviral immune response and host susceptibility to HBV. The host genetic factors are believed to be responsible for clinical outcomes of many infectious diseases including HBV infection.

Epidemiological investigation suggests that there is a strong genetic component to determine the individual susceptibility to infectious pathogens, although to date, no single allele has been associated with HBV persistence or disease severity. Long-term follow-up studies indicated that a small proportion of individuals in high-risk groups never develop the disease. This finding raises the possibility that the host response could be a major factor in determining clinical outcome. HBV-infected individuals display various drug responses to interferon-alpha or lamivudine antiviral therapy. Response to vaccination is also diverse: 85% of healthy subjects can produce the efficient protective anti-HBsAg antibody to the HBV vaccination, while the remaining 15% fail to produce the antibody.

4.1 *HLA* class I and II alleles

The genes for *HLA* class I (*HLA*-A, -B, and -C) and class II (*HLA*-DRB1, DQA1, -DQB1, -DPA1, and DPB1) are located on the short arm of chromosome 6. As the primary effectors of host immune response, *HLA* molecules present foreign antigens to both the CD4+ T lymphocytes and the CD8+ cytolytic T cells, leading to both humoral and cell-mediated immune response. The majority of the human genetic studies associated with HBV infection have focused on *HLA* class II genes. For instance, in a large cohort of pediatric patients from Gambia, an association between DRB1*1302 and viral clearance has been reported[38]. The association between *HLA* class II allele DR13 and a self-limiting course has been confirmed in subsequent studies, suggesting that patients with *HLA*-DR13 can maintain a vigorous CD4+ T cell response to HBV core antigen during acute HBV infection. The beneficial effect of *HLA*-DR13 allele on the outcome of HBV infection may either be the result of more proficient antigen presentation by the *HLA*-DR13 molecules themselves or of a linked polymorphism in a neighboring immunoregulatory gene.

Genetic influence of *HLA* class I alleles has not been reported to be associated with the viral persistence or disease progression in HBV-infected patients until recently. However, a recent study showed that *HLA* class I and class II genetic effects on the outcome of HBV infection in Caucasians,

A*0301 and DRB1*1302 were associated with a 2-fold increase in HBV clearance, whereas B*08 and two B*44 haplotypes were associated with greater than a 1.5-fold increase in viral persistence[38,39].

4.2 Vaccine response

The vaccine incorporating HBsAg generally induces protective antibody following a 3 dose immunization schedule. Detectable anti-HBs antibody level >10 mIU/ml after immunization is protective against HBV. At least 4% of healthy individuals fail to generate significant levels of anti-Hbs antiboides (<10 mIU/ml), and an additional 10% make only low amounts after a standard adult immunization schedule. Milich et al. demonstrated that the level of anti-HBs antibody varied widely in different inbred strains of mice[40]. The response was determined to be a dominant trait related to genes within the class II region of the *MHC*. An antigen-specific T-cell proliferative response occurred in responders but not in nonresponders. In familial studies in humans, the response was dominantly inherited and closely linked to the *MHC*[41]. As in the mice, particular haplotypes (*HLA*-B8, SC01, DR3 and HLA-B44, FC31, DR7) were over-represented in nonresponders. When homozygotes and heterozygotes for the *HLA*-B8, SC01, DR3 haplotypes were prospectively immunized with HBsAg, antibody response was significantly lower in the homozygotes than heterozygotes[42]. However, occasional homozygotes do respond, suggesting that non-*MHC* genes that are subject to (complex) genetic control could also be involved.

5. HOST GENETICS IN SEVERE RSV BRONCHIOLITIS

Respiratory syncytial virus (RSV) is a serious viral respiratory pathogen worldwide. By two years of age, nearly all children are infected and the vast majority manifest mild to moderate upper respiratory disease[43]. However, roughly one quarter of infants and children infected with RSV develop lower respiratory tract disease during initial infection and of these, in 1-2% of infected children younger than 1 year, severe disease develops, often necessitating hospitalization. The basis for the observed spectrum in disease severity among infected children is still not well understood but most likely includes a combination of immune factors that lead to partial obstruction of the tiny infant airways, perhaps, modulated by genetic factors.

The innate defense system of the lung may be particularly important during infections in young infants before acquired immunity has developed. There are two surfactant proteins A1 and A2, which play crucial role in pulmonary defense. Increased circulating levels of total serum IgE and high total eosinophilic counts have been observed in children with persistent wheezing following RSV bronchiolitis compared to children with transient wheezing. Moreover, a history of RSV bronchiolitis, particularly if severe enough to require hospitalization has been associated with subsequent recurrent wheezing. However, there is a growing appreciation that severe RSV disease might be an indicator, rather than a cause, of underlying pulmonary abnormalities involving subtle functional or immunological deficiencies. These observations support the hypothesis that an alteration in Th2 cytokine response in the acute phase of RSV infection predisposes to recurrent wheezing[44]. The available data suggest that in severe RSV disease, a predominant Th2 immune response could be affected by genetic factors, namely SNPs (embedded within common haplotypes) that alter the expression or response of a key cytokine, such as IL-4.

To date, the majority of published studies have examined either pulmonary host defense genes, such as Surfactant proteins A1, A2, and D and the mannose binding lection (MBL2) or the cytokine defense genes interleukin (IL)-4, IL4R, IL8, IL-9, IL-10 and tumor necrosis factor (*TNF*)[45-50]. A strong association, now replicated in several studies is between the IL4 promoter haplotype and severe RSV infection in young children[45,48]. A common haplotype of *IL4*, which contains a common SNP (-589T) known to increase transcriptional activity of IL-4 has been associated with severe hospitalized RSV bronchiolitis in Korean, European and North American Caucasian infants. An association of severe RSV disease with an *IL4* polymorphism, which previously has been shown to be associated with asthma and increased levels of serum IgE, supports the hypothesis that the same genetic susceptibility factors for atopy and hyperreactive airway disease could also play a role in primary acute severe RSV disease.

Severe RSV disease in infants represents a complex disease in which host genetic factors together with viral pathogenetic factors unique to RSV combine to increase or decrease the risk for severe disease. The RSV studies also highlight the importance of population specific genetic factors, for instance, the effect of IL8 in children in the UK is not seen in Korea[45-47]. These results underscore the importance of analyzing variants in well-defined populations of sufficiently large enough size to attain statistical significance.

Since an altered balance between the Th1 and Th2 immune response has been identified as a critical component in severe RSV disease, further genetic association studies should target additional candidate genes drawn from the Th1 and Th2 cytokine pathways as well as novel pathways in innate immunity. It will also be important to investigate pathways, such as interferon-alpha activation or pattern recognition pathways of C-type collectins.

6. CONCLUSIONS

Recent advances in genomics now permit the analysis of germ-line variation and susceptibility to viral infections. So far, studies have investigated variation in genes of innate immunity, but it is clear that we must look at many pathways in host defenses. While the analysis of HLA, chemokines and cytokine defenses have been fruitful, the next steps require careful assessment of pathways to be analyzed because it is likely that sets of SNPs or haplotypes contribute to outcomes in viral infections.

In conclusion, it is now possible to probe the genetic basis of host response to viral pathogens. Rapid progress in the technical and bio-informatic platforms has accelerated the opportunity to examine host-pathogen genomic interactions. Large epidemiological studies are required to adequately study common SNPs in case control or cohort studies addressing viral infections. Since SNP studies generally do identify a highly penetrant variant, follow-up validation studies are critical before implementation in medical care. It can be argued that no SNP study stands alone, but must be replicated to confirm the genetic link between one or more outcomes and the genetic variant(s) of interest.

REFERENCES

1. Risch, N. and Merikangas, K., 1996, The future of genetic studies of complex human diseases. *Science*. **273**:1516-1517
2. Taylor, J.G., Choi, E.H., Foster, C.B. and Chanock, S.J., 2001, Using genetic variation to study human disease. *Trends Mol. Med.* **7**:507-512
3. Foster, C.B. and Chanock, S.J., 2000, Mining variations in genes of innate and phagocytic immunity: current status and future prospects. *Curr. Opin. Hemato.* **7**:9-15
4. Hill, A.V., 1998, The immunogenetics of human infectious diseases. Annu. Rev. Immunol. **16**:593-617
5. Botstein, D. and Risch, N., 2003, Discovering genotypes underlying human phenotypes: past successes for mendelian disease, future approaches for complex disease. Nat. Genet. **33**:228-237
6. Cargill, M., Altshuler, D., Ireland, J., Sklar, P., Ardlie, K., Shaw, N., Lane, C.R., Lim, E.P., Kalyanaraman, N., Nemesh, J., Ziaugra, L., Friedland, L., Rolfe, A., Warrington, J., Lipshutz, R., Daley, G.O. and Lander, E.S., 1999, Characterization of single-nucleotide polymorphisms in coding regions of human genes. Nat. Genet. **22**:231-238
7. Chanock, S.J., Candidate genes and single nucleotide polmorphisms (SNPs) in the study of human disease. *Dis. Markers.* **17**:89-98
8. Kruglyak, L., 1999, Prospects for whole-genome linkage disequilibrium mapping of common disease genes. Nat. Genet. **22**:139-144
9. Daly, M.J., Rioux, J.D., Schaffner, S.F., Hudson, T.J. and Lander, E.S., 2001, High-resolution haplotype structure in the human genome. Nat. Genet. **29**:229-232
10. Gabriel, S.B., Schaffner, S.F., Nguyen, H., Moore, J.M., Roy, J., Blumenstiel, B., Higgins, J., DeFelice, M., Lochner, A., Faggart, M., Liu-Cordero, S.N., Rotimi, C.,

Adeyemo, A., Cooper, R., Ward, R., Lander, E.S., Daly, M.J. and Altshuler, D., 2002, The structure of haplotype blocks in the human genome. Science **296:**2225-2229
11. Excoffier, L. and Slatkin, M., 1995, Maximun-likelihood estimation of molecular haplotype frequencies in a diploid population. Mol. Biol. Evol. **12:**921-927
12. O'Brien S.J. and Nelson, G.W., 2004, Human genes that limit AIDS. *Nat. Genet.* **36:**565-572
13. Zimmerman, P.A., Buckler-White, A., Alkhatib, G., Spalding, T., Kubofcik, J., Combadiere, C., Weissman, D., Cohen, O., Rubbert, A., Lam, G., Vaccarezza, M., Kennedy, P.E., Kumaraswami, V., Giorgi, J.V., Detels, R., Hunter, J., Chopek, M., Berger, E.A., Fauci, A.S., Nutman, T.B. and Murphy, P.M., 1997, Inherited resistance to HIV-1 conferred by an inactivating mutation in CC chemokine receptor 5:studies in populations with contrasting clinical phenotypes, defined racial background, and quantified risk. *Mol. Med.* **3**:23-36
14. Huang, Y., Paxton, W.A., Wolinsky, S.M., Neumann, A.U., Zhang, L., He, T., Kang, S., Ceradini, D., Jin, Z., Yazdanbakhsh, K., Kunstman, K., Erickson, D., Dragon, E., Landau, N.R., Phair, J., Ho, D.D. and Koup, R.A., 1996, The role of a mutant CCR5 allele in HIV-1 transmission and disease progression. *Nat. Med.* **2**:1240-1243
15. Samson, M., Libert, F., Doranz, B.J., Rucker, J., Liesnard, C., Farber, C.M., Saragosti, S., Lapoumeroulie, C., Cognaux, J., Forceille, C., Muyldermans, G., Verhofstede, C., Burtonboy, G., Georges, M., Imai, T., Rana, S., Yi, Y., Smyth, R.J., Collman, R.G., Doms, R.W., Vassart, G. and Parmentier, M., 1996, Resistance to HIV-1 infection in caucasian individuals bearing mutant alleles of the CCR-5 chemokine receptor gene. *Nature.* **382**:722-725
16. Dean, M., Carrington, M., Winkler, C., Huttley, G.A., Smith, M.W., Allikmets, R., Goedert, J.J. Buchbinder, S.P. Vittinghoff, E., Gomperts, E., Donfield, S., Vlahov, D., Kaslow R., Saah, A., Rinaldo, C., Detels, R. and O'Brien, S.J., 1996, Genetic restriction of HIV-1 infection and progression to AIDS by a deletion allele of the CKR5 structuralgene. Hemophilia Growth and Development Study, Multicenter AIDS Cohort Study, Multicenter Hemophilia Cohort Study, San Francisco City Cohort, ALIVE Study. *Science* **273**:1856-1862
17. Bamshad, M.J., Mummidi, S., Gonzalez, E., Ahuja, S.S., Dunn, D.M., Watkins, W.S, Wooding, S., Stone, A.C., Jorde, L.B., Weiss, R.B. and Ahuja, S.K., 2002, A strong signature of balancing selection in the 5' cis-regulatory region of CCR5. *Proc. Natl. Acad. Sci. USA.* **99:**10539-10544
18. An, P., Nelson, G.W., Wang, L., Donfield, S., Goedert, J.J., Phair, J., Vlahov, D., Buchbinder, S., Farrar, W.L., Modi, W., O'Brien, S.J. and Winkler, C.A., 2002, Modulation of HIV infection and disease progression by interacting RANTES variants. *Proc. Natl. Acad. Sci. USA.* **99:**10002-10007
19. Cocchi, F., DeVico, A.L., Garzino-Demo, A., Arya, S.K., Gallo, R.C. and Lusso, P., 1995, Identification of RANTES, MIP-1 alpha, and MIP-1 beta as the major HIV-suppressive factors produced by CD8+ T cells. *Science.* **270**:1811-1815
20. Liu, H., Chao, D., Nakayama, E.E., Taguchi, H., Goto, M., Xin, X., Takamatsu, J.K., Saito, H., Ishikawa, Y., Akaza, T., Juji, T., Takebe, Y., Ohishi, T., Fukutake, K., Maruyama, Y., Yashiki, S., Sonoda., S., Nakamura, T., Nagai, Y., Iwamoto, A. and Shioda, T., 1999, Polymorphism in RANTES chemokine promoter affects HIV-1 disease progression. *Proc. Natl. Acad. Sci. U S A* **96**:4581-4585
21. Gonzalez, E., Dhanda, R., Bamshad, M., Mummidi, S., Geevarghese, R., Catano, G., Anderson, S.A., Walter E.A., Stephan, K.T., Hammer, M.F., Mangano, A., Sen, L., Clark, R.A., Ahuja, S.S., Dolan, M.J. and Ahuja, S.K., 2001, Global survey of genetic variation in CCR5, RANTES, and MIP-1alpha: impact on the epidemiology of the HIV-1 pandemic. *Proc. Natl. Acad. Sci. U S A* **98**:5199-5204
22. McDermott, D.H., Beecroft, M.J., Kleeberger, C.A., Al-Sharif, F.M., Ollier, W.E, Zimmerman, P.A., Boatin, B.A., Leitman, S.F., Detels, R., Hageer, A.H. and Murphy,

P.M., 2000, Chemokine RANTES promoter polymorphism affects risk of both HIV infection and disease progression in the Multicenter AIDS Cohort Study. *AIDS.* **14**:2671-2678

23. Winkler, C., Modi, W., Smith, M.W., Nelson, G.W., Wu, X., Carrington, M., Dean, M., Honjo, T., Tashiro, K., Yabe, D., Buchbinder, S., Vittinghoff, E., Goedert, J.J., O'Brien, T.R., Jacobson, L.P., Detels, R., Donfield, S., Willoughby, A., Gomperts, E., Vlahov, D., Phair, J. and O'Brien, S.J., 1998, Genetic restriction of AIDS pathogenesis by an SDF-1 chemokine gene variant. *Science.* **279:**389-393
24. Smith, M.W., Dean, M., Carrington, M., Winkler, C., Huttley, G.A., Lomb, D.A., Goedert, J.J., O'Brien, T.R., Jacobson, L.P., Kaslow, R., Buchbinder, S., Vittinghoff, E., Vlahov, D., Hoots, K., Hilgartner, M.W. and O'Brien, S.J., 1997, Contrasting genetic influence of CCR2 and CCR5 receptor gene variants on HIV-1 infection and disease progression. *Science.* **277:**959-965
25. Fenyo, E.M., Morfeldt-Manson, L., Chiodi, F., Lind, B., von Gegerfelt, A., Albert, J., Olausson, E. and Asjo, B., 1998, Distinct replicative and cytopathic characteristics of human immunodeficiency virus isolates. *J. Virol.* **62**:4414-4419
26. Shin, H.D., Winkler, C., Stephens, J.C., Bream, J., Young, H., Goedert, J.J., O'Brien, T.R., Vlahov, D., Buchbinder, S., Giorgi, J., Rinaldo, C., Donfield, S., Willoughby, A., O'Brien, S.J. and Smith, M.W., 2000, Genetic restriction of HIV-1 infection and AIDS progression by promoter alleles of interleukin 10. *Proc. Natl. Acad. Sci. USA.* **97:**14467-14472
27. An, P., Vlahov, D., Margolick, J.B., Phair, J., O'Brien, T.R., Lautenberger, J., O'Brien, S.J. and Winkler, C.A., 2003, A TNF-inducible promoter variant of interferon-gamma accelerates CD4 T cell depletion in HIV-1 infected individuals. *J. Infect. Dis.* **188:**228-231
28. Lehrnbecher, T.L., Foster, C.B., Zhu, S., Venzon, D., Steinberg, S.M., Wyvill, K., Metcalf, J.A., Cohen, S.S., Kovacs, J., Yarchoan, R., Blauvelt, A. and Chanock, S.J., 2000, Variant genotypes of FcgammaRIIIA influence the development of Kaposi's sarcoma in HIV-infected men. *Blood.* **95:**2386-2390
29. Gao, X., Nelson, G.W., Karacki, P., Martin, M.P., Phair, J., Kaslow, R., Goedert, J.J., Buchbinder, S., Hoots, K., Vlahov, D., O'Brien, S.J. and Carrington, M., 2001, Effect of a single amino acid change in MHC class I molecules on the rate of progression to AIDS. *N. Engl. J. Med.* **344:**1668-1675
30. Carrington, M., Nelson, G.W., Martin, M.P., Kissner, T., Vlahov, D., Goedert, J.J., Kaslow, R., Buchbinder, S., Hoots, K. and O'Brien, S.J., 1999, HLA and HIV-1: Heterozygote advantage and *B*35-Cw*04* disadvantage. *Science.* **283:**1748-1752
31. Martin, M.P., Gao, X., Lee, J.H., Nelson, G.W., Detels, R., Goedert, J.J., Buchbinder, S., Hoots, K., Vlahov, D., Trowsdale, J., Wilson, M., O'Brien S.J., and Carrington M., 2002, Epistatic interaction between KIR3DS1 and HLA-B delays the progression to AIDS. *Nat. Genet.* **31:**429-434
32. Ahlenstiel, G., Berg, T., Woitas, R.P., Grunhage, F., Iwan, A., Hess, L., Brackman, H.H., Kupfer, B., Schernick, A., Sauerbruch, T. and Spengler U., 2003, Effects of the CCR5-Delta32 mutation on antiviral treatment in chronic hepatitis C. *J. Hepatol.* **39:**245-252
33. Dorak, M.T., Folayan G. O., Niwas, S., van Leeuwen, D.J., Yee, L.J., Tang, J. and Kaslow, R.A., 2002, C-C chemokine receptor 2 and C-C chemokine receptor 5 genotypes in patients treated for chronic hepatitis C virus infection. *Immunol. Res.* **26:**167-175
34. Glas, J., Torok, H.P., Simperl, C., Konig, A., Martin, K., Schmidt, F., Schaefer, M., Schiemann, U. and Folwaczny, C., 2003, The Delta 32 mutation of the chemokine-receptor 5 gene neither is correlated with chronic hepatitis C nor does it predict response to therapy with interferon-alpha and ribavirin. *Clin. Immunol.* **108:**46-50
35. Knapp, S., Hennig, B.J., Frodsham, A.J., Zhang, L., Hellier, S., Wright, M., Goldin, R., Hill, A.V., Thomas, H.C. and Thursz, M.R., 2003, Interleukin-10 promoter

polymorphisms and the outcome of hepatitis C virus infection. *Immunogenetics,* **55:**362-369

36. Yee, L.J., Tang, J., Gibson, A.W., Kimberly, R., Van Leeuwen, D.J. and Kaslow, R.A., 2001, Interleukin 10 polymorphisms as predictors of sustained response in antiviral therapy for chronic hepaptitis C patients. *Hepatology*. **33:**708-712
37. Khakoo SI, Thio CL, Martin MP et al, 2004. HLA and NK cell inhibitory receptor genes in resolving hepatitis c virus infection. *Science* **305**: 872-874
38. Thursz, M.R., Kwiatkowski, D., Allsopp, C.E., Greenwood, B.M., Thomas, H.C. and Hill, A.V., 1995, Association between an MHC class II allele and clearance of hepatitis B virus in the Gambia. *N. Engl. J. Med.* **332**:1065-1069
39. Thio, C.L., Thomas, D.L., Karacki, P., Gao, X., Marti, D., Kaslow, R.A., Goedert, J.J., Hilgartner, M., Strathdee, S.A., Duggal, P., O'Brien, S.J., Astemborski, J. and Carrinton, M., 2003, Comprehensive analysis of class I and class II HLA antigens and chronic hepatitis B virus infection. *J. Virol.* **77:**12083-12087
40. Milich, D.R., 1991, Immune response to hepatitis B virus proteins: relevance of the murine model. *Semin. Liver. Dis.* **11**:93-112
41. Salazar, M., Deulofeut, H., Granja, C., Deulofeut, R., Yunis, D.E., Marcus-Bagley, D., Awdeh, Z., Alper, C.A. and Yunis, E.J., 1995, Normal HBsAg presentation and T-cell defect in the immune response of nonresponders. *Immunogenetics.* **41:**366-74
42. Alper, C.A., Kruskall, M.S., Marcus-Bagley, D., Craven, D.E., Katz, A.J., Brink, S.J., Dienstag, J.L., Awdeh, Z. and Yunis E.J., 1989, Genetic prediction of nonresponse to hepatitis B vaccine. *N. Engl. J. Med.* **321:**708-712
43. Collins, P.L., Chanock, R.M. and Murphy, B.R., 2001, Respiratory Syncytial Virus. *Fields Virology. 4th ed. Philadelphia:* Lippincott-Raven, pp.1443-1486
44. Roman, M., Calhoun, W.J, Hinton K.L., Avendano, L.F., Simon, V., Escobar, A.M., Gaggero, A. and Diaz, P.V., 1997, Respiratory syncytial virus infection in infants is associated with predominant Th-2-like response. *Am. J. Respir. Crit. Care Med.* **156:**190-195
45. Choi, E.H., Lee, H.J., Yoo, T. and Chanock, S.J., 2002, A common haplotype of interleukin-4 gene IL4 is associated with severe respiratory syncytial virus disease in Korean children. *J. Infect. Dis.* **186:**1207-11
46. HullJ, Thomson, Kwiatkowski D Association of respiratory virus bronchiolitis with interleukin 8 gene region in UK families. 2000. *Thorax*. **55**:1023-1027
47. Hull J, Ackerman H, Isles K, Usen S, Pinder M. Thomson A, Kwiatkowski D. Unusual haplotypic structure of IL8, a susceptibility locus for common respiratory virus. 2001. *Am J Hum Genet* **69**: 413-419
48. Hoebee, B., Rietveld, E., Bont, L., Oosten, M., Hodemaekers, H.M., Nagelkerke, N.J., Neijens, H.J., Kimpen, J.L. and Kimman, T.G., 2003, Association of severe respiratory syncytial virus bronchiolitis with interleukin-4 and interleukin-4 receptor alpha polymorphisms. *J. Infect. Dis.* **187:**2-11
49. Hoebee, B., Bont, L., Rietveld, E., van Oosten, M., Hodemaekers, H.M., Nagelkerke, H.J., Kimpen, J.L. and Kimman, T.G., 2004, Influence of promoter variants of interleukin-10, interleukin-9, and tumor necrosis factor-alpa genes on respiratory syncytial virus bronchiolitis. *J. Infect. Dis.* **189:**239-247
50. Lofgren, J., Ramet, M., Renko, M., Marttila, R. and Hallman, M., 2002, Association between Surfactant Protein A Gene Locus and Severe Respiratory Syncytial Virus Infection in Infants. *J. Infect. Dis.* **185:**283-289

Index

1

129/J, 80, 81

2

2A, 270
2BC, 271

3

3A, 271
$3C^{Pro}$, 6

6

6-16, 37

9

9-27, 37

A

A/NS1, 104, 105
A46R, 85, 173, 176, 185, 225, 233, 234
A52R, 85, 173, 176, 185, 225, 233, 234
actinomycin D, 47
ADAR, 3, 37
Adenovirus, 6, 51, 73, 76, 269
adhesin, 183
Akt, 251, 258, 270
ankyrin, 43, 45, 173, 174, 176, 185
antigenic variation, 68
anti-interferon, 4, 10, 115
antiviral, 117, 118, 119, 120, 125, 129, 130, 131, 133, 170, 171, 174, 175, 176, 177, 184, 187, 199, 203, 221, 223, 224, 225, 226, 228, 230, 231, 233, 234, 235, 236
AP-1, 4, 21, 24, 56, 73, 97, 101, 199
APC, 154
apoptosis, 44, 45, 46, 48, 52, 66, 70, 77, 85, 106, 119, 142, 143, 150, 151, 152, 171, 172, 173, 174, 175, 177, 184, 185 211, 257, 265
arsenite, 47
Arterial hyperplasia, 186
Arthritis, 186
Asthma, 186
ATF2/c-jun, 20, 21
atg, 247, 251, 252, 253, 256, 258, 262, 265, 266, 271
autophagosomes, 249, 251, 253, 254, 263, 270, 271, 272
autophagy, 245, 246, 247, 249, 250, 251, 252, 253, 254, 255, 256, 257, 258, 259, 260, 261, 262, 263, 264, 265, 266, 267, 268, 269, 270, 271, 272, 273
autophosphorylation, 46, 47, 48, 49, 51, 180
Avulavirus, 116, 121

B

B/NS1, 104, 105
B18R, 6, 178
bacterial artificial chromosomes (BAC), 168
bafilomycin A, 232
BALB/c, 80, 101, 184, 226, 228, 233, 235
BALB/c mice, 184, 226, 233, 235
Bcl-2, 70, 150, 152, 257, 258, 265, 270
Bcl-x, 152

beclin, 252, 256, 257, 258, 262, 265, 266, 267, 268
bioterrorism agent, 1, 187
Bovine respiratory syncytial virus (BRSV), 116, 118, 119, 131, 132
bronchiolitis, 227, 284, 289, 290
Bunyamwera virus (BUNV), 7, 210, 211

C

C protein, 7, 117, 118, 126, 127, 128, 129, 130, 207, 215
C3H/HeJ mice, 226, 227, 228
C3H/HeN mice, 227
C57BL/6, 80
C57BL/ScCr mice, 226
calcium channels, 79, 83
Canine distemper virus, 116
CARD, 27
cardiomyopathy, 259
caspase, 27, 70, 150, 172, 173, 175, 181, 266
CBP, 21, 105, 143, 182, 183
CCR2, 284, 286, 287
CCR5, 284, 285, 286, 287
CD14, 223, 226, 231
CD150, 229, 230
CD4, 8, 77, 84, 154, 170, 173, 285, 288
CD40, 83, 85, 223
CD80, 83, 85, 223
CD81, 79, 83
CD86, 83, 85
CD94, 78, 79, 81
CD94-NKG2A, 78, 79
cdc2, 145, 146, 147
cdc34, 145
cdk4, 145
cdk6, 145
CENP-A, 144
CENP-C, 144
c-fos, 150
chemokine, 36, 66, 69, 70, 78, 83, 121, 176, 178, 182, 183, 184, 185, 202, 204, 212, 224, 227, 234, 283, 284, 286, 287, 291
chicken embryos, 100
chloroquine, 151, 232, 235
CIITA, 39, 77
c-jun, 21, 123
Class II MHC, 77
CMV, 38, 67, 68, 69, 71, 73, 77, 79, 80, 81, 83, 84, 85, 230, 231
complement control protein, 176, 183
Cowpox, 164
CpG, 44, 200, 225, 231, 232, 233
CPSF, 102
CPV, 164, 169, 173, 177, 179, 181, 182, 184, 186
CrmA, 172, 173, 175
cullin 4a (Cul4a), 123
cyclin D1, 145
cyclin D3, 145
cytokine, 19, 27, 36, 39, 46, 56, 66, 67, 69, 70, 77, 79, 83, 84, 107, 121, 154, 170, 172, 174, 175, 176, 177, 179, 182, 184, 197, 201, 202, 204, 205, 223, 224, 225, 226, 227, 228, 230, 231, 232, 233, 234, 235, 282, 283, 286, 287, 290, 291
cytotoxic T-lymphocytes (CTL), 66, 69, 70, 71, 72, 78, 120, 170, 171, 172

D

DAP12, 80
Daxx, 143
DBA/2, 80
DDB1, 123, 124
DDB2, 123
DEAD-box, 200
delNS1, 10, 99, 100, 101, 103, 205
dendritic cell (DC), 8, 26, 28, 29, 67, 77, 83, 85, 179, 206, 212, 222, 224, 228, 229, 230, 232, 234, 235
dengue fever virus, 6, 199, 212, 213

Dictyostelium, 247, 256, 257
DNA bending, 21, 22
DNA shuffling, 9
DRBP76, 49
double-stranded RNA (dsRNA), 3, 6, 10, 25, 26, 28, 35, 36, 37, 38, 39, 40, 41, 42, 44, 46, 47, 48, 49, 50, 51, 52, 56, 99, 101, 103, 104, 105, 125, 127, 147, 149, 173, 175, 185, 198, 200, 204, 205, 210, 223, 224, 225, 227, 231, 234, 235, 262, 269

E

E1A, 6
E2, 6, 51, 79, 83, 105, 124, 144, 145, 253, 266, 269
E2F1, 85
E3 ubiquitin ligase, 143, 144, 145
E3/19k, 71, 76
E3L, 6, 51, 103, 173, 175, 176, 185
E6, 6
E7, 6
EBER-1, 44
Ebola virus, 7, 103, 199, 201, 202, 203, 204, 205, 206, 214
Ectromelia virus (EV), 164, 169, 173, 175, 177, 179, 181, 185
EGF, 173, 180, 185
eIF2α, 250, 256, 261, 262, 263, 264, 269
electrophoretic mobility shift assays, 204
embryonic stem, 253
EMCV, 45, 46
EMSA, 204
enhanceosome, 21, 22, 23, 29
Env, 84, 225, 229
eosinophilic counts, 290
epidermal growth factor, 180
Epstein Barr virus (EBV), 50, 68, 71, 72, 85, 232, 257, 258, 270
ES, 253, 258, 264, 265, 271

F

Fas, 70, 172, 173
Fc receptor, 66, 70, 236
Filoviridae, 115
filoviruses, 201, 202, 203
formalin-inactivated virus, 226
Fowlpox, 164
fruit bats, 116, 209, 215

G

GADD, 148
GAF, 39, 131
Gamma Activated Factor, 39
gB, 38, 151, 155, 225, 230, 231
GBP I, 37
gC, 150, 151
Gcn, 256, 261, 262
gD, 150, 151
genetic variation, 281, 282, 287
GFP-LC3, 271
gJ, 150, 151, 152
granzyme, 70, 77, 172

H

H5N1, 106
H60, 82
HA, 96
HBx, 270
HCF, 142
hemagglutinin, 68, 96, 98, 229
hemagglutinin-esterase-fusion protein (HEF), 96
Hendra virus (HeV), 7, 116, 129, 130, 207, 208
Henipavirus, 116, 129, 207
Hep2, 132
hepatitis C virus (HCV), 6, 7, 9, 45, 51, 53, 79, 83, 197, 212, 213, 260, 269, 270, 282, 287
herpes simplex virus (HSV), 6, 28, 37, 38, 51, 68, 69, 71, 73, 79, 84, 104,

141, 142, 143, 144, 145, 146, 147, 148, 149, 150, 151, 152, 153, 154, 155, 200, 225, 231, 232, 250, 251, 258, 261, 262, 263, 264, 265, 267, 269
HHV-8, 73, 83
HLA, 74, 75, 79, 81, 155, 284, 287, 288, 289, 291
homeobox, 22
homeostasis, 57, 273
horses, 96, 207, 213
host cell factor, 142
HPIV2, 117, 118, 121, 122, 124, 125
HPIV3, 127
HSV-1, 37, 38, 51, 141, 142, 143, 144, 145, 146, 147, 148, 149, 150, 151, 153, 154, 155, 225, 231, 232, 258, 262, 263, 264, 265, 267, 269
HSV-2, 141, 148, 232
human cytomegalovirus (HCMV), 66, 71, 72, 73, 74, 75, 76, 77, 81, 82, 83, 84, 225, 230
human herpes virus 6, 8, 72
Human Immunodeficiency Virus (HIV), 6, 10, 44, 50, 69, 72, 73, 77, 79, 83, 84, 107, 197, 259, 269, 282, 283, 284, 285, 286, 287
human papilloma virus, 6, 77, 259
human parainfluenza virus, 2, 7, 116, 121
Human respiratory syncytial virus (HRSV), 8, 98, 116, 118, 119, 131, 132, 225, 226, 227, 228, 229, 236, 282, 284, 289, 290
HuP54, 52
HuP58, 52, 53
HuP60, 52, 53
HUVECs, 204, 211, 212
hyperthermia, 150

I

ICP0, 6, 143, 144, 145, 156
ICP22, 145, 146, 147, 154
ICP34.5, 103, 148, 154, 156, 200, 258, 262, 263, 264, 269, 270
ICP47, 71, 153, 154
IEX-1, 150
IFN antagonist, 1, 2, 4, 5, 6, 10, 95, 96, 97, 101, 103, 105, 106, 107, 116, 118, 119, 122, 123, 128, 130, 132, 133, 197 198, 199, 200, 207, 210, 212, 213, 215
IFN receptor, 39, 42, 98, 133, 177
IFN stimulated gene (ISG), 3, 4, 9, 39, 230
IFN viroceptors, 177
IFNAR, 38, 39, 97, 197, 198, 199, 200, 203, 206, 210, 212, 213
IFN-α/β, 2
IFNαβ receptor, 3, 4
IFN-β promoter, 3, 4
IFN-γ, 2
IFNα4, 23
IKK, 24, 25, 28, 29, 39, 97, 209
IKKβ, 24
IL-1β viroceptors, 178
IL-10, 8, 84, 173, 179, 185, 228, 286, 290
IL-11, 37
IL-12, 2, 9, 77, 84, 226, 228, 230
IL-15, 77
IL-18, 77, 173, 179, 185, 233
IL-1β, 204
IL-2, 77, 83, 154, 173
IL-4, 8, 290
IL-4Ralpha, 8
IL-6, 37, 121, 128, 129, 181, 202, 204, 226, 228, 230, 231
ILT, 78
immunoevasive, 65, 67, 68, 69, 70, 73, 81, 82, 85
immunoglobin-like inhibitory receptors, 78
immunomodulation, 66
influenza A, 6, 95, 96, 99, 101, 102, 103, 104, 106, 205
influenza C, 96, 104

innate defense, 184, 289
innate immunity, 1, 51, 66, 80, 126, 215, 246, 247, 250, 252, 254, 261, 262, 265, 267, 268, 272, 273, 281, 283, 290, 291
interferon regulatory factor 3, 3, 198, 199, 205
IP10, 37, 223
IRAK, 28, 40, 223, 225
IRAK-4/TRAF-6, 223
IRES, 53
IRF1, 21, 37
IRF3, 3, 4, 21, 22, 23, 24, 25, 27, 28, 39, 103, 104, 209, 233
IRF7, 3, 21, 22, 23, 24, 28, 29
IRF9, 3, 4, 37, 38, 128, 129, 198, 199
ISG15, 37, 38, 105
ISG54, 37, 38, 52, 53
ISG56, 37, 38, 39, 52, 53
ISG58, 37, 38, 53
ISGF3, 3, 4, 38, 123, 128, 131
ISRE, 3, 38, 39, 204, 206

J

JAK/STAT, 3, 98, 198
Jak1, 39, 127, 129, 156, 199
Jak2, 39, 199
Japanese encephalitis virus, 213

K

K1, 270
K3, 72, 73, 83, 170
K3L, 6, 51, 173, 175, 176, 185, 269
K5, 72, 73, 83, 170
Kaposi, 141, 259, 270, 286
killer-cell immunoglobin-like receptors, 78, 79
KIR, 78, 79
knock-out poxviruses, 166, 167

L

lactacystin lactone, 143
LAMP1, 271
LAT, 142, 150, 152
LC3, 253, 254, 271
Legionella, 271
leukocyte function-associated antigen 1, 78
LFA-1, 78, 83
LMP1, 270
LPS, 24, 28, 52, 98, 200, 222, 223, 224, 226, 227, 234
Lumpy skin disease, 164
Ly49, 78, 80
Ly49D, 78
Ly49I, 79, 81
lymphocytic choriomeningitis virus (LCMV), 225, 235

M

m04/gp34, 71, 73, 81
m06/gp48, 72, 73, 81
M1, 96
M11L, 173, 174, 184, 185
m144, 79, 81
m152/gp60, 73
M153R, 170, 173, 185
m157, 79, 80, 81, 82
macrophages, 28, 66, 67, 77, 83, 107, 174, 179, 183, 202, 204, 222, 224, 226, 227, 230, 232, 234
Malignant fibroma, 164
mannose 6 phosphate, 151
Marburg virus, 201, 202, 203
MC066, 174
MC159L, 172, 173, 175, 176
MC160L, 172
MCF7 breast carcinoma cells, 269
MCMV, 71, 73, 77, 79, 80, 81, 82, 83, 84
MCP-1, 202, 204, 205
MD-2, 223
MDCK, 100

measles virus (MeV), 7, 10, 98, 116, 128, 129, 225, 229
Metapneumovirus, 116
MG132, 143
MHC, 39, 66, 67, 69, 70, 71, 72, 73, 74, 75, 76, 77, 78, 79, 80, 81, 82, 83, 84, 85, 153, 154, 155, 170, 173, 183, 185, 204, 289
MHV, 271, 272
miRNAs, 85
mitochondrial permeability transition, 174
ML, 7, 97, 100, 105, 107
Molluscum contagiosum, 164
Monkeypox, 164
Mononegavirales, 115
Morbillivirus, 116, 128
mouse mammary tumor virus (MMTV), 53, 225, 227, 228, 229
mouse γ-herpesvirus 68, 234
MRC5, 120, 132
MRV, 164, 169
M-T2, 173, 177, 184, 185
M-T4, 173, 174, 185
M-T5, 173, 174, 176, 185
M-T7, 173, 178, 182, 185, 186
MULT-1, 82
mumps virus, 7, 10, 116, 117, 120, 124
MuP49, 52, 53
MuP54, 52, 53
MuP56, 52, 53, 55
Murray Valley, 213
mutation rate, 65
MuV, 116, 117, 118, 120, 121, 122, 124, 129
Mx GTPases, 3
MxA, 8, 37, 119, 131, 198, 211
MyD88, 28, 29, 40, 223, 231, 232, 233, 235
Myt-1, 145
Myxoma virus (MV), 164, 168, 169, 170, 172, 173, 174, 175, 176, 177, 178, 180, 181, 182, 183, 184, 185, 186, 225, 228, 229, 230

N

N protein, 267
N1L, 225, 233, 234
NA, 96
NAP, 24
natural killer cells (NK), 66, 67, 70, 74, 75, 77, 79, 80, 81, 82, 83, 84, 85, 170, 172, 179, 226, 233, 235, 287
ND10, 143, 145
Nef, 72
NEP, 96, 99, 100, 104, 105
neuraminidase, 68, 96
Newcastle disease virus (NDV), 23, 27, 116, 119, 121, 207, 212, 225, 235
NF90, 49
Nipah virus (NiV), 7, 116, 129, 130, 197, 199, 207, 208, 209, 212, 214
NKG2D, 78, 82
NKp30, 78
Nod1, 27
NS1, 1, 5, 6, 7, 10, 51, 96, 97, 99, 100, 101, 102, 103, 104, 105, 106, 107, 116, 131, 132, 200, 205, 269
NS2, 7, 96, 116, 131, 132
NS2A, 212
NS3/4A, 6
NS4A, 212
NS4b, 199
NS4B, 6, 212
NS5A, 6, 45, 51, 269, 270
NSs, 7, 199, 210, 211, 214
nuclear factor κB (NF- κB), 3

O

OAS, 6, 41, 42, 43, 44, 56, 57, 175, 198, 204
OAS1, 8, 37, 41, 42, 43, 44, 45
OAS1b, 44

OAS2, 37, 41, 42, 43, 44
OAS3, 37, 41, 43
OASL, 41, 43, 44
Oct-1, 142
oncolytic agents, 107, 186
Orf, 164
ORF O, 142
ORF P, 142
osteoclast, 119
oxidative stress, 173, 174, 260, 261

P

P48, 38, 53, 55
p53, 70, 85, 143
P56, 3, 39, 40, 41, 52, 53, 57
$p58^{IPK}$, 51
p65, 21, 24, 149, 153
PA, 96, 221
PABII, 102
PACT, 46, 47, 48, 49, 51, 57
pandemics, 68, 96
Paramyxoviridae, 115, 116, 120, 131, 133, 207
Paramyxovirinae, 115, 116, 117, 118, 133
PAS, 249, 252, 253, 271
pathogen-associated molecular patterns (PAMP), 223
PB1, 96
PB1-F2, 96
PB2, 96
PBR, 174
PCI, 53, 55
peptidoglycans, 223
perforins, 70, 77
peripheral benzodiazepine receptor, 174
PERK, 261
phleboviruses, 211
PI3-K, 40, 250, 251, 252, 262, 268, 270
pigs, 96, 207, 215
PKA, 150
PKR (protein kinase R), 3, 6, 7, 8, 25, 26, 37, 39, 40, 41, 44, 46, 47, 48, 49, 50, 51, 53, 57, 103, 104, 105, 106, 119, 142, 147, 148, 149, 152, 156, 175, 176, 198, 200, 204, 212, 250, 256, 261, 262, 263, 264, 265, 269, 272
PKR activation by dsRNA, 47, 48, 49, 51
plasmacytoid dendritic cell, 28, 198, 205, 232
Pneumovirinae, 115, 116, 117, 133
Pneumovirus, 116, 131
poliovirus, 6, 50, 249, 270, 271, 272
poly(I)poly(C), 10, 125, 131, 132, 234
polyglutamine, 260, 261
poultry, 96
pRB, 143, 145
programmed cell death, 66, 70, 150, 252, 254, 266, 267, 268
promyelocytic leukemia protein (PML), 3, 6, 143, 144, 156
Proteasome, COP9-signalosome, Initiation factor, 53
protein kinase A, 150
Pseudocowpox, 164
PT, 174
PTEN, 250, 258

R

R protein, 267
R3616, 148
Rabbitpox, 164
rabies virus, 131
RACK1, 124, 129
RAE-1, 79, 82
RanGAP1, 144
RANTES, 204, 284, 285
Rapamycin, 260
reovirus, 6, 50
respiratory syncytial virus (RSV), 8, 98, 116, 118, 119, 131, 132, 225, 226, 227, 228, 229, 236, 282, 284, 289, 290
Respirovirus, 116, 117, 126, 127
reverse genetics, 126, 206
Rhabdoviridae, 115

ribonucleoprotein, 25
Rift valley fever virus (RVFV), 7, 199, 210, 211, 214
RIG-I, 27, 28, 99, 200
rinderpest virus, 116
RING4, 37
RIP2, 27
RNA editing, 207
RNA helicase, 27, 29, 99, 200
RNA polymerase, 26, 65, 68, 96, 207, 211
RNase L, 3, 6, 40, 41, 42, 43, 44, 45, 46, 56
RNP, 25, 26, 28
rRV, 131
rubella, 10, 67
Rubulavirus, 116, 121, 122, 124, 125, 129

S

SCID-mice, 80
SDF-1, 286
semaphorins, 181
Sendai, 7, 8, 40, 101, 116, 117, 126, 206, 209, 236
serine kinase, 24
serine phosphorylation, 20, 24
serine protease inhibitor, 172
SERP-1, 181, 182
SERP-2, 172
serpin, 172, 181, 185
Sheep pox, 164
Shope fibroma virus (SFV), 164, 169, 171, 175, 177
simian B virus, 148
Simian virus 5 (SV5) 7, 116, 117, 118, 119, 120, 121, 122, 123, 124, 125
Sindbis, 214, 253, 262, 264, 265, 266, 267
single nucleotide polymorphisms, 7, 282
SLAM, 229, 230
SNPs, 8, 282, 285, 286, 290, 291
sorbitol, 150, 153
Sp100, 143, 144
Sp140, 143
STAT1, 3, 4, 5, 7, 37, 38, 40, 98, 101, 117, 120, 121, 122, 123, 124, 125, 126, 127, 128, 129, 130, 131, 198, 199, 200, 203, 208, 213
STAT2, 3, 4, 38, 117, 121, 122, 124, 127, 128, 129, 130, 198, 199, 208
STAT3, 117, 120, 121, 122, 124, 128, 129
Swinepox, 164

T

T2K, 24
TAK1, 28
Tanapox, 164
TAP, 71, 72, 74, 76, 153, 154
TAR, 6, 44, 51
Tat, 6, 50, 79, 83, 269
TATA, 22
TBK1, 8, 24, 25, 28, 29, 40, 97, 209
tetratricopeptides, 52
thapsigargin, 47
Thogoto, 7, 97, 100, 104, 105, 107
threonine, 72, 145, 251
TICAM, 223, 231, 234
TIR, 27, 28, 163, 176, 209, 222, 223, 233
TLR1, 222, 223, 228, 231
TLR2, 8, 27, 98, 223, 228, 229, 230, 231
TLR3, 8, 26, 28, 39, 51, 56, 98, 200, 209, 222, 223, 224, 227, 231, 234, 235
TLR4, 28, 98, 200, 209, 222, 224, 225, 226, 227, 228, 229, 230, 234
TLR6, 223, 231
TLR7, 26, 28, 98, 227, 231, 234, 235, 236
TLR8, 98
TLR9, 26, 28, 29, 200, 224, 231, 232, 233, 234, 235, 236
TMV, 267, 268

TNF, 24, 52, 70, 77, 85, 106, 150, 153, 172, 173, 174, 176, 177, 185, 202, 204, 223, 226, 228, 290
TNF receptors, 176
TNF-R, 70, 173, 185
Toll-like receptor, 8, 38, 85, 98, 200
topoisomerase, 146, 147
TPR, 52, 53, 57
TRAF6, 28, 29, 40, 225, 233
TRAM, 28
transgenic mice, 24, 228, 230, 254, 260
DAP12-deficient, 80
IFNAR -/-, 98, 200, 203
MyD88 knockout, 232
PKR-/-, 51, 103, 153, 263
RNase L -/-, 46
$STAT1^{-/-}$, 5, 101
$TICAM^{-/-}$, 235
TLR3-/-, 235
translation initiation factor, 25, 40, 46, 53, 147
Transplant rejection, 186
transporter associated with antigen presentation, 71, 72
TRIF, 8, 28, 40, 209, 223
Tula virus (TULV), 211
tumor suppression, 143, 251, 257, 258
tumorigenesis, 227, 258
Tupaia virus, 116
Turkey rhinotracheitis, 116
Tyk2, 199
type I IFNs, 2, 38, 223, 232, 233
tyrosine phosphorylation, 3, 39, 121, 126, 127, 128

U

U21, 72, 73
UbcH5a, 144
UbcH8, 105
Ube1L/E1, 105
ubiquitin, 29, 45, 72, 123, 124, 143, 144, 145, 153, 251, 253
U_L13, 145, 146
UL18, 79, 81
U_L38, 146
U_L41, 141, 146
U_L42, 146, 147
U_S 6, 150
U_S 7, 150
U_S1.5, 142
US10, 74
Us11, 51, 269
US11, 6, 71, 74, 75, 81, 146, 148, 149
U_S12, 142, 153
US2, 71, 74, 75, 76, 77, 81
US3, 71, 74, 76, 77, 81, 150
US6, 71, 74, 81
US8, 74

V

V protein, 104, 117, 118, 120, 121, 122, 123, 124, 125, 126, 127, 128, 129, 130, 133, 208, 209
VA RNAs, 6
vaccines, 8, 9, 10, 107, 119, 120
vaccinia virus (VV), 6, 51, 85, 103, 164, 168, 169, 170, 173, 175, 176, 177, 178, 179, 180, 181, 182, 183, 185, 186, 200, 225, 233, 235, 269
variola, 164, 187
vBcl-2, 258
VCP, 173, 183, 184, 186
VDC, 124
VEGF, 173, 180, 185
VEGFR-2, 180
Venezuelan equine encephalitis virus (VEEV), 213, 214
Vero, 100, 128, 146, 202
vesicular stomatitis virus (VSV), 143, 225, 235
vhs, 69, 141, 142, 150, 153, 154, 155
Viral IL-10, 179
Viral IL-18 binding proteins, 179
viral stress-inducible genes, 36

vIRF, 6, 51
vIRF-2, 51
viroceptors, 170, 176, 178, 184
virokines, 170, 176
viromimetics, 170
virostealth proteins, 170
virotransducers, 170, 184
virus host shut-off, 69
VP35, 7, 103, 199, 205, 206
VPS34, 252, 262, 268
VSIGs, 36, 37, 39
vTNFR, 177
vTNFRs, 176, 177

W

Wee-1, 145

X

X-linked lymphoproliferative syndrome, 281

Y

Yaba monkey tumor, 164
yeast, 148, 211, 247, 249, 250, 251, 252, 253, 254, 256, 257, 259, 262, 265, 268, 269

Z

Zaire Ebola, 201, 202, 204, 205

Α

α-synuclein, 260

Γ

$\gamma_1$34.5, 6

Σ

σ3, 6, 269

FSC
www.fsc.org
MIX
Papier | Fördert gute Waldnutzung
FSC® C083411

Kolibri 360 GmbH
Ferdinand-Jühlke-Straße 7
99095 Erfurt, Deutschland
produktsicherheit@kolibri360.de